VOLUME FIFTY EIGHT

Advances in
ECOLOGICAL RESEARCH
Next Generation Biomonitoring: Part 1

ADVANCES IN ECOLOGICAL RESEARCH

Series Editors

DAVID A. BOHAN
Directeur de Recherche
UMR 1347 Agroécologie
AgroSup/UB/INRA
Pôle GESTAD, Dijon, France

ALEX J. DUMBRELL
School of Biological Sciences
University of Essex
Wivenhoe Park, Colchester
Essex, United Kingdom

VOLUME FIFTY EIGHT

ADVANCES IN
ECOLOGICAL RESEARCH
Next Generation Biomonitoring: Part 1

Edited by

DAVID A. BOHAN
Directeur de Recherche
UMR 1347 Agroécologie
AgroSup/UB/INRA
Pôle GESTAD, Dijon, France

ALEX J. DUMBRELL
School of Biological Sciences
University of Essex
Wivenhoe Park, Colchester, Essex,
United Kingdom

GUY WOODWARD
Imperial College London, Ascot, Berkshire,
United Kingdom

MICHELLE JACKSON
Imperial College London, Ascot, Berkshire,
United Kingdom

ACADEMIC PRESS

An imprint of Elsevier

Academic Press is an imprint of Elsevier
The Boulevard, Langford Lane, Kidlington, Oxford OX5 1GB, United Kingdom
125 London Wall, London EC2Y 5AS, United Kingdom
50 Hampshire Street, 5th Floor, Cambridge, MA 02139, United States
525 B Street, Suite 1800, San Diego, CA 92101-4495, United States

First edition 2018

Notices
Knowledge and best practice in this field are constantly changing. As new research and experience broaden our understanding, changes in research methods, professional practices, or medical treatment may become necessary.

Practitioners and researchers must always rely on their own experience and knowledge in evaluating and using any information, methods, compounds, or experiments described herein. In using such information or methods they should be mindful of their own safety and the safety of others, including parties for whom they have a professional responsibility.

To the fullest extent of the law, neither the Publisher nor the authors, contributors, or editors, assume any liability for any injury and/or damage to persons or property as a matter of products liability, negligence or otherwise, or from any use or operation of any methods, products, instructions, or ideas contained in the material herein.

ISBN: 978-0-12-813949-3
ISSN: 0065-2504

For information on all Academic Press publications
visit our website at https://www.elsevier.com/books-and-journals

Working together
to grow libraries in
developing countries

www.elsevier.com • www.bookaid.org

Publisher: Zoe Kruze
Acquisition Editor: Jason Mitchell
Editorial Project Manager: Joanna Collett
Production Project Manager: Abdulla Sait
Cover Designer: Mark Rogers

Typeset by SPi Global, India

CONTENTS

3. Advances in Monitoring and Modelling Climate at Ecologically Relevant Scales   101

Isobel Bramer, Barbara J. Anderson, Jonathan Bennie, Andrew J. Bladon,
Pieter De Frenne, Deborah Hemming, Ross A. Hill, Michael R. Kearney,
Christian Körner, Amanda H. Korstjens, Jonathan Lenoir, Ilya M.D. Maclean,
Christopher D. Marsh, Michael D. Morecroft, Ralf Ohlemüller, Helen D. Slater,
Andrew J. Suggitt, Florian Zellweger, and Phillipa K. Gillingham

4. Challenges With Inferring How Land-Use Affects Terrestrial Biodiversity: Study Design, Time, Space and Synthesis   163

Adriana De Palma, Katia Sanchez-Ortiz, Philip A. Martin, Amy Chadwick,
Guillermo Gilbert, Amanda E. Bates, Luca Börger, Sara Contu,
Samantha L.L. Hill, and Andy Purvis

CONTRIBUTORS

Kessy Abarenkov
University of Tartu, Tartu, Estonia

Florian Altermatt
Eawag, Dübendorf; University of Zurich, Zürich, Switzerland

Barbara J. Anderson
Manaaki Whenua Landcare Research, Biodiversity and Conservation Team, Dunedin, New Zealand

Amanda E. Bates
Ocean and Earth Science, National Oceanography Centre Southampton, University of Southampton, Southampton, United Kingdom

Jonathan Bennie
College of Life and Environmental Sciences, University of Exeter, Penryn, Cornwall, United Kingdom

Andrew J. Bladon
RSPB Centre for Conservation Science, The Lodge, Sandy, Bedfordshire, United Kingdom

David A. Bohan
Agroécologie, AgroSup Dijon, INRA, University of Bourgogne Franche-Comté, Dijon, France

Luca Börger
College of Science, Swansea University, Swansea, United Kingdom

Ángel Borja
AZTI, Pasaia, Spain

Agnès Bouchez
INRA UMR CARRTEL, Thonon-les-bains, France

Isobel Bramer
Faculty of Science and Technology, Bournemouth University, Poole, Dorset, United Kingdom

Kat Bruce
NatureMetrics, CABI Site, Surrey, United Kingdom

Amy Chadwick
University College London, London, United Kingdom

Fedor Čiampor
Zoology Lab, Plant Science and Biodiversity Center, Slovak Academy of Sciences, Bratislava, Slovakia

Zuzana Čiamporová-Zaťovičová
Zoology Lab, Plant Science and Biodiversity Center, Slovak Academy of Sciences, Bratislava, Slovakia

Sara Contu
Natural History Museum, London, United Kingdom

Filipe O. Costa
Centre of Molecular and Environmental Biology (CBMA), University of Minho, Braga, Portugal

Pieter De Frenne
Forest and Nature Lab, Ghent University, Ghent, Belgium

Adriana De Palma
Natural History Museum, London, United Kingdom

Stéphane A.P. Derocles
Agroécologie, AgroSup Dijon, INRA, University of Bourgogne Franche-Comté, Dijon, France

Sofia Duarte
Centre of Molecular and Environmental Biology (CBMA), University of Minho, Braga, Portugal

Alex J. Dumbrell
School of Biological Sciences, University of Essex, Colchester, United Kingdom

Torbjørn Ekrem
Norwegian University of Science and Technology, Trondheim, Norway

Vasco Elbrecht
Aquatic Ecosystem Research, University of Duisburg-Essen, Essen, Germany; Centre for Biodiversity Genomics, University of Guelph, Guelph, ON, Canada

Darren M. Evans
School of Natural and Environmental Sciences, Newcastle University, Newcastle upon Tyne, United Kingdom

Diego Fontaneto
National Research Council of Italy, Institute of Ecosystem Study, Verbania Pallanza, Italy

Alain Franc
BIOGECO, INRA, Univ. Bordeaux, Cestas, and Pleiade Team, INRIA Sud-Ouest, Talence, France

Matthias F. Geiger
Zoologisches Forschungsmuseum Alexander Koenig, Leibniz Institute for Animal Biodiversity, Bonn, Germany

Ilse R. Geijzendorffer
Tour du Valat, Research Institute for the Conservation of Mediterranean Wetlands, Arles, France

Guillermo Gilbert
Natural History Museum, London, United Kingdom

Phillipa K. Gillingham
Faculty of Science and Technology, Bournemouth University, Poole, Dorset, United Kingdom

Anis Guelmami
Tour du Valat, Research Institute for the Conservation of Mediterranean Wetlands, Arles, France

Deborah Hemming
Met Office Hadley Centre, Exeter, Devon, United Kingdom; Birmingham Institute of Forest Research, Birmingham University, Birmingham, United Kingdom

Daniel Hering
Aquatic Ecology; Center of Water and Environmental Research (ZWU), University of Duisburg-Essen, Essen, Germany

Ross A. Hill
Faculty of Science and Technology, Bournemouth University, Poole, Dorset, United Kingdom

Samantha L.L. Hill
Natural History Museum, London; UN Environment World Conservation Monitoring Centre, Cambridge, United Kingdom

Lawrence N. Hudson
Natural History Museum, London, United Kingdom

Maria Kahlert
Swedish University of Agricultural Sciences, Uppsala, Sweden

Belma Kalamujić Stroil
University of Sarajevo—Institute for Genetic Engineering and Biotechnology, Sarajevo, Bosnia and Herzegovina

Michael R. Kearney
School of BioSciences, The University of Melbourne, Melbourne, Australia

Martyn Kelly
Bowburn Consultancy, Durham, United Kingdom

Emre Keskin
Evolutionary Genetics Laboratory (eGL), Ankara University Agricultural Faculty, Ankara, Turkey

James J.N. Kitson
School of Natural and Environmental Sciences, Newcastle University, Newcastle upon Tyne, United Kingdom

Christian Körner
Institute of Botany, University of Basel, Basel, Switzerland

Amanda H. Korstjens
Faculty of Science and Technology, Bournemouth University, Poole, Dorset, United Kingdom

Florian Leese
Aquatic Ecosystem Research; Center of Water and Environmental Research (ZWU), University of Duisburg-Essen, Essen, Germany

Jonathan Lenoir
UR "Ecologie et dynamique des systèmes anthropisés" (EDYSAN, UMR 7058 CNRS-UPJV), Université de Picardie Jules Verne, Amiens, France

Igor Liska
ICPDR Permanent Secretariat, Vienna International Centre, Vienna, Austria

Igor Lysenko
Grand Challenges in Ecosystems and the Environment, Imperial College London, Ascot, United Kingdom

Ilya M.D. Maclean
College of Life and Environmental Sciences, University of Exeter, Penryn, Cornwall, United Kingdom

Christopher D. Marsh
Faculty of Science and Technology, Bournemouth University, Poole, Dorset, United Kingdom

Philip A. Martin
Conservation Science Group, University of Cambridge, Cambridge, United Kingdom

François Massol
CNRS, UMR 8198 Evo-Eco-Paleo, Université de Lille, SPICI group, Lille, France

Kristian Meissner
Finnish Environment Institute, General Director's Office, Jyväskylä, Finland

Patricia Mergen
Botanic Garden Meise, Meise; Royal Museum for Central Africa, Tervuren, Belgium

Michael D. Morecroft
Natural England c/o Mail Hub, County Hall, Worcester, Worcestershire, United Kingdom

Tim Newbold
Centre for Biodiversity and Environment Research, University College London, London, United Kingdom

Ralf Ohlemüller
University of Otago, Dunedin, New Zealand

Marc Paganini
European Space Agency, Frascati, Italy

Charlie Pauvert
BIOGECO, INRA, Univ. Bordeaux, Pessac, France

Jan Pawlowski
University of Geneva, Geneva, Switzerland

Lyubomir Penev
Pensoft Publishers, Sofia, Bulgaria

Christian Perennou
Tour du Valat, Research Institute for the Conservation of Mediterranean Wetlands, Arles, France

Petra Philipson
Brockmann Geomatics Sweden AB, Stockholm, Sweden

Helen R.P. Phillips
German Centre for Integrative Biodiversity Research (iDiv) Halle-Jena-Leipzig, Leipzig, Germany

Manuel Plantegenest
UMR 1349 IGEPP, INRA, Agrocampus-Ouest, Université de Rennes 1, Rennes Cedex, France

Brigitte Poulin
Tour du Valat, Research Institute for the Conservation of Mediterranean Wetlands, Arles, France

Andy Purvis
Natural History Museum, London; Grand Challenges in Ecosystems and the Environment, Imperial College London, Ascot, United Kingdom

Yorick Reyjol
AFB, The French Agency for Biodiversity, Direction de la Recherche, Vincennes, France

Ana Rotter
National Institute of Biology, Ljubljana, Slovenia

Katia Sanchez-Ortiz
Natural History Museum, London; Grand Challenges in Ecosystems and the Environment, Imperial College London, Ascot, United Kingdom

Jörn P.W. Scharlemann
School of Life Sciences, University of Sussex, Brighton; UN Environment World Conservation Monitoring Centre, Cambridge, United Kingdom

Helen D. Slater
Faculty of Science and Technology, Bournemouth University, Poole, Dorset, United Kingdom

Dirk Steinke
Centre for Biodiversity Genomics, University of Guelph; University of Guelph, Guelph, ON, Canada

Adrian Strauch
University of Bonn, Center for Remote Sensing of Land Surfaces (ZFL), Bonn, Germany

Andrew J. Suggitt
University of York, York, Yorkshire, United Kingdom

Christian Tottrup
DHI GRAS, Hoersholm, Denmark

John Truckenbrodt
Friedrich-Schiller-University Jena, Institute of Geography, Jena, Germany

Corinne Vacher
BIOGECO, INRA, Univ. Bordeaux, Pessac, France

Bas van der Wal
STOWA, Stichting Toegepast Onderzoek Waterbeheer, Amersfoort, The Netherlands

Simon Vitecek
University of Vienna, Vienna, Austria; Senckenberg Research Institute and Natural History Museum, Frankfurt am Main, Germany

Alexander M. Weigand
Aquatic Ecosystem Research; Center of Water and Environmental Research (ZWU), University of Duisburg-Essen, Essen, Germany; Musée National d'Histoire Naturelle de Luxembourg, Luxembourg, Luxembourg

Florian Zellweger
Forest Ecology and Conservation Group, University of Cambridge, Cambridge, Cambridgeshire, United Kingdom; Swiss Federal Research Institute WSL, Birmensdorf, Switzerland

Jonas Zimmermann
Botanic Garden and Botanical Museum, Freie Universität Berlin, Berlin, Germany

PREFACE

Biomonitoring the Earth's ecosystems and their attendant communities, functions and ecoservices underpins decision making in many areas of policy and can have considerable value for the public, particularly in the case of species with high conservation value. In almost all cases, however, current biomonitoring approaches suffer from problems of accuracy, high costs that restrict coverage and limited generality. Biomonitoring schemes are also based upon methods developed in the early or middle part of the last century and have largely ignored subsequent advances in ecological theory and techniques, especially those derived from molecular ecology, remote sensing, network science and ecoinformatics. Consequently, the full diversity of functions and species in an ecosystem has rarely been evaluated. This is problematic because it only provides a partial view of the greater whole and cannot account for—or predict—the "ecological surprises" that commonly arise through indirect food web effects in nature. In this two-volume Thematic Issue of *Advances in Ecological Research* focusing on Ecological Biomonitoring, we showcase some of the new biomonitoring approaches that have begun to appear in the last 15 years and that have started to tackle these problems directly; to generate the more sophisticated Next-Generation Biomonitoring (NGB) approaches, we will need to cope with our rapidly changing environment. Potentially, NGB could, even within the next decade, revolutionise our understanding of the functioning of Earth's major ecosystems, allowing us to both measure and predict the effects of a range of abiotic stressors as well as those from the biotic sphere (e.g. species invasion and extinction), which will lead to better-informed and more effective management. Moreover, as they are often rooted in standardised, functional metrics, these approaches could potentially be applied at local to global scales, both accurately and cheaply.

The first couple of papers of this two-volume Thematic Issue consider the role that new DNA-based approaches might play in the future of NGB. Derocles et al. (this issue) examine the potential that Next-Generation Sequencing (NGS) of environmental samples of DNA has to provide the means to rapidly build highly resolved species interaction networks across multiple trophic levels. Their paper details how the analysis of multilayer ecological networks, constructed from NGS data, could be used to characterise the ecological mechanisms that underpin ecosystem functioning and ecosystem service provision within future NGB frameworks. The authors propose that the future of network ecology and biomonitoring is extremely exciting given that the tools needed to build highly resolved multilayer networks are now finally within reach.

In the subsequent paper, Leese et al. (this issue) place the current start of the art in environmental DNA sampling for NGB of freshwaters within the historical context of the limitations and strengths of traditional biomonitoring methods. The authors use a new research consortium, DNA-Aquanet, recently established and supported by the European Union, as the lens through which to view the development of the novel approaches that will augment—and ultimately supersede—current practices. They emphasise the fundamental differences in the traditional and NGB approaches, as well as highlighting some of the key areas of common ground, especially where there is scope for the "handshaking" and cross-calibration that is needed to form the bridge between the old and the new, thus preserving the value of the vast store of historical data that have already been amassed. The increase in capacity and a decrease in costs of molecular tools are discussed in relation to the far slower development of traditional methods. Unresolved issues the authors highlight include those that are still holding the field back, such as bioinformatics database errors, amplification bias and problems of estimating relative abundance across taxa from DNA data. These are discussed against the backdrop of end-user community inertia due to past investment in older biomonitoring approaches—a classic example of the "sunk cost fallacy". Leese et al. then focus on the key advances that are now being made in NGB and how the DNA-Aquanet consortium is helping to drive those changes, in both the scientific and non-academic spheres. The paper has a strong applied focus, with a strong link to EU legislative frameworks, but this is complemented by the consideration of the role these new approaches could play in addressing fundamental questions in ecology, reshaping not just our current view of the world but also the questions we will be able to ask in the future.

The final four papers of this volume are more explicitly practical in tone, emphasising the application of ecological approaches to biomonitoring and the measurement of ecosystem change. The paper by Bramer et al. (this issue) examines an important, but often overlooked, component of biomonitoring—the local microclimate that supports the focal organisms in the ecosystem of interest. Based on recent discussions from a British Ecological Society Open Workshop (organised by the Climate Change Ecology Special Interest Group), they provide a broad overview of recent advances in microclimate monitoring and modelling, highlighting some of the key research challenges and solutions in this field, and scan the horizon for future developments. Ultimately, the spatiotemporal distribution of all organisms is largely controlled by their physiological tolerances to environmental conditions. Most research to date examines where and when species exist as a function of broad climatic envelopes operating over many kilometres. However, within these areas, local microclimates can vary immensely, even approaching the physiological limits of life for short periods or in particular patches within an otherwise seemingly benign landscape. Without understanding microclimatic variability, our understanding of the controls of species distributions is limited, as is our ability to predict how climatic changes may reshape them. Bramer et al. (this issue) tackle these problems directly and provide recommendations for improving NGB and our understanding of the controls on species distributions.

In the next paper, De Palma et al. (this issue) explore the strengths and weaknesses of the different study designs that are commonly used in biomonitoring in relation to land-use change, including space–for–time substitution, time series, and before–after–control–impact design. Comparisons of data from different types of studies can be problematic, and different designs may even detect different trends in biodiversity change. The authors discuss how new syntheses can incorporate multiple study types to provide a new and more holistic perspective in NGB. To develop more realistic future projections of biodiversity change, they stress the need for a better understanding of temporal dynamics. In conclusion, De Palma et al. call for more studies using a before–after–control–impact design, which are relevant for the widest range of questions related to NGB. These studies are still surprisingly rare, but disproportionately important because they can be used to validate or correct inferences from simpler designs.

The paper by Purvis et al. (this issue) gives a detailed account of the project "Projecting Responses of Ecological Diversity In Changing Terrestrial Systems (PREDICTS)". Since 2012, PREDICTS has collated abundance and occurrence data from thousands of sites facing different land-use pressures across the globe and now covers over 50,000 species in nearly 100 countries. In their paper, the authors discuss key design decisions for making predictions for biological diversity, including using space–for–time substitution, and detail the modelling approaches they have used. The project focuses on site-level biodiversity data because many ecosystem functions and services depend on the local, rather than global, state of biodiversity. They emphasise how the PREDICTS database can be used to improve global biodiversity assessments, which often rely on expert opinion or data from species representing only a small fraction of total global biodiversity (e.g. vertebrates). For instance, PREDICTS has implemented a version of the Biodiversity Intactness Index (BII) that is based on objective primary biodiversity data, rather than subjective expert judgement. This PREDICTS paper gives the most detailed overview of this large project to date and illustrates the value that tools, models, indicators and projections will have for biomonitoring and predicting change in global biodiversity.

In the final paper of this issue, Perennou et al. (this issue) describe developments and approaches to improve current space-borne remote sensing of ecosystems, using a case study from wetlands in the Mediterranean biodiversity hot spot. Given current challenges that affect wetlands, but which also have corollaries in the remote sensing of all ecosystems, of delineating and separating habitat types, mapping of the internal environmental dynamics and the detection of trends over time that need to be disentangled from natural background variability, Perrenou et al. argue that the solutions to improving current remote sensing approaches will only be achieved by allying the rapidly developing methodologies of remote sensing to ecological understanding of the ecosystems being monitored.

These two volumes present a snapshot of some of the work currently being done in biomonitoring. The combination of papers across them reveals the huge value in using novel NGS, sensing and informatics approaches and better fusions of pure and applied disciplines to monitor and model how natural ecosystems will respond to the accelerating rates and increasing magnitude of environmental change we are already seeing across the globe. There is clearly plenty of exciting and challenging work still to be done, but this Thematic Issue illustrates some of the most important steps being taken towards developing the NGB approaches we will need to achieve a more sustainable future.

ALEX J. DUMBRELL
GUY WOODWARD
MICHELLE C. JACKSON
DAVID A. BOHAN

ACKNOWLEDGEMENTS

David A. Bohan would like to acknowledge the support of the French Agence Nationale de la Recherche project *NGB* (ANR-17-CE32-0011) and FACCE SURPLUS project *PREAR* (ANR-15-SUSF-0002-03).

CHAPTER ONE

Biomonitoring for the 21st Century: Integrating Next-Generation Sequencing Into Ecological Network Analysis

Stéphane A.P. Derocles*,[1], David A. Bohan*, Alex J. Dumbrell[†],
James J.N. Kitson[‡], François Massol[§], Charlie Pauvert[¶],
Manuel Plantegenest[‖], Corinne Vacher[¶], Darren M. Evans[‡]

*Agroécologie, AgroSup Dijon, INRA, University of Bourgogne Franche-Comté, Dijon, France
[†]School of Biological Sciences, University of Essex, Colchester, United Kingdom
[‡]School of Natural and Environmental Sciences, Newcastle University, Newcastle upon Tyne, United Kingdom
[§]CNRS, UMR 8198 Evo-Eco-Paleo, Université de Lille, SPICI group, Lille, France
[¶]BIOGECO, INRA, Univ. Bordeaux, Pessac, France
[‖]UMR 1349 IGEPP, INRA, Agrocampus-Ouest, Université de Rennes 1, Rennes Cedex, France
[1]Corresponding author: e-mail address: stephane.derocles@inra.fr

Contents

Advances in Ecological Research, Volume 58
ISSN 0065-2504
https://doi.org/10.1016/bs.aecr.2017.12.001

Abstract

Ecological network analysis (ENA) provides a mechanistic framework for describing complex species interactions, quantifying ecosystem services, and examining the impacts of environmental change on ecosystems. In this chapter, we highlight the importance and potential of ENA in future biomonitoring programs, as current biomonitoring indicators (e.g. species richness, population abundances of targeted species) are mostly descriptive and unable to characterize the mechanisms that underpin ecosystem functioning. Measuring the robustness of multilayer networks in the long term is one way of integrating ecological metrics more generally into biomonitoring schemes to better measure biodiversity and ecosystem functioning. Ecological networks are nevertheless difficult and labour-intensive to construct using conventional approaches, especially when building multilayer networks in poorly studied ecosystems (i.e. many tropical regions). Next-generation sequencing (NGS) provides unprecedented opportunities to rapidly build highly resolved species interaction networks across multiple trophic levels, but are yet to be fully exploited. We highlight the impediments to ecologists wishing to build DNA-based ecological networks and discuss some possible solutions. Machine learning and better data sharing between ecologists represent very important areas for advances in NGS-based networks. The future of network ecology is very exciting as all the tools necessary to build highly resolved multilayer networks are now within ecologists reach.

1. INTRODUCTION

Traditionally, community ecology tends to focus on patterns of species richness and community composition, while ecosystem ecology focuses on fluxes of energy and materials. Ecological networks (sometimes called food webs for trophic interactions), however, provide a quantitative framework to combine these approaches and unify the study of biodiversity and ecosystem function (Thompson et al., 2012). Ecological networks, which describe which species are interacting with which (i.e. qualitative networks) as well as the strength of their interactions (i.e. quantitative networks), are now routinely used to understand ecosystem 'robustness' to species extinctions (Evans et al., 2013; Säterberg et al., 2013), quantify ecosystem services (Derocles et al., 2014a; Macfadyen et al., 2009) or examine the impacts of environmental change (Morris et al., 2015; Thompson and Gonzalez, 2017; Tylianakis et al., 2007). By using a burgeoning range of metrics to describe network structure, complexity and stability (see Arnoldi et al., 2016; Bersier et al., 2002; Donohue et al., 2013; Dunne et al., 2002a,b), ENA is considerably improving our understanding of ecology and evolution, with a growing number of applications for biomonitoring (Bohan et al., 2017; Gray et al., 2014). Indeed, ENA is increasingly being used to assess ecosystem response to environmental changes (e.g. climate change, pollution, invasive species; Aizen et al., 2008; Blanchard, 2015; Bohan et al., 2017; Thompson et al., 2016). There is consequently a growing shift in biodiversity monitoring away from conventional species and community-level descriptions towards a more comprehensive and mechanistic approach using species interaction networks (Bohan et al., 2013; Derocles et al., 2014a; Evans et al., 2013; Fontaine et al., 2011; Gray et al., 2014; Ings et al., 2009; Kéfi et al., 2012; Macfadyen et al., 2009; Pocock et al., 2012; Wirta et al., 2014). Nevertheless, ecological networks can be difficult to construct with conventional approaches and suffer some major pitfalls mainly centred on sampling issues, taxonomic misidentification and/or incorrect species interactions (Evans et al., 2016; Gibson et al., 2011). Major errors occurring in either of these steps could ultimately affect network-level structural metrics and thus our understanding of ecosystem functioning (Novak et al., 2011). DNA-based methods (based on combined taxonomic identification and interaction data from DNA sequences) have the potential to overcome many of these issues, providing large, highly resolved, phylogenetically structured networks suitable for rapid and reliable biomonitoring (Bohan et al., 2017; Evans et al., 2016; Vacher et al., 2016; Valentini et al., 2009b).

Today, next-generation sequencing (NGS) or high-throughput sequencing (see Goodwin et al., 2016 for a review) can rapidly generate millions of DNA sequences. Sequences can describe, very precisely, not only the biodiversity present within an ecosystem, but also species interactions, the data from which can then be used to construct ecological networks (Evans et al., 2016). Recently, ecological network studies have taken advantage of NGS to successfully construct networks (e.g. Toju et al., 2014 for a plant–fungus network). Advances in statistical modelling and machine learning approaches bring a new opportunity to predict species interactions and rapidly build multilayer ecological networks from DNA sequences data generated with NGS (Vacher et al., 2016).

Despite species identification from DNA sequences commonly being seen as a universal way to identify species (Hebert et al., 2003), the NGS technology to build food webs is not applied uniformly in network ecology. Experimental designs (field sampling and molecular protocols) and the construction of ecological networks are heavily dependent on the ecosystem studied, and particularly on the type of interactions (see Box 1). Here, we

BOX 1 Species interactions in ecological networks

Species interactions are a major component of ecosystem functioning. In ecological communities, a wide range of interactions can be described and visualized as ecological networks. These include direct and indirect interactions. Direct interactions relate to cases where a species directly affects another (i.e. species A impacts species B). Indirect interactions refer to cases where the impact of a species on another is mediated or transmitted by a third species (i.e. a first species A affects a second species B through an intermediary species C).

In ecological networks, direct interactions are usually described and collectively shape the structure of the networks. However, this does not mean that indirect interactions are ignored as ecological networks are also used as a framework to study indirect interactions such as resource competition (Tilman, 1982), apparent competition (interactions through shared natural enemies; Derocles et al., 2014a; Holt, 1997; Morris et al., 2004; van Veen et al., 2006) or trophic cascades (Hairston et al., 1960; Oksanen et al., 1981). Indeed, indirect interactions result from the cooccurrence of several direct interactions. Hence, because the purpose of ecological networks is to describe the set of (direct) species interactions in an ecosystem, building networks constitutes a powerful approach to identify potential indirect interactions. Within a network it is, for example, possible to detect shared natural enemies when searching for cases of apparent competition—a particular instance of three-node network motifs (Stouffer et al., 2007).

Table 1 summarizes a general classification of direct ecological interactions. Although a wide range of interactions occur in nature, ecological network

BOX 1 Species interactions in ecological networks—cont'd

Table 1 Direct Interactions Between Two Species (According to Lidicker, 1979; see Faust and Raes, 2012)

Type of Interaction	Effect on Species A	Effect on Species B	Nature of Interaction	Examples in Ecological Networks
Mutualism	**Positive**	**Positive**	**Mutual benefits of the species**	Plant–pollinator, plant–ant, plant–seed disperser, plant–fungi
Interference competition	Negative	Negative	Species have negative effect on each other	
Trophic/ predation	**Positive**	**Negative**	**Predator gains at the expense of the prey, which is killed. We include here prey–predator, plant–consumer and host–parasitoid interactions**	Host–parasitoid, prey–predator, plant–herbivore
Parasitism	**Positive**	**Negative**	**Parasite develops at the expense of the host, which is not killed**	Host–parasite, host–pathogen
Commensalism	Positive	Null	Species A is benefited, species B is not affected	
Amensalism	Null	Negative	Species A has a negative effect on species B, but species A is not affected	
Neutralism	Null	Null	Neither species is affected	

The interaction types studied in depth in network ecology are in *bold*.

Continued

BOX 1 Species interactions in ecological networks—cont'd
studies to date have tended to focus on three types of interactions: parasitism, mutualism and trophic interactions. Other types of interactions have been studied (see Allesina and Levine, 2011; Coyte et al., 2015; Mougi, 2016 for networks with competitive interactions), but they are relatively rare in comparison with the large majority of studies dealing with trophic and mutualist networks. A complementary classification was established by Pantel et al. (2017) accounting for the degree of interaction immediacy: whether the interaction takes place over a short or long part of an organism's life cycle. This distinguishes, for example, parasitism from predation, scramble competition from contest competition and mutualistic symbiosis from external mutualism. However, when discussing species interactions in this chapter, we will be referring to parasitism, mutualism and trophic interactions.

distinguish two cases in particular. First, NGS can be directly used to build quantitative ecological interactions between organisms by resolving species interactions (e.g. Evans et al., 2016; Kitson et al., 2016; Piñol et al., 2014; Toju et al., 2013, 2014). This use of NGS data is, however, only possible in ecosystems in which relationships between organisms can clearly be established, such as host–parasitoid interactions where the parasitoid can be detected within the host (Derocles et al., 2014a, 2015; Wirta et al., 2014), prey–predator interactions by detecting prey in gut contents (e.g. Piñol et al., 2014; Tiede et al., 2016) or faeces (Clare et al., 2014; Zeale et al., 2011; see Symondson and Harwood, 2014) and plant–pollinator interactions by using high-throughput sequencing to identify the pollen carried (Bell et al., 2017; Galimberti et al., 2014; Pornon et al., 2016; Sickel et al., 2015). Second, there are systems in which it is impossible (or logistically very problematic) to detect interactions between organisms and assessing whether these interactions are positive or negative, such as those within microbial (Jakuschkin et al., 2016) or planktonic communities (Lima-Mendez et al., 2015). For these systems, NGS approaches can only identify cooccurring species and their relative abundance. NGS data then need to be combined with theoretical approaches, including statistical modelling (Faust and Raes, 2012) or machine learning (Bohan et al., 2011a), to predict species interactions from their abundance patterns and finally to build ecological networks (Bohan et al., 2017; Kamenova et al., 2017; Vacher et al., 2016). These two ways of building ecological networks have their own specificities and challenges to overcome but also share common problems. These problems are related to (1) the qualitative and quantitative reliability of NGS data (i.e. polymerase chain reaction (PCR) bias and errors, sequencing bias and estimation of species abundances and frequency of interactions with number of NGS reads; Sommeria-Klein et al., 2016); (2) the identification of nodes and interactions in the network (inferring species interactions with statistical models when interactions are not directly resolved by molecular tools); (3) the costs of the sequencing technology and the expertise needed to process the data (Toju et al., 2013, 2014; Vacher et al., 2016).

Here, we bring new insights on how to integrate NGS and ENA into biomonitoring (Fig. 1). We first consider why ecological networks provide a suitable framework for a better understanding of biodiversity and ecosystem functioning and how they can be used to complement or supersede conventional biomonitoring approaches. Second, we underline the challenges that ecologists face in building ecological networks when DNA-based tools are not available (which represent the vast majority of food web studies in the literature). Third, we demonstrate how molecular methods, NGS in particular, can overcome (at least partially) the numerous constraints inherent in conventional network construction methodologies (e.g. taxonomic identification, insect rearing, fieldwork issues), while considering the challenges of using NGS tools for building networks. Fourth, we give insights on how to overcome NGS data issues and efficiently build networks through machine learning and data sharing. Finally, we discuss new areas of research and development centred on ENA of multilayer networks to ultimately create more resilient ecosystems.

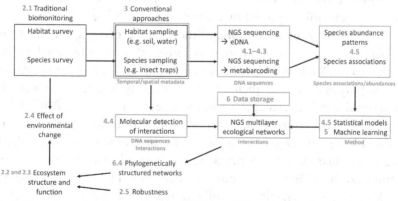

Fig. 1 A road map to integrate NGS and ENA into biomonitoring. The successive steps are discussed in this chapter, and the corresponding sections are indicated in *blue*. Data stored and shared for an efficient biomonitoring are indicated in *green*.

2. HOW ARE ECOLOGICAL NETWORKS USEFUL FOR BIOMONITORING?

2.1 Traditional Biomonitoring Is Typically Descriptive and Rarely Provides an Understanding of the Underlying Mechanisms Behind Ecosystem Functions

Biomonitoring of change lies at the core of ecosystem conservation, management and restoration. As biomonitoring is an obligation today, biomonitoring programs are framed by government organizations (e.g. European Commission, Joint Nature Conservation Committee in the United Kingdom). In its simplest form, biomonitoring consists of recording species diversity and abundances across different locations and times using a range of ecological census techniques and taxonomic identification. Most biomonitoring sampling methodologies were developed in the middle of the 20th century (Bohan et al., 2017) and were selected for entirely pragmatic reasons that reflected the current state of knowledge, simplicity and cost. Indicators are sampled to evaluate risks to human health and the environment for communication to the public or government policy makers. These include pesticide residues, elements and metabolites as pollution indicators, while abundances of target species or community descriptors are used to assess the ecological condition of ecosystems. However, these established methodologies are often of low generality. They are also often limited to particular ecosystems of species and communities of study and may not allow comparison between different systems. The evaluation of the myriad of changes in ecosystems that can occur is simply too costly, time-intensive and not necessarily captured by current biomonitoring indicators.

Consequently, biomonitoring of the full diversity of species and their interactions within an ecosystem is rarely, if ever, attempted (Bohan et al., 2017). While traditional biomonitoring is useful for simple conservation purposes such as identifying biodiversity hot spots or mapping 'functional gaps' in ecological communities (Forest et al., 2007; Myers et al., 2000; Raxworthy et al., 2003), such an approach is clearly not suited to the task of predicting the consequences of human actions that specifically target particular species or habitats. This is due to the fact that these human actions can have unintended consequences that spread through the network of species interactions at different spatial and temporal scales (Estes et al., 1998; Polis et al., 1997). For instance, traditional biomonitoring schemes have repeatedly failed at predicting the consequences of species introductions and have only just begun to look for guidance in interaction network approaches (David et al., 2017; Médoc et al., 2017; Pantel et al., 2017).

2.2 Ecological Networks Provide a Framework to Describe and Monitor Ecological Processes and Ecosystem Functions

Networks have become a prominent tool for studying community and ecosystem ecology, as they serve as a generic, conceptual framework for undertaking research across a broad range of ecological systems. Ecological networks, famously described by Darwin as the 'tangled bank', describe the interactions between species, the underlying structure of communities and the function and stability of ecosystems (Montoya et al., 2006). Historically, the ecologist Charles Elton pioneered the concepts of food chains and food webs, organizing species into functional groups (Elton, 1927; see also Cousins, 1987; Polis, 1991). These concepts formed the basis for ecologist Raymond Lindeman's classic and landmark paper on trophic dynamics (Lindeman, 1942). The examination of networks has then been spurred by now classic studies such as the keystone predation experiments and theory (Paine, 1966, 1969, 1974), the complexity–stability debate (Gilpin, 1975; MacArthur, 1955; May, 1972, 1973a,b) and the search for invariant patterns linking, for example, species diversity with the number of links in food webs (Briand and Cohen, 1984; Cohen and Briand, 1984; Cohen and Newman, 1985; Cohen et al., 1990; Pimm, 1980; Stenseth, 1985; Williams and Martinez, 2000, 2004). The past decade in particular has seen significant advances in the theoretical understanding, construction, analysis and application of complex species interactions networks (see Fontaine et al., 2011; Kéfi et al., 2012 for reviews). This area of ecology has been marked by two trends: (i) the building of more sophisticated models aimed at predicting and/or explaining the structure of ecological networks based on a variety of mechanisms (e.g. Allesina et al., 2008; Bascompte et al., 2003; Canard et al., 2012; Dalla Riva and Stouffer, 2016; Eklöf et al., 2013; Jordano, 1987; Jordano et al., 2003; Lewis and Law, 2007; Rohr et al., 2016; Williams and Martinez, 2000; reviewed in Kamenova et al., 2017) and (ii) the search for more precise data (in particular, taxonomic identification), and practical methods to obtain them, that has chiefly been done to counteract the tendency to lump together insufficiently described species that reduces the ability to make predictions and identify food web invariants (Novak et al., 2011; Solow and Beet, 1998; Yodzis, 1998). More recently,

Thompson et al. (2012) proposed using ecological networks as a conceptual framework to reconcile biodiversity and ecosystem function studies.

A network approach can be built on current biomonitoring schemes: if interaction data is collected alongside conventional monitoring of biodiversity, then it is possible to start monitoring both biodiversity and ecosystem functioning (see Mulder et al., 2006 for an example for soil microbial communities). For example, plant surveys could be complemented with insect flower visitation data to create plant–flower–visitor networks. Conversely, when pollinators are targeted by biomonitoring programs, the pollen carried by the species could be identified and used to create pollen-transport networks. These complementary approaches could be implemented in traditional biomonitoring methodologies and would give a better understanding of ecological processes through the construction of networks. Taking a step further, a combination of NGS and ENA together could provide a radically new approach to understand how environmental change affects ecosystems.

2.3 Ecological Network Structure Characterizes Ecosystem Properties

To measure changes in ecosystems, a wide range of metrics have been developed to encapsulate the emergent architecture of the networks (see Bersier et al., 2002). ENA relies on a wide range of network descriptors to assess the effect of environmental changes on ecosystem function. Ma et al. (2017) discuss descriptors of network complexity, such as connectance (a measure of network complexity), modularity (representing compartmentalization within the network) and nestedness (i.e. nodes with few connections linked to a subset of nodes interacting with more connected nodes) and their importance for detecting changes occurring in ecosystems (see Fortuna et al., 2010; Poisot and Gravel, 2014 for a critical view of some network metrics). Metrics of consumer–prey asymmetries are, in addition, very important to consider. The effect of environmental changes may vary across a food web (Thompson et al., 2012), and a change in an ecosystem may go undetected using measures of network complexity but nevertheless can affect consumer–prey asymmetries, with consequences on ecosystem function and services. Such asymmetries can be described as 'vulnerability' and 'generality' introduced by Schoener (1989) as, respectively, the mean number of consumers per prey and the mean number of prey per consumer within a food web. These consumer–prey asymmetry metrics are particularly well suited to the study of host–parasitoid networks (Derocles et al.,

2014a; Wirta et al., 2014). Other metrics of ecological network structure have also been proposed as determinants of ecosystem properties, such as the existence of fast and slow energy channels (Rooney et al., 2006), negative relations between interaction strength and the length (Neutel et al., 2002) of the trophic loop it is part of, or the frequency of network motifs (Stouffer et al., 2007).

Ecological processes such as pollination, pest control and seed dispersal are historically and still currently well studied in network ecology. These processes rely on mutualist and antagonist interactions with structural properties that can be characterized with network descriptors. Mutualist networks are, for example, often described as nested structures (Bascompte et al., 2003; Thébault and Fontaine, 2010). Network structure thus constitutes an efficient indicator of pollination quality (Kaiser-Bunbury et al., 2017). Similarly, a compartmentalized (or modular) structure often emerges from antagonist networks (Derocles et al., 2014a; Ma et al., 2017). Compartmentalized networks have important implications for natural pest control as they suggest a high specificity between the pest species and their natural enemies. With the current threat to food security (Godfray et al., 2010), ENA could help in our understanding of the underlying mechanisms involved in pest control and provide indicators to help agroecosystem management.

Nevertheless, characterizing ecosystem properties through ENA must be done with caution. Most networks metrics are highly dependent on sampling completeness (see Blüthgen et al., 2006; Jordano, 2016; Rivera-Hutinel et al., 2012). Consequently, the effort spent to sample and characterize an environment may directly affect the structure highlighted. Since DNA is ubiquitous in ecosystems, NGS constitutes a promising way to overcome the sampling completeness issues in ENA.

2.4 Knowledge of Ecological Networks Helps to Assess the Effect(s) of Environmental Changes on Ecosystem Processes and Associated Services

Ecological networks are increasingly (but not systematically) used to assess the effects of environmental changes on ecosystems as they provide a more complete description of ecological processes than conventional community or species-oriented approaches. For instance, Tylianakis et al. (2007) demonstrated that habitat modification altered the structure of networks of cavity-nesting bees, wasps and their parasitoids. The altered network structure had effects on parasitism rate, with consequences on ecosystem services such as pollination and biological control. A striking result from this study was that, despite only little observed variation in species richness, marked changes arose in network structure. Evans et al. (2013) demonstrated in an organic farm model system that two particular seminatural habitats (representing less than 5% of total area of the farm) were disproportionately important to maintain the integrity of the overall network, and thus of the associated ecosystem services (i.e. natural pest control, pollination). More recently, Kaiser-Bunbury et al. (2017) showed that ecosystem restoration in mountaintop communities affects the network structure in a positive way with a higher functional redundancy in restored communities. This modification of network architecture had direct and positive effects on the reproductive performance of the most abundant plant species. Thus, the development and application of ENA represent a paradigm shift in the biomonitoring of ecosystems (Kaiser-Bunbury and Blüthgen, 2015). However, empirical studies of this sort are still relatively rare in the literature, mainly because of the underlying network construction process. In particular, theoretical links between network structure and ecological function need to be better established. In this context, ecological network modelling has made some impressive progress in the understanding of ecosystem functioning. For example, the allometric food web model designed by Schneider et al. (2016) established the link between the diversity of animal communities and primary productivity. They demonstrated that diverse animal communities are more exploitative on plants but do not reduce plant biomass because this communities are composed of energetically more efficient plant and animal species. Network modelling such as the allometric food web model can therefore complement empirical studies. Consequently, more collaborative research between empirical and theoretical network ecologists is urgently needed and could be especially useful in helping to address a number global challenges, such as climate change, biodiversity loss and food security.

2.5 The Robustness of Networks of Ecological Networks: Applications for Understanding Species and Habitat Loss, Restoration and Building Ecosystem Resilience

The study of network 'robustness' (Dunne et al., 2002a,b; Memmott et al., 2004) has grown rapidly in recent years, partly driven by advances in computational modelling (Kaiser-Bunbury et al., 2010; Staniczenko et al., 2010), but mostly by the objective of understanding the threat of biodiversity loss to ecosystem services and functioning (Astegiano et al., 2015; Pocock et al., 2012). Studies have progressed from simple qualitative, bipartite mutualistic networks (Memmott et al., 2004), to investigations of patterns across eco-systems (Srinivasan et al., 2007) and to current quantitative approaches that take into account species abundance (Kaiser-Bunbury et al., 2010).

Pocock et al. (2012) constructed and analysed a 'network of ecological networks' (i.e. 11 groups of animals interacting with shared plants on farm-land), providing new analytical tools for understanding both the conse-quences of species extinctions across multiple animal groups, and the potential for ecological restoration. The study provided a method to calcu-late the relative importance of plants, and thus identified some plants that were disproportionately important in the network of networks (i.e. com-mon agricultural plants such as clover *Trifolium* and thistle *Cirsium* spp.). Although yet to be tested empirically, one application of this approach is that important plants could be targets for conservation and restoration that would benefit multiple animal groups. By examining the robustness of the joined networks, the study found that animal groups varied in their robustness to sequences of plant extinction, with the plant–pollinator network exhibiting much lower robustness than the seed-feeding bird network. Therefore, using a network approach, it should be possible to identify more sensitive groups for targeted conservation effort and/or assessment for biomonitoring rather than spending limited funds on charismatic species. Evans et al. (2013) developed this approach further by modelling the cascading effects of habitat loss, driven by plant extinctions, on the robustness of multiple animal groups. Habitat robustness analysis identified two seminatural habitats (i.e. waste ground and hedgerows together comprising <5% of the total area of the farm) as disproportionately important to the integrity of the overall network. This provides another tool for directing the management of multiple-habitat sites and landscape restoration, although it is yet to be tested empirically. Field and landscape-scale manipulations are required to both test and improve robustness models as a way of increasing the resilience of ecosystems.

More recently, Pilosof et al. (2017) demonstrated that the multilayer net-work from Pocock et al. (2012) provides much more realistic information on the stability and robustness of ecological communities than the examination of a single disconnected monolayer network (e.g. a bipartite host–parasitoid network). Parasitoid extinctions (representing a major aspect for the natural pest control) differ between scenarios purely based on the plant–parasitoid network and more comprehensive scenarios considering a multilayer net-work of both plant–parasitoid and plant–flower–visitor interactions. As flower visitors are involved in plant pollination, pollinator extinctions lead to secondary plant extinctions and tertiary parasitoid loss. This demonstrates that the biomonitoring of ecosystems cannot be realized reliably without considering the myriad of interactions occurring between organisms, as everything is connected in an ecosystem (Evans et al., 2017).

The robustness of interactions calculated from multilayer networks represents a powerful indicator of the ecological condition of an ecosystem and should therefore be developed further in the context of biomonitoring programs. Multilayer network approaches allow the long-term monitoring of the fragility of key components of ecological processes and ecosystem services such as plant–flower visitor networks (i.e. pollination) or insect pest–parasitoid networks (i.e. natural pest control) across spatial scales. With the development of ENA and the availability of NGS, we foresee a complementary use of traditional biomonitoring indicators (i.e. species richness, population surveys) with new indicators based on the architecture of ecological networks (in particular, the robustness) which are ultimately much more intimately linked to ecological processes.

3. ECOLOGICAL NETWORKS CAN BE CHALLENGING TO BUILD USING CONVENTIONAL APPROACHES

Despite their proven value in ecological research, networks are nevertheless limited by the difficulties of building them. These difficulties are centred around three major issues: (i) the sampling effort required to capture a significant range of species interactions; (ii) the reliable identification of specimens; and (iii) the adequate description of interactions between the organisms (see Box 1).

First, detecting the majority of species and their interactions within a network requires monumental effort. The challenges increase with the species richness in the ecosystem, the spatial scale of the habitat/ecosystem of interest and the temporal scale over which interactions are being considered. For example, the biodiversity of tropical ecosystems is much more difficult to assess accurately than its equivalent in arctic environments or temperate agroecosystems (Lewinsohn and Roslin, 2008; Morris et al., 2004), even if the latter is not trivial to study either (Derocles et al., 2014a, 2015; Evans et al., 2013; Macfadyen et al., 2009; Pocock et al., 2012; Wirta et al., 2014). Moreover, quantifying any aspect of species diversity in order to monitor environmental changes in biodiverse regions runs into major issues of scale-dependency (e.g. Dumbrell et al., 2008), raising further logistical challenges associated with repeatedly monitoring species diversity, while environmental changes modifying the spatial (and most likely temporal) scaling properties of species within these systems. These problems all arise from the sampling effort and associated logistical constraints required to detect a representative and significant proportion of the species living in the ecosystem (Gotelli and Chao, 2013; Gotelli and Colwell, 2011; Jordano, 2016). Furthermore, all species are not sampled equitably (and some of them simply cannot be sampled at all; Valentini et al., 2009b). For example, temporally transient species (e.g. due to migration or phenology), which when present may have a disproportional influence on network interactions, are almost always ignored. Thus, these issues all lead to a biased view of biodiversity, which favours reporting the presence of species that are the most conspicuous and easiest to sample. As sampling effort and completeness greatly impact the inferred structure of ecological networks, it is now usual to quantify network sampling completeness (see Costa et al., 2016) using estimators such as Chao 2 (Chao, 1984; Colwell and Coddington, 1994). This approach partially alleviates the sampling issues (assuming high sampling completeness is attained), but ecologists still need new tools for a more exhaustive detection of species and interactions.

Second, accurate species identification remains a major challenge, with two separate but related issues. The first issue is that accurate and reliable species identification requires specific taxonomic expertise for the studied group (Derocles et al., 2012a; Evans et al., 2016). Consequently, if multiple taxonomic groups are studied, many taxonomists may be required to assess the biodiversity within a network (Valentini et al., 2009a). Hence, reliable morphological identification may not always be possible for all taxa. For example, the existence of cryptic species (i.e. hard-to-identify species using morphological criteria) may lead to an underestimation of the species richness within ecosystems, resulting in biases at the network level (Derocles et al., 2016; Hebert et al., 2004; Kaartinen et al., 2010; Smith et al., 2006, 2007, 2008) and inaccurate model predictions (Novak et al., 2011). The second issue is that numerous taxa cannot be identified in situ (e.g. microbes) and require additional laboratory processing that is often limited due to financial constraints. In the case of microbial species, this is further hindered by the need to culture them in order to provide sufficient numbers for identification. This provides a major identification bias as most microbial taxa are not readily cultivable in the laboratory. In network ecology, misidentifications can be very problematic as they may bias the structure and distribution of interactions. As species may interact with numerous other organisms at different network levels, each identification error is compounded with each interaction across the network. Consequently, sufficient sampling and accurate identification are crucial steps in the construction of highly resolved ecological networks.

Third, exhaustively describing the possible range of species interactions that structure ecosystems is an onerous task (Bohan et al., 2013) and sampling all interactions of even a single type is conditioned by the number of observations (Blüthgen et al., 2008). While most species interactions are hard to identify in the field, some of them simply cannot be detected or observed with traditional sampling methods (see Jordano, 2016). As discussed by Gotelli and Colwell (2011), sampling biodiversity is very labour-intensive and often fails to detect most of the species in an ecosystem. For example, the construction of mutualist networks (e.g. plant–pollinator and plant–flower visitor interactions in Pocock et al., 2012) requires laborious and time-consuming field observations: it is therefore very hard to exhaustively capture mutualist interactions. Similarly in food webs built from host–parasitoid interactions (Derocles et al., 2012b; Gariepy et al., 2008), specimen sampling and rearing in the laboratory are imperfect for most taxonomic groups (Derocles et al., 2012b, 2015; Evans et al., 2016). For instance, rearing hosts sampled in the field until the emergence of adult parasitoids is a very challenging task. Indeed, both hosts and parasitoids have a high risk of dying during the rearing process, hence compromising the identification of host–parasitoid interactions. In webs based on prey–predator interactions, a 'Russian doll' effect may lead to the detection of false interactions from morphological (or molecular) identification of gut contents (Woodward et al., 2012). In this latter case, the prey is not directly consumed by the predator from which the gut content was analysed, but in the gut content of an intermediate consumer present in the focal predators gut. Consequently, overlooking these cases of secondary predation may lead to unrepresentative ecological networks.

Finally, some interactions cannot realistically be observed in the field despite providing valuable information on ecosystem services, such as seed–ground beetle interactions associated with weed regulation by Carabids (Bohan et al., 2011b), or belowground plant–microbe interactions, such as arbuscular mycorrhizae that influence terrestrial ecosystem productivity (Fitter et al., 2005). When relying purely on classic approaches (i.e. field observations, specimen rearing, morphological identification) to build ecological networks, the construction of networks becomes risky.

In order to limit the complexities and costs of describing complete ecological networks for a given ecosystem, most ecological network studies to date have assessed ecosystem function and services by studying a subsample of the network and focusing on particular types of interactions (e.g. mutualist or trophic interactions). The choice of subsampling (according to the question addressed) makes sampling, field observation, specimen rearing and taxonomic identification logistically possible. Focussing on a subset of interactions also illustrates the a priori expectations ecologists have on the underlying role of some taxonomic groups on ecosystem functions and services. Many ecological networks studies to date may better reflect the interactions which are easy to study or that ecologists think more important, rather than an actual representation of ecosystem functioning. This can lead to key aspects of networks being overlooked: there are still vast numbers of as yet 'unknown' interactions that need to be described and their role in ecosystem function evaluated. For example, in an agricultural network, a machine learning approach discovered an unexpected role for predatory spiders as prey (Bohan et al., 2011a; Tamaddoni-Nezhad et al., 2013), a finding confirmed by subsequent gut content analyses (Davey et al., 2013), giving a new mechanistic insight into the role of spiders in agroecosystems. The application of combined approaches, such as machine learning and NGS, that are less limited by the a priori expectations and assumptions of ecologists could greatly expand and speed up the discovery of links to build a more holistic and exhaustive view of ecological networks (Bohan et al., 2017).

Pocock et al. (2012) were among the first to assess multiple types of interactions that were pooled in a 'network of ecological networks' in the context of farmland ecosystem services and functioning, providing new insights into the robustness of these interconnected networks (Evans et al., 2013). These networks of networks were built using conventional methodologies that rely on field observations or rearing specimens followed by morphological identification by taxonomists. Although species interactions were highly resolved and well quantified for many of the subnetworks (e.g. plant–insect pollinators), others were potentially subject to bias (e.g. plant–leafminer–parasitoids) because of the limitations of taxonomically selective rearing success and the reliance on accurate morphological identification. Given that the construction of such networks is labour-intensive, building larger, highly resolved ecological networks in a wide range of ecosystems is likely to be hindered until more cost-effective methodologies can be developed. The application of NGS technology is one such method that is likely to revolutionize network ecology.

Advances in DNA-sequencing technologies are answering previously intractable questions in functional and taxonomic biodiversity and provide

enormous potential to determine hitherto difficult to observe species interactions. Thus, DNA-based approaches, NGS in particular, hold the potential to provide many of the solutions to the problems described earlier
(Bohan et al., 2017; Evans et al., 2016; Vacher et al., 2016). Combining
DNA-barcoding technologies with ENA offers important new opportunities for understanding large-scale ecological and evolutionary processes (such
as invasive species, see Kamenova et al., 2017), as well as providing powerful
tools for building ecosystems that are resilient to environmental change (see
Evans et al., 2016 for a conceptual framework). Until recently, ecological
networks represented therefore a powerful but challenging approach to
establish and were consequently difficult to integrate in biomonitoring.
As discussed in the next section, NGS technologies together with the prediction of interactions with statistical modelling and machine learning represent now an exciting opportunity to include ENA more systematically
into biomonitoring.

4. COMBINING NGS WITH ENA: OPPORTUNITIES AND CHALLENGES

4.1 Using NGS to Construct Ecological Networks

Currently, ecological networks constructed using DNA-based approaches
are not used to regularly monitor ecosystems. This may be partially due
to the historical reliance on classic field survey methods in network ecology,
which rely on observation, specimen sampling, laboratory rearing and morphological identification to construct bipartite networks. Recent work has
demonstrated that NGS can be rapid, universal and relatively cheap, in comparison to conventional (i.e. 'traditional' taxonomy based) approaches to
assess biodiversity (Beng et al., 2016; Ji et al., 2013; Liu et al., 2013). Beyond
the characterization of biodiversity, NGS can also be used to efficiently build
ecological networks (see Evans et al., 2016; Toju et al., 2014; Vacher et al.,
2016). First, NGS has the potential to directly establish species interactions
(Evans et al., 2016; Kitson et al., 2016; Toju et al., 2014). Second, with
metabarcoding and eDNA approaches, NGS can also generate millions of
DNA sequences which then can be processed and used in statistical models
to construct ecological networks (Vacher et al., 2016).

However, molecular approaches and NGS in particular are yet to be
widely used to build ecological networks. Non-NGS molecular approaches
such as diagnostic PCRs (using taxon-specific primers to detect targeted
species in samples) and Sanger sequencing approaches (see Table 2)

Table 2 Comparison of the Main Sequencing Technologies

Sequencing Platform	Sequencing Generation	Amplification Method	Sequencing Method	Read Length (bp)	Error Rate (%)	Error Type	Number of Reads Per Run	Time Per Run (Hours)	Cost Per Million Bases (USD)
Sanger ABI 3730xl	1	PCR	Dideoxy chain termination	600–1000	0.001	Indel–Substitution	96	0.5–3	500
Ion Torrent	2	PCR	Polymerase synthesis	200	1	Indel	8.2×10^7	2–4	0.10
454 Roche GS FLX+	2	PCR	Pyrosequencing	700	1	Indel	1×10^6	23	8.57
Illumina HiSeq 2500; high output	2	PCR	Synthesis	2×125	0.1	Substitution	8×10^9 (paired)	7–60	0.03
Illumina HiSeq 2500; rapid run	2	PCR	Synthesis	2×250	0.1	Substitution	1.2×10^9 (paired)	24–144	0.04
Illumina MiSeq v3	2	PCR	Synthesis	2×300	0.1	Substitution	3×10^8	27	0.15
SOLiD 5500xl	2	PCR	Ligation	2×60	5	Substitution	8×10^8	144	0.11
PacBio RS II: P6-C4	3	Real-time single-molecule template	Synthesis	~10,000–15,000	13	Indel	$3.5–7.5 \times 10^4$	0.5–4	0.40–0.80
Oxford Nanopore MinION	3	None	Nanopore	~2000–5000	38	Indel–Substitution	$1.1–4.7 \times 10^4$	50	6.44–17.90

Based on Schendure, J., Ji, H., 2008. Next-generation DNA sequencing. Nat. Biotechnol. 26, 1135–1445; Glenn, T.C., 2011. Field guide to next-generation DNA sequencers. Mol. Ecol. Resour. 11, 759–769; Niedringhaus, T.P., Milanova, D., Kerby, M.B., Snyder, M.P., Barron, A.E., 2011. Landscape of next-generation sequencing technologies. Anal Chem. 83, 4327–4341; Liu, L., Li, Y., Li, S., Hu, N., He, Y., Pong, R., Lin, D., Lu, L., Law, M., 2012. Comparison of next-generation sequencing systems. J. Biomed. Biotechnol. 2012, 251364; Escobar-Zepeda, A., Vera-Ponce de León, A., Sanchez-Flores, A., 2015. The road to metagenomics: from microbiology to DNA sequencing technologies and bioinformatics. Front. Genet. 6, 348; Rhoads, A., Au, K.F., 2015. PacBio sequencing and its applications. Genomics Proteomics Bioinformatics 13, 278–289; Weirather, J.L., de Cesare, M., Wang, Y., Piazza, P., Sebastiano, V., Wang, X.-J., Buck, D., Au, K.F., 2017. Comprehensive comparison of Pacific Biosciences and Oxford Nanopore Technologies and their applications to transcriptome analysis. F1000Res. 6, 100.

remain more intuitive and easier for network ecologists to understand than NGS (Derocles et al., 2014a, 2015; Traugott et al., 2008; Wirta et al., 2014). These two explanations focus on why network ecologists have yet to fully embrace NGS approaches over more traditional methods. The flip side to this argument is that molecular ecologists using NGS for metabarcoding studies have yet to fully realize the potential of the data they generate. The vast majority of NGS studies quantifying the diversity of ecological communities have heavily relied on descriptive statistics based on classical measures of community diversity, and/or changes in species composition between samples. However, data are often collected in such a way that (ecological) networks could be constructed, but are not, and the vast potential of the NGS data thus remains unrealized.

Conventional approaches to ecological network construction have some major drawbacks that could make them inefficient in the biomonitoring of ecosystems. Visual species identification (with or without microscopy) can sometimes be slow and labour-intensive at best, or unreliable at worst. Diagnostic PCRs need very good prior knowledge of the species composition of the ecosystem monitored, because they require the design of multiple PCR primers to detect the full range of species. For Sanger sequencing, costs increase linearly with experiment size and quickly become too expensive for large-scale biomonitoring.

In contrast, NGS metabarcoding may scale more efficiently to large samples compared with microscopy, diagnostic PCRs and Sanger sequencing, providing opportunities for much more intensive sampling of species interaction networks than has previously been possible. The investment in time and materials goes 'per plate' (96, 384 or 1536 samples) rather than 'per sample', although for large numbers of samples, additional sequencing runs may be required, which increases the cost. However, the number of samples that can be processed with a single sequencing run varies widely depending on a range of factors, including sequencing technology, chemistry techniques, and the quality of DNA extraction and amplification procedures (see Table 2). While in principle highly useful, it is these technical, and some of the theoretical issues linked to the use of NGS to quantify interactions that may have limited their adoption by researchers. We discuss ways of overcoming these limitations below and foresee that the construction of ecological networks using NGS will soon become commonplace and be integrated into biomonitoring schemes.

4.2 PCR Bias and Abundance Estimation in NGS Community Analyses

Understanding the limitations of molecular approaches in detecting species interactions is fundamentally important when correctly designing a sequencing approach and interpreting the produced network. Primer and amplification biases are well-known phenomena in the PCR. Mismatches between primer and binding site sequences or structural and compositional variation in the DNA strand can lead to variation in PCR efficiency (Polz and Cavanaugh, 1998), causing two distinct issues, respectively: (1) the inability to detect species present within the sample as the primer mismatches exclude their detection, and (2) the preferential amplification of DNA from some species at the expense of amplifying DNA from others (a cloning step is added to separate mixed DNA sample; e.g. Dunshea, 2009). However, for metabarcoding where the aim is to parallel sequence an entire community (or to identify two parties in an ecological interaction), this becomes more critical as differing PCR efficiencies among species can result in a final PCR product composition that is not representative of the input DNA composition (although some authors have reported broad correlations e.g. Elbrecht and Leese, 2015; Leray and Knowlton, 2017; Razgour et al., 2011). In practical terms, biases can lead to false negatives and read depths that are of no use for determining quantitative or even relative community composition (Elbrecht and Leese, 2015; Leray and Knowlton, 2017; Piñol et al., 2014). The most common approach to dealing with this is to develop PCR primers that are as general and unbiased as possible (Elbrecht and Leese, 2015; Leray et al., 2013), but even these are prone to the above-mentioned biases and a certain degree of PCR-induced bias is now commonly acknowledged in PCR-based metabarcoding studies (Leray and Knowlton, 2017). That said, in some instances by using carefully designed primers and targeting genes which vary in base-pair composition but not structural properties among species, elimination of PCR amplification biases is entirely possible (Cotton et al., 2014), and researchers should continue to pursue development and validation of these unbiased approaches.

Some authors have attempted to circumvent this by using a metagenomic approach (e.g. Tang et al., 2015) where they sequence all DNA present in their extraction and then filter the resulting sequences to only retain the data of use for identifying species. In theory, with no PCR step, there is no amplification bias, so read depths are more representative of the input DNA composition. In practice, the relationship between community composition and metagenomic read depth is not so simple. The availability of mitochondrial DNA (mtDNA) for extraction varies significantly with tissue mass and metabolic activity (e.g. there is a significant nonlinear increase in mitochondrial count in developing oocytes; Cotterill et al., 2013), and this relationship can be further modified by tissue type and age (Barazzoni et al., 2000). This bias concerns mtDNA (mainly used to identify animals), but similar issues surround plastid DNA (used for plant identification), and the overall metagenome is often swamped with 16S ribosomal DNA gene reads (used for microbe identification), which may mask the presence of rarer higher taxa. Even if it were possible to know how read counts vary with tissue mass/type/age for all the organisms in our community (e.g. the contribution of multicellular organisms to eDNA has been modelled by this chapter; Sommeria-Klein et al., 2016), relative read counts can be further skewed by differences in ease of DNA extraction across taxa (Schiebelhut et al., 2017) and the extraction method used to obtain the DNA (Deiner et al., 2015; Vesty et al., 2017). Taken together we are forced to conclude that, as currently performed, metabarcoding is not generally suitable for estimating tissue biomass from sequence data (Clare, 2014) and thus any such metabarcoding-based estimations would have to be idiosyncratically calibrated using conventional abundance surveys (as in Tang et al., 2015). Estimation of abundances is also problematic except for single-celled organisms for which they can be assessed accurately by targeting genes with minimum amplification biases (see Fischer et al., 2017). For higher taxa, however, only relative abundances can be recovered from NGS data. Relative abundances may nevertheless have a limited use in a conservative biology perspective and thus in biomonitoring as well (Clare, 2014).

4.3 NGS Without a Prior PCR Step

A significant drawback of the metagenomic approach is cost (see Table 2). By sequencing all DNA in an extraction, researchers greatly limit the sample size per sequencing run and thereby either increase the sequencing costs for the study or, in fixed cost studies, reduce the statistical power of the study dramatically. They also discard much of the available read depth when they filter reads to only those of direct interest. One general approach to solve this is to enrich the DNA to be sequenced for a specific genomic region without PCR, avoiding PCR bias. This can be achieved in several ways, but the most common is to use an hybridization approach that employs sets of degenerate probes that can bind to target DNA and then themselves be bound to magnetic beads (Gnirke et al., 2009) or a solid substrate (Albert et al., 2007) with nontarget DNA simply being washed away. The enriched DNA is biased towards useful genomic regions so a smaller proportion of the sequencing reads are discarded and more samples can be included in a sequencing run (see 'sequencing coverage' in the glossary). Variations of this approach exist, which range from very simple centrifugation-based approaches (Macher et al., 2017) to much more complex methodologies using isothermal DNA replication (Dapprich et al., 2016) that can allow researchers to enrich for extremely long genomic regions suitable for the latest sequencing technologies such as Pacific Bioscience (PacBio) Single Molecule, Real-Time (SMRT) sequencing and Oxford Nanopore MinION (see Table 2). Depending on the laboratory, these approaches can be highly scalable, but their utility for community assessment is yet to be proven.

4.4 Detection of Species Interactions Using Molecular Tools

As an alternative to attempting to infer relative abundances via read depths of bulk extracted communities, it is possible to simply analyse individual organisms separately and link the metadata for each sample (i.e. individual). In this situation, the number of samples for each species is a proxy measure of relative abundance. If the sample contains multiple DNA templates arising from a species interaction (e.g. predator gut contents, host/parasite systems or plant–pollinator systems), then it is possible to use molecular tools to detect these species interactions in a quantitative or semiquantitative manner. Prior to the advent of parallel sequencing technologies, this would have been achieved by one of the following two broad approaches. First, PCR diagnostic approaches use sets of primer pairs that each produce species-specific bands of different lengths (or with different attached fluorophores as in microsatellite genotyping) that can then be separated by gel or capillary electrophoresis (e.g. aphid/parasitoid interactions Traugott et al., 2008; predatory beetle gut contents King et al., 2011). Second, PCR amplification of all DNA in a sample using general primers, separating PCR products via cloning (e.g. Dunshea, 2009) or gel electrophoresis, followed by Sanger sequencing (e.g. Kitson et al., 2013). Third, the design of primers specific to a taxonomic group (e.g. a parasitoid family or a genus of prey) amplifying a short but variable region (such as a fragment of COI) is another approach to resolve species interactions (Derocles et al., 2012b; Fayle et al., 2015). This method allows identification of an interaction that is occurring by relying first on a PCR diagnostic (e.g. a parasitoid within a host, a prey with a gut content) and then to identify the nature of the interaction by sequencing the organisms detected (Derocles et al., 2012b; Rougerie et al., 2011). Thus, this approach was successfully applied to build ecological networks in a farmland system (Derocles et al., 2014a) and an arctic system (Wirta et al., 2014). However, because of the linear cost of the Sanger sequencing and/or the time to process samples with these molecular tools, these approaches allow to examine a relative limited number of organisms (e.g. a low number of hosts).

The advent of NGS technologies allows researchers to parallelize this process and work more effectively on a larger scale. The use of PCR primers tagged with known sequences to track samples is well established in NGS (Binladen et al., 2007), and this has been shown to be effective not only for community metabarcoding (Yu et al., 2012) but can also be used to build webs. Toju et al. (2013, 2014) were one of the first to use NGS (454 pyrosequencing) to resolve species interactions between trees and arbuscular mycorrhizal fungi and then build ecological networks from that data. One step further, Shokralla et al. (2015) and Cruaud et al. (2017) showed that a 'nested tagging' approach for amplicons involving two rounds of PCR permits (see nested PCR in the glossary) extensive multiplexing to increase throughput of barcoding programs, and Evans et al. (2016) have proposed this as an approach to building larger, replicated networks in ecological studies. In the future, it is likely that these sorts of nested tagging approaches will be combined with PCR-free approaches to sequencing to allow quantified networks to be produced while reducing concerns over PCR bias and missing interactions caused by false negatives.

All the molecular approaches described earlier represent tools able to rapidly characterize the biodiversity of ecosystems or describe species interactions. There is no doubt that this area of research is expending very rapidly as that new advances must be expected, pushing the limits of the description of biodiversity and the understanding of ecosystems. In the future, we believe that NGS will be fully integrated by ecologists to build networks and will be a usual approach of biomonitoring programs.

4.5 How to Deal With Interactions Not Directly Resolved by NGS: Are Species Association Networks Species Interaction Networks? The Case of Microorganisms

For several decades, ecological networks have been constructed from the observations of both the species and their interactions (Ings et al., 2009; Poisot et al., 2016b). Databases of observation-based ecological networks, such as plant–pollinator and predator–prey interaction networks, have been compiled (e.g. Interaction web database IWDB; https://www.nceas.ucsb.edu/interactionweb/) and used to describe the architecture of biodiversity (Bascompte and Jordano, 2007; Bascompte et al., 2003; Jordano et al., 2003; Lewinsohn et al., 2006; Olesen et al., 2007), to understand how species have assembled through time (Peralta, 2016; Rezende et al., 2007; Vacher et al., 2008), to elucidate the network properties that sustain species coexistence (Allesina and Tang, 2012; Tang et al., 2014; Thébault and Fontaine, 2010) and to predict the resistance and resilience of ecosystem functions to environmental change (Memmott et al., 2004; Schleuning et al., 2016; Vanbergen et al., 2017). These observation-based networks, however, often remain incomplete (Chacoff et al., 2012; Jordano, 2016), despite intense efforts to make them as realistic as possible by merging interaction types (Fontaine et al., 2011; Genrich et al., 2017; Kéfi et al., 2012, 2016; Melián et al., 2009; Pocock et al., 2012), by linking networks occurring at different times and in different sites ('multilayer networks' or 'metawebs'; Pilosof et al., 2017; Poisot et al., 2012) and by linking them to the functioning of human societies (The QUINTESSENCE Consortium, 2016). Sampling effort has been shown to significantly influence some whole-network properties (Blüthgen et al., 2008; Costa et al., 2016), while the integration of new sets of species, such as parasites, can completely disrupt the architecture of ecological networks (Hudson et al., 2006; Lafferty et al., 2008). This lack of integration of small organisms (e.g. bacteria, fungi) is a shortcoming of the ecological network literature, since these small organisms represent a major part of the Earth biodiversity in terms of species number and biomass (Dobson et al., 2008; Hawksworth, 2001; van der Heijden et al., 2008; Whitman et al., 1998).

The emergence of metabarcoding and NGS in the past decade has given us a chance to fill this gap but raised new issues (Bálint et al., 2016). Metabarcoding approaches revolutionized the field of microbial ecology because they gave us access to the composition of whole microbial communities, including noncultivable microorganisms (Hibbett et al., 2009; Peay et al., 2008). However, metabarcoding approaches only detect molecular species named operational taxonomic units (OTUs). They do not detect ecological interactions. A current challenge is therefore to reconstruct species interactions networks from species (relative) abundance data (Fig. 2), such as those obtained from metabarcoding techniques (Abreu and Taga, 2016; Biswas et al., 2016; Faust and Raes, 2012; Layeghifard et al., 2017; Vacher et al., 2016). New theoretical frameworks have been developed by community ecologists to tackle this issue (Cazelles et al., 2016;

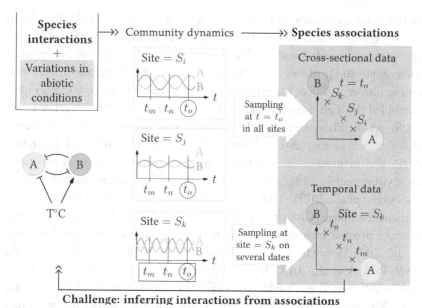

Challenge: inferring interactions from associations

Fig. 2 Challenges to overcome in order to reconstruct species interaction networks from species abundance patterns. The structure of species interaction networks and the variations in the abiotic environment trigger spatial and temporal variations in species abundances. These variations can be captured by sampling several geographical sites at one sampling date (*circled*), by sampling one site at multiple dates (*framed*) or by combining both sampling designs (not shown). The challenge is to reconstruct the species interaction network (often unknown) from the variables and relationships that can be measured (*in grey*), including species association statistics and abiotic environmental variables. Knowledge only of the species associations is not sufficient, as illustrated in Fig. 3.

Ovaskainen et al., 2016, 2017a,b), and software adapted to metabarcoding data has been developed concomitantly (Bucci et al., 2016; Deng et al., 2012; Faust et al., 2015a; Friedman and Alm, 2012; Kurtz et al., 2015; Li et al., 2016; Shang et al., 2017; Weiss et al., 2016). These methods all produce species association networks, where a link between two species represents a significant statistical association between their abundances. This is a critical issue, because species association networks differ from species interaction networks (Fig. 3). Species association networks, which are now very popular in the field of microbial ecology, usually have two type of links, positive and negative (Agler et al., 2016; Faust et al., 2015b; Jakuschkin et al., 2016). Realistic species interaction networks have multiple types of links, including mutualism, commensalism, parasitism, predation, amensalism and competition (Faust and Raes, 2012).

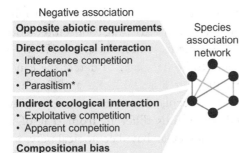

Fig. 3 Why species association networks are not species interaction networks. Significant associations between species abundances can be triggered by ecological interactions, but also by species abiotic requirements and methodological biases. Moreover, linking statistical association sign with ecological interaction type is not straightforward (*the cases of predation and parasitism are discussed in the text). These issues must be overcome to reconstruct species interaction networks from species abundance data, such as those obtained from metabarcoding approaches.

The relationship between statistical associations and ecological interactions is not straightforward (Fig. 3). For instance, parasitism is rarely captured as a significant link in microbial association networks and when it is, it is retrieved as a positive link despite the detrimental effect of the parasite on its host (Weiss et al., 2016). This might be because the copresence of the host species and the parasite species is necessary for the interaction to occur: analysing together samples where neither the host nor the parasite are present, and samples where they both interact, tends to create a positive relationship between the abundances of the host and the parasite species in cross-sectional datasets. The same reasoning holds for predator–prey interactions (Fig. 3). Predation and parasitism are also expected to trigger various types of association in the case of time series. Indeed, in resource–consumer interactions, species evolution can alter the dynamics of the interacting populations (Bengfort et al., 2017; Hiltunen et al., 2014). It can therefore modulate the statistical association between species abundances, especially in the case of fast-evolving species such as microorganisms. Eco-evolutionary models suggest that when the resource (i.e. the prey or the host) evolves faster than its interacting partner, the two-species dynamics exhibit antiphase cycles (Hiltunen et al., 2014). Such dynamics are expected to yield a negative association between species abundances in time series. In contrast, no association is expected if none of the species evolve or if both evolve at the same rate, because their dynamics would then be lagged by a quarter-period (Hiltunen et al., 2014). In real systems, both species dynamics and evolution are actually shaped by the complex network of trophic and nontrophic interactions. Current methods generally fail at retrieving robust interaction signals from these complex dynamic systems (Sander et al., 2017; Weiss et al., 2016). New frameworks seem to be necessary to detect the causes of real ecosystems dynamics (Sugihara et al., 2012; Ye and Sugihara, 2016). The meaning of a link in a molecular-based ecological network reconstructed using current methods is thus different from that of an observation-based network: links of molecular-based ecological networks represent statistical associations (that may be triggered by biotic interactions), but they do not represent true species interactions. Because of this conceptual difference, the knowledge derived from observation-based interaction networks hardly applies to molecular-based association networks. A new line of research, using molecular-based association networks as indicators of plant, animal and ecosystem health, has therefore emerged (Bohan et al., 2017; Karimi et al., 2017; Poudel et al., 2016). To get a more complete view of ecosystem structure and function, future research should strengthen the synergies between observation-based interaction networks (involving macroorganisms) and molecular-based association networks (involving microorganisms).

Finally, even though molecular-based association networks concern predominantly microorganisms, advances in this area of research go well beyond purely microbial networks. In a biomonitoring scheme building networks based on eDNA, the issues related to the link between species association networks and species interaction networks are valid for any kind of network constructed. Consequently, recovering accurate species interaction networks from species association networks represents a major issue today in order to rapidly construct networks from monitored environments.

5. MACHINE LEARNING AS A WAY TO RAPIDLY BUILD MOLECULAR ECOLOGICAL NETWORKS IN A RAPID AND RELIABLE WAY?

5.1 Learning Ecological Networks From Data

A basic premise of ecology is that the variation in the observed data, through sampling ecosystems, contains information about past and current ecological interactions between species in the ecosystem. Thus, the abundance and variation of any one species are in part a consequence of past interactions between the individuals of previous generations, such as sexual reproduction, and of the current generation, such as competition, cannibalism or migration. Across a community the observed abundance, phenotypic variation and diversity of species are further determined by ecological processes and interactions between species, including trophic and competition interactions. In analysing variation in sample data, ecologists aim to recover some information about these interactions, often using data sampled at particular times or in manipulative experiments that target specific interactions.

More recently, these ideas have been extended to simultaneously building whole ecological networks of interactions from data using statistical or logic-based machine learning. The idea behind these machine learning methods is simply that already used by ecologists that embedded in a dataset is the imprint of the recent processes and interactions that created the data and this information can be recovered to reconstruct networks. The underlying hypothesis of machine learning for network reconstruction is therefore that ecological interactions produce correlations and relational patterns in the abundance of species that can be recovered. In statistical machine learning, the variation in the sample is treated statistically, typically using Bayesian approaches (Jakuschkin et al., 2016; Vacher et al., 2016). Significant correlations between any given pair of species within the data are then considered as potential network edges. Logic-based machine learning treats relational patterns rather like the structure of grammar in a language (Muggleton, 1991; Tamaddoni-Nezhad et al., 2006).

In both statistical and logic-based machine learning, the trick is to sort, from the list of edges (i.e. interactions) hypothesized by the learning algorithms, those links that are ecologically meaningful from those that are artefacts. This selection approach is done differently in the two approaches. In logic-based machine learning, the grammar for an interaction can be coded as background information. In the agroecological network learnt by Bohan et al. (2011a) and Tamaddoni-Nezhad et al. (2013), trophic interactions were selected by background information that was a set of grammar rules (a model) for a trophic interaction whereby the predator and prey species must cooccur in the same samples and predators should be larger than their prey (in this case big things eat small things, but that is not always the case in nature). A trophic interaction between two species was only identified if this grammar rule was realized. In statistical approaches, links are selected using environmental factors or species functional trait covariates integrated into the modelling (Cazelles et al., 2016; Jakuschkin et al., 2016; Ovaskainen et al., 2016; Vacher et al., 2016).

Learning networks is currently limited by our background information rules. Mechanistic rules for trophic interactions, based upon body size or gape size, allow the reconstruction of food webs. Where ecological networks are structured by processes for which we have no general mechanistic explanation, there is no background information that can be employed, and machine learning is of little value for reconstructing networks. However, recent developments in logical machine learning are now allowing background information rules to be discovered from data. Tamaddoni-Nezhad et al. (2015) showed using simple subnetworks that the trophic interaction rule 'big things eat small things' can be recovered. Developments of this work are now extending this possibility of rule learning to larger and noisier data sets (Dai et al., 2015).

5.2 Exploiting eDNA-Derived Information as a Source for Network Data

While machine learning could greatly speed up the process of constructing ecological networks, the abundance data used for the reconstruction are currently something of a bottleneck. Highly replicated and taxonomically resolved data sets, such as the farm-scale evaluations data used by Bohan et al. (2011a) and Tamaddoni-Nezhad et al. (2013) to reconstruct an agroecological network, are few and costly to create. One solution will be to move towards assessing the presence of species, and potentially their relative abundance and variation, using highly replicated samples of DNA taken from the environment. These eDNA samples, which in principle could be sampled (relatively) easily and cheaply, could then be used to identify the taxa present at a sampling point using NGS approaches. NGS describes a number of similar technologies for generating large numbers of nucleic acid sequences for the identification of species (OTUs) and functions. The great beauty of these methods is that the nucleic acids with which they work are common to all life forms and ubiquitous. The fact that NGS might be applied to the identification of OTUs and functions in environmental samples from any biome, habitat and environment and any source material with minimal change in protocol has driven interest in eDNA as a generic source of data (Barnes and Turner, 2015; Evans et al., 2016; Thomsen and Willerslev, 2015). Coupling machine learning and NGS data could greatly speed up the reconstruction of networks in all ecosystems.

The raw OTU information that would be produced by eDNA sampling would contain all the interactions that structure the data. Treating this complexity of information has been highly problematic and difficult to date, and thus network researchers have tended to use DNA data in which many of the interactions are filtered out. Probably the best example is gut content data in which the sample is effectively an individual predator, and the data are the OTUs contained in the predator's gut (e.g. Fayle et al., 2015). Adopting this predator-derived data effectively selects for realized links, without a process of learning, and allows trophic ecological networks to be constructed directly. While learning approaches might be applied to these data to learn something about the processes underscoring trophic interactions, such as the traits that make some species predators of particular prey via background information, these gut content DNA samples are essentially limited to describing food webs which alone cannot explain community-level species cooccurrence and ecosystem structure (Bohan et al., 2013; Pocock et al., 2012). DNA analysis of gut content samples (or other type of DNA samples allowing a direct highlight of interactions such as host parasitized or faeces) is nevertheless mandatory as it allows to confirm the predictions from machine learning. In biomonitoring using NGS networks established with learning approach (or statistical models), it would be necessary that new interactions discovered by machine learning are systematically tested (with molecular tools or with experiments) for an accurate understanding of ecosystems.

The challenge for the future is to use appropriate machine learning and background information to tackle the problem of the many interactions that in combination have created the eDNA data we observe. While the best machine learning approaches have yet to be determined (Bohan et al., 2017), it is clear that background information will play a pivotal role here. The background information 'model' proposed by Bohan et al. (2011a) and Tamaddoni-Nezhad et al. (2013) for their agroecological network was essentially a model for a hypothetical trophic interaction and consequently selected interactions that conformed to this—trophic interactions. Subsequent work showed that a logic-based machine learning approach, called metainterpretative learning (MIL) (Muggleton et al., 2014, 2015), could discover background information directly from data. MIL was demonstrated discovering the rule used by Bohan et al. (2011a) and Tamaddoni-Nezhad et al. (2013) that 'big things eat small things' directly from a simulated, synthetic food web (Tamaddoni-Nezhad et al., 2015). This holds out the exciting possibility that reconstruction of an ever greater number of ecological networks, from eDNA, will drive a rapid improvement in our understanding of ecosystem structure and function because it will require the discovery of the background information models—the generic rules—that describe/determine the ecological interactions that structure all ecosystems.

6. NGS NETWORK DATA SHARING

With NGS approaches, network ecologists will be able to rapidly generate a large amount of data, including both DNA sequences and putative species interactions (interactions derived from machine learning must be validated through observation and experiment, see Lima-Mendez et al., 2015). This means that data curation and sharing between researchers must be done systematically for both DNA sequences and species interactions. Currently, sequences data are available publicly to researchers through databases such as Genbank, BOLD (Ratnasingham and Hebert, 2007), European Nucleotide Archive (for raw NGS data, see Leinonen et al., 2011; Nilsson et al., 2011). In network construction, reliable species identifications are directly dependent on the quality of these databases (Sonet et al., 2013; Wells and Stevens, 2008). First, databases of DNA sequences must ideally be representative and comprehensive to cover the identification of all the species entangled in the ecosystem studied. This is particularly important as relying only on OTUs would certainly allow to build networks and to describe the architecture of the ecosystems, but would also miss the important ecological functions of the organisms found interacting. For example, this was the case in the plant–fungi network built by Toju et al. (2014). In this very impressive network gathering 33 plant species and 387 fungal OTUs, only 26.4% of the OTUs were identified to the genus level and the ecological function remained to be determined for an important part of the fungi detected (Toju et al., 2013, 2014). Although, as more complete genomes become described and deposited in databases, it may be possible to infer ecological functions from OTU data. While still in its infancy, this approach has been realized via the PICRUSt algorithm, which infers complete genome data from 16S rRNA gene sequences used to define bacterial OTUs, linking the functional data held within genomes to their corresponding 16S sequences (Langille et al., 2013). Other approaches, which use trait-based databases to map taxonomic identities inferred from OTU data to that taxon's function, also exist (e.g. FUNGuild; Nguyen et al., 2016), and these will become increasingly important as the quality and quantity held within expands. We emphasize that the completeness of databases will improve over time through the increasing use of sequencing technologies and will therefore enhance the reliability of species identifications and their function. Second, the validity of NGS networks also depends on the frequency of misidentifications and sequencing errors in databases (Wells and Sperling, 2000) because misidentifications from public libraries can spread in network construction. Taxonomic databases such as BOLD (Ratnasingham and Hebert, 2007) nevertheless limit these risks since DNA sequences (i.e. DNA barcodes in BOLD) are associated with specimen vouchers. This practice allows reexamination of specimens after DNA sequencing and also gives information such as sampling location, photography, sequencing laboratory or specimen depository. Despite their imperfections, public libraries are improving over time and there is a general consensus to share DNA-sequencing data which improve DNA-based research in a wide range of areas.

In ecological network research, data sharing is, however, not a systematic practice. Ecologists generate together massive high-value data through individual projects. However, data curation and sharing are still an important issue to overcome to enable ecologists to take full advantage of existing data as well as future generated data (see Hampton et al., 2013; Poisot et al., 2016a,b). IWDB is an example of data repository where nearly a hundred of webs are available today, half of them being plant–pollinator networks. Mangal (Poisot et al., 2016a,b) has been designed (with an R package [rmangal]) to access, curate and deposit data on ecological interactions. The Global Biotic Interactions (GloBI) database (Poelen et al., 2014) is an open structure to share and analyse species interactions in a structured data repository. Ecological data, networks included, can also be found on non-specific repositories such as Dryad or Figshare. One important challenge for these data is to conform to some common presentation standards, despite attempts at providing a common format for open access ecological network data (Poisot et al., 2016a,b).

With the rapid increase of studies using NGS to construct ecological networks, need of a common way for molecular network ecologists to store and share their highly valuable data, which includes both DNA sequences and species interactions (or at least species cooccurrence), has never been of greater importance.

6.1 The Importance of a Dedicated NGS Network Database: Linking DNA Sequences and Ecological Interactions to Limit Species Identification Errors

While it is still possible to rely on existing public repositories for DNA sequences (e.g. Genbank) and ecological data (e.g. Dryad or IWDB) to store and share NGS networks, a common way to register NGS networks would be considerably beneficial for network ecologists. Indeed, there is a need to be able to recover directly, easily and reliably the links existing between

DNA sequences and species interactions (or at least cooccurrence) within NGS networks. To date, there is no single environment providing these features. The field could build upon the standard file format (e.g. BIOM) to efficiently store OTU table associated with metadata concerning samples (e.g. treatment) or species (e.g. taxonomy). Public data-sharing initiatives have sprouted both from microbial ecology (e.g. QIITA platform; qiita. microbio.me.) and ecology (e.g. mangal; Poisot et al., 2016a,b), but NGS networks would still need to connect both fields. Taking inspiration from current databases, a systematic storage of DNA sequences and ecological interactions together would allow network ecologists to rapidly recreate the NGS networks published, compare them with their own ecological networks and make further analysis possible.

In particular, this would allow the reexamination of existing networks. Because accurate identification of species and interactions directly affect the network structure (e.g. Wirta et al., 2014), it seems important in the future to be able to reassign the species or at least the OTUs in the NGS networks created. Because species delimitation methods are improving (Kekkonen and Hebert, 2014; Puillandre et al., 2012; Zhang et al., 2013; Zorita et al., 2015) as well as public databases with the increasing number of sequencing programs, species identification or OTUs delimitation will certainly follow this trend. Consequently, the ability to reexamine existing NGS networks and reassigning species and OTUs thanks to a systematic practice of data storage will improve the precision of the networks described which may be today biased by current database completeness and species delimitation methods.

6.2 Reconstructing Ecological Networks With Different Predicting Methods of Species Interactions

As already discussed, in some systems in which interactions cannot be directly determined, such as networks of microbes (see Vacher et al., 2016), only species cooccurrence and/or abundance patterns are available, and thus statistical methods (Kurtz et al., 2015) or machine learning (Bohan et al., 2011a,b) are needed to predict species interaction. Despite their ability to rapidly build networks from NGS data, these methods are still in their infancy (Bohan et al., 2017). It is clear that this research area is still young, and future developments are expected, which will affect the way ecological networks are created and how accurate they are. As a NGS network database would allow the reassignment of species and OTUs paired with the improvement of public databases and identification methods, it would also allow the reconstruction of preexisting networks based on cooccurrence patterns by benefiting from the arrival of new statistical inference methods and machine learning approaches. Moreover, since machine learning is very reliant on available data and information on the biological traits of species (e.g. body size), a centralized way of data storage for NGS networks and a systematic data-sharing practice between network ecologists would considerably improve the efficiency of network construction with learning approaches.

This means that not only DNA sequences and species interactions or cooccurrence must be stored and shared, but also the bioinformatics pipeline that created the networks (e.g. the mathematical and bioinformatics framework 'molecular ecological networks' to construct ecological association networks developed by Deng et al., 2012). Reproducibility of bioinformatic pipelines is an important topic in molecular phylogeny research where new phylogenetic trees are constantly built (Szitenberg et al., 2015). Network ecology should follow this example, and a NGS network database could be a first step in that direction.

6.3 Do Only Sequences and Species Interactions/ Cooccurrences Matter in a NGS Network Database?

As demonstrated previously, being able to track DNA sequences together with species interactions or abundance patterns is important. However, there is much more to ecological data than simply interactions and cooccurrences. In particular, temporal and spatial data are also highly valuable. Indeed, it is tempting to use databases to merge data from different studies in order to build large-scale ecological networks and to picture a broader view of ecosystem functioning. For instance, this has been done by Derocles et al. (2014a) for an aphid–parasitoid food web by pooling all interactions described in Europe between these organisms. However, merging data without caution would also create unrealistic ecological networks. Ecological networks cannot be created from all kinds of data: covariates (i.e. metadata) are needed. Indeed, the structure of ecological communities revealed by eDNA does not only depend on biotic interactions but also on abiotic filtering, dispersal limitation, ecological drift and historical contingency (Ovaskainen et al., 2017a,b). Temporal and spatial metadata would be the minimum information required to prevent building ecological networks with interactions or species not present in the same location (e.g. species A is eating species B in America and species C in Europe, can a network be built with the three species?)

or during the same period of the year (e.g. species A is eating species B in the winter, but not in summer). Here, BOLD is a very good example to follow since the deposited specimens (i.e. voucher) are systematically assigned with collection data (sampling date, sampling location, name of the collector).

6.4 An Example Output From a NGS Network Database: Phylogenetically Structured Networks

In this chapter, the use of DNA sequences has been considered in species identification or OTU delimitation. However, DNA sequences are too valuable to limit them only to the creation of nodes in ecological networks. Recently, DNA sequences have also been used for the phylogenetic signal they can provide in a network context (Elias et al., 2013; Hadfield et al., 2014; Rafferty and Ives, 2013). The interactions detected, and hence the network structure of ecosystems, reflect both the ecological processes and the evolutionary history of the species (e.g. Pilosof et al., 2014). Consequently, the observed network structures are necessarily constrained by the coevolutionary processes between species, as evidenced by recent studies predicting invasive species interactions from native species phylogenies (Charlery de la Masselière et al., 2017; Pearse and Altermatt, 2013). It seems very important therefore to start to account for phylogenetic signals within networks (Ives and Godfray, 2006), as pioneered by host–parasite cophylogeny studies (e.g. Banks and Paterson, 2005) and now advocated in the holobiont literature (Brooks et al., 2016; Martinson et al., 2017). Hopefully, NGS networks have the ability to easily explore this area of research through the extensive DNA sequences and species interactions they generate. A NGS database would constitute a turnkey solution to build phylogenetically structured ecological networks (see Evans et al., 2016).

As an example here, aphid–parasitoid interactions found in an agroecosystem from Western France by Derocles et al. (2014a,b) together with cytochrome C oxidase I (COI) mined from Genbank was used to build a phylogenetically structured network (Fig. 4). With this approach, the phylogenetic conservatism hypothesis that closely related parasitoid wasps parasitize closely related aphid species was tested (Cavender-Bares et al., 2009; Webb et al., 2002). This hypothesis seems true in some cases: for example, the genus *Lysiphlebus* attacks the genus *Aphis*. In the same way, the genus *Trioxys* exclusively parasitizes the genus *Myszocallis*.

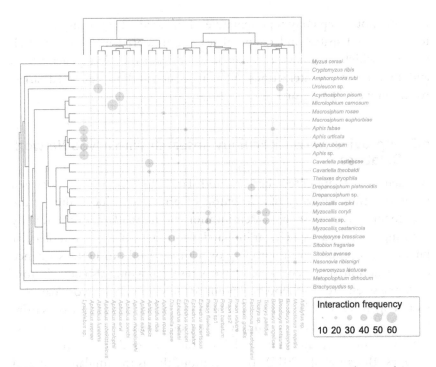

Fig. 4 A phylogenetically structured aphid–parasitoid network created using the networks from Derocles et al. (2014a,b). Cytochrome oxidase subunit I (COI) sequences for the detected species were extracted from Genbank and the Barcode of Life Database version 3 (BOLD v3). Trees were created using the default settings in RAxML version 8.2.10 (Stamatakis, 2014), and the figure was produced in R (R Development Core Team, 2017) using the library dualingTrees as used by Chaudhary et al. (2016). *Circles* represent interactions found in an agroecosystem (Derocles et al., 2014a,b). The radius of the *circle* represents the strength of the interaction. The aphids are in *rows*, the parasitoids in *columns*.

A particular case contradicts this theory: *Sitobion avenae* is parasitized by parasitoid species phylogenetically distant (*Aphidius* spp., *Ephedrus plagiator* or *Praon volucre*).

However, because the network in Derocles et al. (2014a) was based on morphological identifications for aphids and 16S sequence identifications for parasitoids, Fig. 4 must be considered with caution: COI sequences used to build the phylogenetic trees in Fig. 4 did not come from the specimens sampled by Derocles et al. (2014a). Moreover, for some species not identified to the species level (e.g. *Lysiphlebus* sp. or *Aphis* sp.), COI sequences mined from Genbank from a closely related species were used (e.g. COI of *Lysiphlebus fabarum*). Finally, the interactions used here are certainly not completely representative of the studied agroecosystem due to the limited spatial and temporal scales considered in Derocles et al. (2014a), which may have consequences both on the phylogenetic trees and the structure of interactions. A database such as those described earlier would prevent these issues.

6.5 Improving Network Ecology Research With a NGS Network Database

A database as presented here would facilitate comparison between networks, thus allowing the detection of errors introduced during the sequencing process, species identification or network construction by pointing out unusual species or interactions. In the same way as molecular phylogenetic studies are improving over time, thanks to DNA sequence curation and sharing, ecological network research would also benefit from such practices. From a theoretical point of view, a NGS network database would open ways for large-scale comparisons of ecological networks from various ecosystems, and thus improve our understanding of how ecosystems vary in space and time and how they respond to environmental changes (Bohan et al., 2017). By definition, this is the fundamental goal of biomonitoring programs.

Nevertheless, ecologists still need to be cautious when using preexisting data to build or compare ecological networks (including phylogenetically structured networks) and particularly when making ecological interpretations. Patterns derived from associations between community ecology and phylogeny can be interpreted in very different ways, and phylogenetic patterns do not necessary reflect community assembly mechanisms (Gerhold et al., 2015; Mayfield and Levine, 2010). As demonstrated earlier, interpreting coabundance pattern-based networks is not straightforward as complex oscillatory dynamics, indirect interactions or trophic cascades may alter the structure of cooccurrence networks. Consequently, the efficient use of databases to build and compare networks must be considered as a complementary approach. Experimental evidences based on field or laboratory experiments are still mandatory to confirm the ecological patterns highlighted and the networks constructed. To sum up, if used with appropriate caution, a NGS-based network database would constitute a way to start a virtuous circle that would improve the methodology of network construction, the fundamental research in ecology and finally the biomonitoring of ecosystems.

7. CONCLUSION: TOWARDS THE CONSTRUCTION OF MULTILAYER NETWORKS IN ECOLOGY USING NGS

7.1 Towards Larger, Highly Resolved Networks

The recent conceptual revolution in ecology brought forth by the development of metacommunity and metacosystem theory (Holyoak et al., 2005; Leibold et al., 2004; Loreau et al., 2003) has stressed the importance of spatial processes for the functioning, diversity, complexity and dynamics of ecological systems (Massol et al., 2011). Recent theoretical findings have highlighted the role of the spatial structure and scales of communities, food webs and ecosystems to understand properties such as ecosystem productivity and nutrient fluxes (Gravel et al., 2010; Loreau and Holt, 2004), ecosystem stability (Gravel et al., 2016; Mougi and Kondoh, 2016; Wang et al., 2017), species coexistence (Amarasekare et al., 2004; Haegeman and Loreau, 2015; Mouquet and Loreau, 2002), food-chain length (Calcagno et al., 2011; Holt, 1997), food web complexity (Bolchoun et al., 2017; Pillai et al., 2011), alternate steady states of ecosystems (Gounand et al., 2014), nutrient colimitation (Marleau et al., 2015; Mouquet et al., 2006) and the existence of 'keystone' ecosystems (Mouquet et al., 2013). From a practical, applied viewpoint, these models have also fostered change in the understanding of human-managed ecosystems such as agroecosystems (Bohan et al., 2013; Gaba et al., 2010; Massol and Petit, 2013) or marine protected areas (Andrello et al., 2013; Parravicini et al., 2014). However, our knowledge of 'networks of networks', (e.g. multiple animal groups interacting with shared plants; Pocock et al., 2012), is still scant and poorly defined. In part, this is due to the difficulty of obtaining necessary data using traditional empirical tools (i.e. direct observation, gut contents, stable isotopes, etc.). As discussed earlier, even if it is more efficient and reliable than other approaches, NGS is not the key to every lock as it might not see some species (PCR issues for instance), misidentify them (incomplete databanks) or miscalculate the links. However, merging NGS with ENA will provide the ecological research community with the tools it needs to create multilayer networks with increased ease in the coming decades, thus improving the power to detect many kinds of spatial food web effects. Increasing network coverage (i.e. more species identified, more links identified) through networks of ecological networks is a step forward for different reasons:

(1) It provides more precise metrics on network properties (connectance, modularity, degree distribution, etc.; see Ma et al., 2017);

(2) It better connects species demographics with community and food web dynamics through decoupling the population dynamics and interaction effects of cryptic species;

(3) It improves precision regarding indirect measures of species interaction strengths, which is crucial to better understand ecosystem stability theory and test its predictions (see Jacquet et al., 2016).

7.2 NGS Networks to Link Above- and Belowground Ecosystems, as Well as Eukaryotes and Prokaryotes

Currently, ecological network research is mainly split between studies focusing either on aboveground or belowground interactions, with very few linking the two (Mulder et al., 2013; Rodríguez-Echeverría and Traveset, 2015). However, the structure of ecological networks is inevitably affected by the range of organisms considered (Hudson et al., 2006; Lafferty et al., 2008). NGS is a promising way to start to link all organisms, from above- and belowground, prokaryotes and eukaryotes, in the same network. This approach has been attempted recently with a combination of observations and DNA-based identifications to infer a multiorganism network including spiders, earthworms, Enchytraeids, nematodes, plants, protists and other microorganisms (Morriën et al., 2017). By combining existing NGS techniques, it is now possible to take this approach even further and rapidly construct networks much more complex than those currently established. With a nest-tagging NGS tool (Cruaud et al., 2017; Evans et al., 2016; Shokralla et al., 2015), it would be possible to rapidly construct both aboveground (e.g. flower visitation based on pollen sequencing carried by pollinating insects and animal DNA found in plant nectaries) and belowground networks (with arbuscular mycorrhizal fungi, as previously done by Toju et al., 2014) where all organisms are linked by the shared plants they interact with. In parallel, metabarcoding data from microbial communities and machine learning inference would enable the inclusion of microorganisms in a network of networks, and understand how they affect the network structure and how they constrain interactions.

7.3 Biomonitoring of Ecosystems With Multilayer Phylogenetically Structured Networks

NGS methods provide an unprecedented opportunity to rapidly build complex multilayer ecological networks. A combination of sequencing technologies and machine learning approaches would open new opportunities for describing networks at various spatial and temporal scales

(Bohan et al., 2017). Such 'next-generation biomonitoring' through the construction of networks of networks paves the way for an exciting new area of research and a comprehensive understanding of the response of ecosystems to environmental changes. From our point of view, we now have access to the tools needed to respond to a number of contemporary global challenges.

However, more complex networks in large-scale biomonitoring programs require new analytical tools to disentangle actual patterns of change in network structure from noise. Indeed, the construction of multilayer networks is only a step (albeit a major one) when assessing the impacts of environmental changes on ecosystems. The identification of consistent patterns within networks is another crucial step that cannot be neglected. Most of the networks analysed to date are still relatively simple as they do not integrate a wide range of organisms or different kinds of interactions. In more complex networks, such as the network of ecological networks established by Pocock et al. (2012), all species are linked by the plants. While giving a more complete view of the architecture of interactions, this approach still neglects some links such as intraguild predation, which play a role in the stability of ecosystems (see Nakazawa and Yamamura, 2006). For example, integrating the full range of interactions mediated by the ground beetles (Coleoptera: Carabidae) in the type of network of networks built by Pocock et al. (2012) would be currently problematic. Indeed, these organisms can interact with plants, herbivorous species (e.g. aphids), as well as natural enemies of herbivorous species (e.g. parasitoids). Describing the structure and the robustness of very complex ecological networks, integrating a wide range of interactions as well as prokaryotes and eukaryotes, therefore requires new statistical methods.

Combining phylogenetic information with ecological networks is also a new area of research that remains to be more systematically explored (Evans et al., 2016). NGS is generating an increasing amount of new data which give the opportunity to investigate how the structure of species interactions and the phylogenetic signals are linked. This approach would certainly be an added value to the current biomonitoring programs as ecosystem condition and coevolutionary processes could be monitored together. More importantly, in a changing world, phylogenetic signals may be key to predicting ecological interactions (Elias et al., 2013; Ives and Godfray, 2006; Rafferty and Ives, 2013; Rezende et al., 2007).

The future of network ecology is exciting, with a great opportunity to biomonitoring and to improve our understanding of the ecosystem functioning and services as well as finding ways to mitigate the impact of environmental changes or to restore ecosystems.

ACKNOWLEDGEMENTS

We are grateful to Frédéric Barraquand (BIOGECO, INRA, Univ. Bordeaux, Pessac, France) for his contribution to the manuscript and the figures. We acknowledge the support of the projects FACCE SURPLUS PREAR and ANR-17-CE32-0011. We thank Michael J.O. Pocock (Centre for Ecology & Hydrology, Wallingford, United Kingdom) for his help in defining the direction of the manuscript. We thank Eleanor Collinson (School of Natural and Environmental Sciences, Newcastle University, United Kingdom) for her careful reading and thoughtful comments on this chapter. We are grateful to Mattias Jonsson (SLU, Uppsala, Sweden) and Athen Ma (Queen Mary University of London, United Kingdom) for their very helpful review of the manuscript.

GLOSSARY

Molecular Biology

Amplicon a DNA fragment amplified by primers during the polymerase chain reaction (PCR).

Diagnostic PCR specific primer pairs are designed for each targeted species or higher taxonomic group to produce amplicons of different sizes. These primers are then used in a multiplex and/or several singleplex PCRs for taxonomic identification. Identification of taxonomic groups in samples is then based on the presence/absence of bands on an electrophoresis gel (taxonomic group is present or not) as well as the position of each band (which taxonomic group is present). Quantitative PCR (qPCR) or digital droplet PCR (ddPCR) could then be used to get a precise estimate of the abundance of a species/group. This approach does not require DNA sequencing but requires a prior knowledge of the studied organisms in order to design specific primers.

DNA barcoding in a broad sense, a taxonomic method to identify organisms (ideally to the species level) using DNA sequences.

Environmental DNA (eDNA) remnant DNA in the environment; cells or tissues left behind by organisms (e.g. faeces, hair, epithelial cells).

Gel electrophoresis method used to visualize DNA fragments according to their size usually on an agarose gel.

Metabarcoding approach where several millions of DNA sequences of a specific genomic region are generated (e.g. eDNA sample) to characterize the taxonomic diversity present in a sample.

Metagenomics often confused with metabarcoding, metagenomics sequencing does not target specific genomic regions. Instead, it provides insights into entire genomes from multiple organisms in a particular ecosystem.

Molecular network/molecular ecological networks in the context of this chapter, molecular networks refer to ecological networks constructed with DNA-based techniques (Sanger sequencing, PCR diagnostics, NGS).

Multiplex PCR PCR where more than one primer pair is employed to simultaneously amplify several PCR fragments within a single reaction.

Nested PCR PCR comprising two successive steps. In the second step, the PCR product obtained in the first step is amplified. Nested PCR can improve the amplification of recalcitrant target gene regions. In NGS metabarcoding protocols, a nested PCR allows the addition of a second step for tagging ('nested tagging method', see Evans et al., 2016; Kitson et al., 2016).

Next-generation sequencing (NGS) or high-throughput sequencing sequencing technologies designed to simultaneously generate up to several millions of sequences (see Table 2 for more details).

Operational taxonomic units/OTUs specimens/samples assignment to a taxonomic group sharing similar sequences, but without naming actual species.

Primer short strand of DNA (or RNA) between generally 15 and 25 base pairs. A primer is the starting point used by the DNA polymerase to synthetize a complementary DNA brand.

Sanger sequencing 'first-generation' sequencing method where the output is a single DNA sequence per sample and PCR (see Table 2 for more details).

Sequencing coverage the depth of coverage refers to the number of times a region has been sequenced by independent reads. The breadth of coverage refers to the percentage of bases of a region covered with a given depth (e.g. 90% of the targeted region was covered three times). The number of samples in a NGS run depends mainly on the sequencing coverage. The number of samples in a sequencing run does not depend on labwear units (i.e. 96- or 384-well plate).

Sequence reads DNA sequences produced by NGS.

Sequencing run step during which the DNA is sequenced (see Table 2 for more details).

Singleplex PCR where one pair of primers is used to amplify one specific PCR fragment.

Specimen voucher a sample of an organism deposited and stored in a facility. A voucher is generally associated with a scientific species name, the collector's name, the expert who identified the specimen, the date and location (with GPS coordinates) of collection. Researchers may request the permission to examine a voucher for further studies (see Culley, 2013).

Tag unique short sequence (usually 8–12 bp) added to the 5' end of a primer. Samples with unique tags can be pooled and sequenced in the same sequencing run. Sequences are later assigned to samples using bioinformatic pipelines.

Ecology and Ecological Network Analysis

Bipartite networks a bipartite network can be divided into two disjoint sets. The nodes (i.e. species) of these two sets are connected with links (i.e. interactions). Plant–pollinator networks and prey–predator networks are two examples of bipartite networks.

Compartmentalization the degree to which an ecological network is divided into weakly connected subwebs.

Connectance the proportion of observed links over all possible links. A simple measure of connectance is given by the ratio number of link/number of species^2. More complex measures of connectance account for the frequency of interactions (see Bersier et al., 2002).

Conventional approach/method in the context of this chapter, this refers to 'traditional' taxonomy-based approaches to identify species. These methods do not rely on DNA to identify species.

Ecosystem functions/ecosystem functioning ecological processes that control the fluxes of energy and matter between trophic levels and between the organisms and the environment.

Ecosystem services humankind benefits from ecosystem functioning. Ecosystem services are often grouped into four categories: (1) provisioning: e.g., food and water production; (2) regulation: e.g., control of climate, pests and diseases; (3) supporting: e.g., nutrient cycles, pollinations; (4) cultural: e.g., historical, recreational, educational and therapeutic.

Generality mean number of consumers per prey. In a trophic network, a simple measure of generality is the mean number of prey species per predator species. More complex measures of generality account for the frequency of consumer–prey interactions.

Indicators in biomonitoring, indicators are used to assess risks to human health and environment. These indicators include pollution indicators (pesticides, elements, metabolites) and ecological indicators (species, population, community, behaviour). Indicators are used for communication to the general public and policy makers.

Modularity a measure of compartmentalization within a network. In ecological networks, it refers to a group of species interacting more frequently with themselves than with other species. A wide range of methods have been developed to assess the modularity of a network (see Poisot, 2013).

Multilayer networks set of ecological networks built at different times and/or in different sites and or/with different interaction types (see Pilosof et al., 2017). For example, the 'network of ecological networks' built by Pocock et al. (2012) is a multilayer network gathering different interaction types (mainly mutualist and trophic).

Nestedness tendency of nodes with few connections to be linked to a subset of nodes interacting with more connected nodes. In nested ecological networks, a core group of generalist species interact with each other and with specialist species, while specialist species interact only with the core group of generalist species. A wide range of methods have been developed to assess the nestedness of a network (see Almeida–Neto et al., 2008).

Node in ecological networks, nodes represent taxonomic units (generally species, but it can be other taxonomic levels or OTUs) connected by links (i.e. interactions).

Robustness a measure of the tolerance of ecological networks to species extinctions (Dunne et al., 2002b; Memmott et al., 2004).

Specimen rearing specimens sampled in the field are reared in the laboratory after collection under controlled conditions (e.g. temperature, humidity and photoperiod). Specimen rearing is typically used when the development stage of specimens collected in the field does not allow a reliable taxonomic identification. Specimen rearing is also commonly used for parasitoid identification. In this case, parasitoid hosts are collected in the field and reared in the laboratory until the emergence of adult parasitoids.

Vulnerability mean number of prey per consumer. In a trophic network, a simple measure of vulnerability is the mean number of predator species per prey species. As for the generality, more complex measures of vulnerability account for the frequency of prey–consumers interactions.

REFERENCES

Abreu, N.A., Taga, M.E., 2016. Decoding molecular interactions in microbial communities. FEMS Microbiol. Rev. 40, 648–663. https://doi.org/10.1093/femsre/fuw019.

Agler, M.T., Ruhe, J., Kroll, S., Morhenn, C., Kim, S.-T., Weigel, D., Kemen, E.M., 2016. Microbial hub taxa link host and abiotic factors to plant microbiome variation. PLoS Biol. 14, e1002352. https://doi.org/10.1371/journal.pbio.1002352.

Aizen, M.A., Morales, C.L., Morales, J.M., 2008. Invasive mutualists erode native pollination webs. PLoS Biol. 6, e31.

Albert, T.J., Molla, M.N., Muzny, D.M., Nazareth, L., Wheeler, D., Song, X., Richmond, T.A., Middle, C.M., Rodesch, M.J., Packard, C.J., Weinstock, G.M., Gibbs, R.A., 2007. Direct selection of human genomic loci by microarray hybridization. Nat. Methods 4, 903–905.

Allesina, S., Levine, J.M., 2011. A competitive network theory of species diversity. Proc. Natl. Acad. Sci. U.S.A. 108, 5638–5642.

Allesina, S., Tang, S., 2012. Stability criteria for complex ecosystems. Nature 483, 205–208. https://doi.org/10.1038/nature10832.

Allesina, S., Alonso, D., Pascual, M., 2008. A general model for food web structure. Science 320, 658–661.

Almeida-Neto, M., Guimaraes, P., Guimarães, P.R., Loyola, R.D., Ulrich, W., 2008. A consistent metric for nestedness analysis in ecological systems: reconciling concept and measurement. Oikos 117, 1227–1239.

Amarasekare, P., Hoopes, M.F., Mouquet, N., Holyoak, M., 2004. Mechanisms of coexistence in competitive metacommunities. Am. Nat. 164, 310–326.

Andrello, M., Mouillot, D., Beuvier, J., Albouy, C., Thuiller, W., Manel, S., 2013. Low connectivity between Mediterranean marine protected areas: a biophysical modeling approach for the Dusky Grouper Epinephelus marginatus. PLoS One 8, e68564.

Arnoldi, J.F., Loreau, M., Haegeman, B., 2016. Resilience, reactivity and variability: a mathematical comparison of ecological stability measures. J. Theor. Biol. 389, 47–59.

Astegiano, J., Massol, F., Vidal, M.M., Cheptou, P.-O., Guimarães Jr., P.R., 2015. The robustness of plant-pollinator assemblages: linking plant interaction patterns and sensitivity to pollinator loss. PLoS One 10, e0117243.

Bálint, M., Bahram, M., Eren, A.M., Faust, K., Fuhrman, J.A., Lindahl, B., O'Hara, R.B., Öpik, M., Sogin, M.L., Unterseher, M., Tedersoo, L., Abarenkov, K., Tedersoo, L., Nilsson, R., Ainsworth, T., Krause, L., Bridge, T., Aitchison, J., Amend, A., Seifert, K., Bruns, T., Anderson, M., Anderson, M., Anderson, M., Walsh, D., Antoninka, A., Wolf, J., Bowker, M., Ashelford, K., Chuzhanova, N., Fry, J., Bálint, M., Bartha, L., O'Hara, R., Bálint, M., Schmidt, P.-A., Sharma, R., Ban, Y., An, L., Jiang, H., Baselga, A., Baselga, A., Orme, C., Bik, H., Fournier, D., Sung, W., Bik, H., Thomas, W., Bohannan, B., Hughes, J., Boker, S., Neale, M., Maes, H., Breiman, L., Brodin, J., Mild, M., Hedskog, C., Brown, S., Veach, A., Rigdon-Huss, A., Bullock, H., Harlow, L., Mulaik, S., Burbank, D., Anderson, R., Buttigieg, P., Ramette, A., Byrne, B., Caporaso, J., Kuczynski, J., Stombaugh, J., Carlsen, T., Aas, A., Lindner, D., Chao, A., Colwell, R., Lin, C.-W., Chao, A., Jost, L., Chase, J., Kraft, N., Smith, K., Chow, C.-E., Kim, D., Sachdeva, R., Clarke, K., Gorley, R., Clemmensen, K., Finlay, R., Dahlberg, A., Cole, J., Wang, Q., Cardenas, E., Colwell, R., Chao, A., Gotelli, N., Colwell, R., Mao, C., Chang, J., Csárdi, G., Nepusz, T., De Cáceres, M., Legendre, P., Wiser, S., Delmont, T., Malandain, C., Prestat, E., Deng, Y., Jiang, Y., Yang, Y., DeSantis, T., Hugenholtz, P., Larsen, N., Doolittle, W., Zhaxybayeva, O., Dormann, C., Elith, J., Bacher, S., Dormann, C., Frueund, J., Bluethgen, N., Dormann, C., Gruber, B., Fruend, J., Durno, W., Hanson, N., Konwar, K., Eickbush, T., Eickbush, D., Eiler, A., Heinrich, F., Bertilsson, S., Epstein, S., Eren, A., Maignien, L., Sul, W., Eren, A., Morrison, H., Lescault, P., Eren, A., Sogin, M., Morrison, H., Falkowski, P., Fenchel, T., Delong, E., Fang, H., Huang, C., Zhao, H., Faust, K., Lahti, L., Gonze, D., Faust, K., Raes, J., Favreau, J., Drew, C., Hess, G., Ficetola, G., Coissac, E., Zundel, S., Fierer, N., Bradford, M., Jackson, R., Fisher, C., Mehta, P., Flores, C., Poisot, T., Valverde, S., Friedman, J., Alm, E., Fujita, M., Leaché, A., Burbrink, F., Gevers, D., Cohan, F., Lawrence, J., Gilbert, J., Steele, J., Caporaso, J., Goldstein, H., Spiegelhalter, D., Gotelli, N., Colwell, R., Grace, J., Keeley, J., Johnson, D., Granger, C., Guillot, G., Rousset, F., Gwinn, D., Allen, M., Bonvechio, K., Hadi, A., Ling, R., Haegeman, B., Hamelin, J., Moriarty, J., Hamady, M., Lozupone, C., Knight, R., Hao, X., Jiang, R., Chen, T., He, Y., Caporaso, J., Jiang, X.-T., Hekstra, D., Leibler, S., Hill, M., Hinchliff, C.,

Smith, S., Allman, J., Hong, S.-H., Bunge, J., Jeon, S.-O., Hu, L., Bentler, P., Hughes, J., Hellmann, J., Ricketts, T., Hui, F., Taskinen, S., Pledger, S., Huse, S., Welch, D., Voorhis, A., Huse, S., Welch, D., Morrison, H., Huson, D., Mitra, S., Ruscheweyh, H.-J., Ihrmark, K., Bödeker, I., Cruz-Martinez, K., Jansson, J., Prosser, J., Koeppel, A., Wu, M., Kõljalg, U., Nilsson, R., Abarenkov, K., Kuczynski, J., Liu, Z., Lozupone, C., Kunin, V., Engelbrektson, A., Ochman, H., Kurtz, Z., Müller, C., Miraldi, E., Legendre, P., Fortin, M.-J., Borcard, D., Legendre, P., Legendre, L., Leininger, S., Urich, T., Schloter, M., Lima-Mendez, G., Faust, K., Henry, N., Lindahl, B., Kuske, C., Locey, K., Lennon, J., Love, M., Huber, W., Anders, S., Lozupone, C., Knight, R., Lozupone, C., Hamady, M., Kelley, S., Lütkepohl, H., McArdle, B., Anderson, M., McGill, B., Etienne, R., Gray, J., McMurdie, P., Holmes, S., Magurran, A., Magurran, A., Magurran, A., Dornelas, M., Moyes, F., Mahé, F., Rognes, T., Quince, C., Mandal, S., Van Treuren, W., White, R., Manly, B., Alberto, J., Manoharan, L., Kushwaha, S., Hedlund, K., Margesin, R., Miteva, V., Marino, S., Baxter, N., Huffnagle, G., Matsen, F., Evans, S., Meiser, A., Bálint, M., Schmitt, I., Miki, T., Yokokawa, T., Matsui, K., Mounier, J., Monnet, C., Vallaeys, T., Mutshinda, C., O'Hara, R., Woiwod, I., Nguyen, N., Song, Z., Bates, S., Nguyen, N.-P., Warnow, T., Pop, M., Noss, R., O'Hara, R., O'Hara, R., Kotze, D., Oksanen, J., Blanchet, F., Kindt, R., Oliver, A., Brown, S., Callaham, M., Öpik, M., Davison, J., Moora, M., Pace, N., Parada, A., Needham, D., Fuhrman, J., Patin, N., Kunin, V., Lidström, U., Peres-Neto, P., Jackson, D., Perisin, M., Vetter, M., Gilbert, J., Peršoh, D., Preheim, S., Perrotta, A., Martin-Platero, A., Prosser, J., Bohannan, B., Curtis, T., Pruesse, E., Quast, C., Knittel, K., Quince, C., Curtis, T., Sloan, W., Quince, C., Lanzen, A., Curtis, T., Quinn, G., Keough, M., Team, R.C., Reshef, D., Reshef, Y., Finucane, H., Risso, D., Ngai, J., Speed, T., Robinson, M., McCarthy, D., Smyth, G., Robinson, M., Oshlack, A., Rolstad, J., Gjerde, I., Gundersen, V., Rosling, A., Cox, F., Cruz-Martinez, K., Ruan, Q., Dutta, D., Schwalbach, M., Ryberg, M., Sanli, K., Bengtsson-Palme, J., Nilsson, R., Scherber, C., Eisenhauer, N., Weisser, W., Schloss, P., Westcott, S., Ryabin, T., Schmidt, T., Rodrigues, J.M., von Mering, C., Schnell, I., Bohmann, K., Gilbert, M., Shannon, P., Markiel, A., Ozier, O., Sipos, R., Székely, A., Palatinszky, M., Sloan, W., Lunn, M., Woodcock, S., Smets, W., Leff, J., Bradford, M., Smilauer, P., Lepš, J., Sogin, M., Morrison, H., Huber, J., Steele, J., Countway, P., Xia, L., Stegen, J., Lin, X., Fredrickson, J., Stein, R., Bucci, V., Toussaint, N., Stingl, U., Cho, J.-C., Foo, W., Sugihara, G., May, R., Ye, H., Taylor, L., Tedersoo, L., Anslan, S., Bahram, M., Tedersoo, L., Bahram, M., Põlme, S., Tibshirani, R., Tikhonov, M., Leach, R., Wingreen, N., Unterseher, M., Schnittler, M., Dormann, C., Vamosi, S., Heard, S., Vamosi, J., van der Heijden, M., Bardgett, R., van Straalen, N., van der Heijden, M., Klironomos, J., Ursic, M., Wang, G., Wang, Y., Wang, Y., Naumann, U., Wright, S., Warton, D., Blanchet, F., O'Hara, R., Warton, D., Foster, S., De'ath, G., Warton, D., Wright, S., Wang, Y., Weiss, S., Van Treuren, W., Lozupone, C., Weiss, S., Xu, Z., Amir, A., Weitz, J., Poisot, T., Meyer, J., Westcott, S., Schloss, P., Whittingham, M., Stephens, P., Bradbury, R., Xia, L., Ai, D., Cram, J., Xia, L., Steele, J., Cram, J., Xiang, D., Verbruggen, E., Hu, Y., Yilmaz, P., Kottmann, R., Field, D., Zhang, J., Kapli, P., Pavlidis, P., Zuur, A., Fryer, R., Jolliffe, I., 2016. Millions of reads, thousands of taxa: microbial community structure and associations analyzed via marker genes. FEMS Microbiol. Rev. 6, 189–196. https://doi.org/10.1093/femsre/fuw017.

Banks, J.C., Paterson, A.M., 2005. Multi-host parasite species in cophylogenetic studies. Int. J. Parasitol. 35, 741–746.

Barazzoni, R., Short, K.R., Nair, K.S., 2000. Effects of aging on mitochondrial DNA copy number and cytochrome C oxidase gene expression in rat skeletal muscle, liver, and heart. J. Biol. Chem. 275, 3343–3347.

Barnes, M.A., Turner, C.R., 2015. The ecology of environmental DNA and implications for conservation genetics. Conserv. Genet. 17, 1–17.

Bascompte, J., Jordano, P., 2007. Plant-animal mutualistic networks: the architecture of biodiversity. Annu. Rev. Ecol. Evol. Syst. 38, 567–593.

Bascompte, J., Jordano, P., Melian, C.J., Olesen, J.M., 2003. The nested assembly of plant-animal mutualistic networks. Proc. Natl. Acad. Sci. U.S.A. 100, 9383–9387.

Bell, K.L., Fowler, J., Burgess, K.S., Dobbs, E.K., Gruenewald, D., Lawley, B., Morozumi, C., Brosi, B.J., 2017. Applying pollen DNA metabarcoding to the study of plant–pollinator interactions. Appl. Plant Sci. 5, 1600124.

Beng, K.C., Tomlinson, K.W., Shen, X.H., Surget-Groba, Y., Hughes, A.C., Corlett, R.T., Slik, J.W.F., 2016. The utility of DNA metabarcoding for studying the response of arthropod diversity and composition to land-use change in the tropics. Sci. Rep. 6, 24965.

Bengfort, M., van Velzen, E., Gaedke, U., 2017. Slight phenotypic variation in predators and prey causes complex predator-prey oscillations. Ecol. Complex. 31, 115–124. https://doi.org/10.1016/j.ecocom.2017.06.003.

Bersier, L.F., Banasek-Richter, C., Cattin, M.F., 2002. Quantitative descriptors of food-web matrices. Ecology 83, 2394–2407. https://doi.org/10.1890/0012-9658(2002)083[2394:QDOFWM]2.0.CO;2.

Binladen, J., Gilbert, M.T.P., Bollback, J.P., Panitz, F., Bendixen, C., Nielsen, R., Willerslev, E., 2007. The use of coded PCR primers enables high-throughput sequencing of multiple homolog amplification products by 454 parallel sequencing. PLoS One 2, e197.

Biswas, S., Mcdonald, M., Lundberg, D.S., Dangl, J.L., Jolic, V., 2016. Learning microbial interaction networks from metagenomic count data. J. Comput. Biol. 23, 526–535.

Blanchard, J.L., 2015. Climate change: a rewired food web. Nature 527, 173–174.

Blüthgen, N., Menzel, F., Blüthgen, N., 2006. Measuring spaecialization in species interaction networks. BMC Ecol. 6, 9.

Blüthgen, N., Fründ, J., Vázquez, D.P., Menzel, F., Bluthgen, N., Frund, J., Vazquez, D.P., 2008. What do interaction network metrics tell us about specialization and biological traits? Ecology 89, 3387–3399.

Bohan, D.A., Caron-Lormier, G., Muggleton, S., Raybould, A., Tamaddoni-Nezhad, A., 2011a. Automated discovery of food webs from ecological data using logic-based machine learning. PLoS One 6, e29028.

Bohan, D.A., Boursault, A., Brooks, D.R., Petit, S., 2011b. National-scale regulation of the weed seedbank by carabid predators. J. Appl. Ecol. 48, 888–898. https://doi.org/10.1111/j.1365-2664.2011.02008.x.

Bohan, D.A., Raybould, A., Mulder, C., Woodward, G., Tamaddoni-Nezhad, A., Bluthgen, N., Pocock, M.J.O., Muggleton, S., Evans, D.M., Astegiano, J., Massol, F., Loeuille, N., Petit, S., Macfadyen, S., 2013. Networking agroecology: integrating the diversity of agroecosystem interactions. In: Woodward, G., Bohan, D.A. (Eds.), Ecological Networks in an Agricultural World, pp. 1–67.

Bohan, D.A., Vacher, C., Tamaddoni-Nezhad, A., Raybould, A., Dumbrell, A.J., Woodward, G., 2017. Next-generation global biomonitoring: large-scale, automated reconstruction of ecological networks. Trends Ecol. Evol. 32, 477–487. https://doi.org/10.1016/j.tree.2017.03.001.

Bolchoun, L., Drossel, B., Allhoff, K.T., 2017. Spatial topologies affect local food web structure and diversity in evolutionary metacommunities. Sci. Rep. 7, 1818.

Briand, F., Cohen, J.E., 1984. Community food webs have scale-invariant structure. Nature 307, 264–267.

Brooks, A.W., Kohl, K.D., Brucker, R.M., van Opstal, E.J., Bordenstein, S.R., 2016. Phylosymbiosis: relationships and functional effects of microbial communities across host evolutionary history. PLoS Biol. 14, e2000225.

Bucci, V., Tzen, B., Li, N., Simmons, M., Tanoue, T., Bogart, E., Deng, L., Yeliseyev, V., Delaney, M.L., Liu, Q., Olle, B., Stein, R.R., Honda, K., Bry, L., Gerber, G.K., 2016. MDSINE: Microbial Dynamical Systems INference Engine for microbiome time-series analyses. Genome Biol. 17, 121. https://doi.org/10.1186/s13059-016-0980-6.

Calcagno, V., Massol, F., Mouquet, N., Jarne, P., David, P., 2011. Constraints on food chain length arising from regional metacommunity dynamics. Proc. R. Soc. B 278, 3042–3049.

Canard, E., Mouquet, N., Marescot, L., Gaston, K.J., Gravel, D., Mouillot, D., 2012. Emergence of structural patterns in neutral trophic networks. PLoS One 7, e38295.

Cavender-Bares, J., Kozak, K.H., Fine, P.V.A., Kembel, S.W., 2009. The merging of community ecology and phylogenetic biology. Ecol. Lett. 12, 693–715.

Cazelles, K., Araújo, M.B., Mouquet, N., Gravel, D., 2016. A theory for species co-occurrence in interaction networks. Theor. Ecol. 9, 39–48. https://doi.org/10.1007/s12080-015-0281-9.

Chacoff, N.P., Vázquez, D.P., Lomáscolo, S.B., Stevani, E.L., Dorado, J., Padrón, B., 2012. Evaluating sampling completeness in a desert plant-pollinator network. J. Anim. Ecol. 81, 190–200. https://doi.org/10.1111/j.1365-2656.2011.01883.x.

Chao, A., 1984. Nonparametric estimation of the number of classes in a population. Scand. J. Stat. 11, 265–270.

Charlery de la Masselière, M., Ravigné, V., Facon, B., Lefeuvre, P., Massol, F., Quilici, S., Duyck, P.-F., 2017. Changes in phytophagous insect host ranges following the invasion of their community: long-term data for fruit flies. Ecol. Evol. 7, 5181–5190.

Chaudhary, V.B., Rúa, M.A., Antoninka, A., Bever, J.D., Cannon, J., Craig, A., Duchicela, J., Frame, A., Gardes, M., Gehring, C., Ha, M., Hart, M., Hopkins, J., Ji, B., Collins Johnson, N., Kaonongbua, W., Karst, J., Koide, R.T., Lamit, L.J., Meadow, J., Milligan, B.G., Moore, J.C., Pendergast IV, T.H., Piculell, B., Ramsby, B., Simard, S., Shrestha, S., Umbanhowar, J., Viechtbauer, W., Waters, L., Wilson, G.W.T., Zee, P.C., Hoeksema, J.D., 2016. MycoDB, a global database of plant response to mycorrhizal fungi. Sci. Data 3, 160028.

Clare, E.L., 2014. Molecular detection of trophic interactions: emerging trends, distinct advantages, significant considerations and conservation applications. Evol. Appl. 7, 1144–1157.

Clare, E.L., Symondson, W.O.C., Fenton, M.B., 2014. An inordinate fondness for beetles? Variation in seasonal dietary preferences of night-roosting big brown bats (Eptesicus fuscus). Mol. Ecol. 23, 3633–3647.

Cohen, J.E., Briand, F., 1984. Trophic links of community food webs. Proc. Natl. Acad. Sci. U.S.A. 81, 4105–4109.

Cohen, J.E., Newman, C.M., 1985. A stochastic theory of community food webs: I. Models and aggregated data. Proc. R. Soc. B 224, 421–448.

Cohen, J., Briand, F., Newman, C., 1990. Community Food Webs: Data and Theory. Springer Science & Business Media.

Colwell, R.K., Coddington, J.A., 1994. Estimating terrestrial biodiversity through extrapolation. Philos. Trans. R. Soc. B 345, 101–118.

Costa, J.M., da Silva, L.P., Ramos, J.A., Heleno, R.H., 2016. Sampling completeness in seed dispersal networks: when enough is-enough. Basic Appl. Ecol. 17, 155–164. https://doi.org/10.1016/j.baae.2015.09.008.

Cotterill, M., Harris, S.E., Collado Fernandez, E., Huntriss, J.D., Campbell, B.K., Picton, H.M., 2013. The activity and copy number of mitochondrial DNA in ovine oocytes throughout oogenesis in vivo and during oocyte maturation in vitro. Mol. Hum. Reprod. 19, 444–450.

Cotton, T.E.A., Dumbrell, A.J., Helgason, T., 2014. What goes in must come out: testing for biases in molecular analysis of arbuscular mycorrhizal fungal communities. PLoS One 9, e109234.

Cousins, S., 1987. The decline of the trophic level concept. Trends Ecol. Evol. 2, 312–316.

Coyte, K.Z., Schluter, J., Foster, K.R., 2015. The ecology of microbiome: networks, competition, stability. Science 350, 663–666.

Cruaud, P., Rasplus, J.-Y., Rodriguez, L.J., Cruaud, A., 2017. High-throughput sequencing of multiple amplicons for barcoding and integrative taxonomy. Sci. Rep. 7, 41948.

Culley, T.M., 2013. Why vouchers matter in botanical research. Appl. Plant Sci. 1. apps. 1300076.

Dai, W.-Z., Muggleton, S., Zhou, Z.-H., 2015. In: Logical vision: meta-interpretive learning for simple geometrical concepts. Late Breaking Paper Proceedings of the 25th International Conference on Inductive Logic Programming. CEUR, pp. 1–16. http://ceur-ws.org/Vol-1636.

Dalla Riva, G.V., Stouffer, D.B., 2016. Exploring the evolutionary signature of food webs' backbones using functional traits. Oikos 125, 446–456.

Dapprich, J., Ferriola, D., Mackiewicz, K., Clark, P.M., Rappaport, E., D'Arcy, M., Sasson, A., Gai, X., Schug, J., Kaestner, K.H., Monos, D., 2016. The next generation of target capture technologies—large DNA fragment enrichment and sequencing determines regional genomic variation of high complexity. BMC Genomics 17, 486.

Davey, J.S., Vaughan, I.P., King, R.A., Bell, J.R., Bohan, D.A., Bruford, M.W., Holland, J.M., Symondson, W.O.C., 2013. Intraguild predation in winter wheat: prey choice by a common epigeal carabid consuming spiders. J. Appl. Ecol. 50, 271–279.

David, P., Thébault, E., Anneville, O., Duyck, P.F., Chapuis, E., Loeuille, N., 2017. Impacts of invasive species on food webs: a review of empirical data. In: Bohan, D.A., Dumbrell, A.J., Massol, F. (Eds.), Advances in Ecological Research. Academic Press, pp. 1–60.

Deiner, K., Walser, J.-C., Mächler, E., Altermatt, F., 2015. Choice of capture and extraction methods affect detection of freshwater biodiversity from environmental DNA. Biol. Conserv. 183, 53–63.

Deng, Y., Jiang, Y.-H., Yang, Y., He, Z., Luo, F., Zhou, J., 2012. Molecular ecological network analyses. BMC Bioinformatics 13, 113. https://doi.org/10.1186/1471-2105-13-113.

Derocles, S.A.P., Le Ralec, A., Plantegenest, M., Chaubet, B., Cruaud, C., Cruaud, A., Rasplus, J.-Y., 2012a. Identification of molecular markers for DNA barcoding in the Aphidiinae (Hym. Braconidae). Mol. Ecol. Resour. 12, 197–208.

Derocles, S.A.P., Plantegenest, M., Simon, J.-C., Taberlet, P., Le Ralec, A., 2012b. A universal method for the detection and identification of Aphidiinae parasitoids within their aphid hosts. Mol. Ecol. Resour. 12, 634–645.

Derocles, S.A.P., Le Ralec, A., Besson, M.M., Maret, M., Walton, A., Evans, D.M., Plantegenest, M., 2014a. Molecular analysis reveals high compartmentalization in aphid-primary parasitoid networks and low parasitoid sharing between crop and noncrop habitats. Mol. Ecol. 23, 3900–3911. https://doi.org/10.1111/mec.12701.

Derocles, S.A.P., Le Ralec, A., Besson, M.M., Maret, M., Walton, A., Evans, D.M., Plantegenest, M., 2014b. Data from: molecular analysis reveals high compartmentalization in aphid-primary parasitoid networks and low parasitoid sharing between crop and noncrop habitats. Dryad Digital Repository https://doi.org/10.5061/dryad.v1q8j.

Derocles, S.A.P., Evans, D.M., Nichols, P.C., Evans, S.A., Lunt, D.H., 2015. Determining plant-leaf miner-parasitoid interactions: a DNA barcoding approach. PLoS One 10, e0117872.

Derocles, S.A.P., Plantegenest, M., Rasplus, J.Y., Marie, A., Evans, D.M., Lunt, D.H., Le Ralec, A., 2016. Are generalist Aphidiinae (Hym. Braconidae) mostly cryptic species complexes? Syst. Entomol. 41, 379–391. https://doi.org/10.1111/syen.12160.

Dobson, A., Lafferty, K.D., Kuris, A.M., Hechinger, R.F., Jetz, W., 2008. Homage to Linnaeus: how many parasites? How many hosts? Proc. Natl. Acad. Sci. U.S.A. 105, 11482–11489.

Donohue, I., Petchey, O.L., Montoya, J.M., Jackson, A.L., McNally, L., Viana, M., Healy, K., Lurgi, M., O'Connor, N.E., Emmerson, M.C., 2013. On the dimensionality of ecological stability. Ecol. Lett. 16, 421–429.

Dumbrell, A.J., Clarke, E.J., Frost, G.A., Randell, T.E., Pitchford, J.W., Hill, J.K., 2008. Estimated changes in species' diversity following habitat disturbance are dependent on spatial scale: theoretical and empirical evidence. J. Appl. Ecol. 45, 1469–1477.

Dunne, J.A., Williams, R.J., Martinez, N.D., 2002a. Network structure and biodiversity loss in food webs: robustness increases with connectance. Ecol. Lett. 5, 558–567.

Dunne, J.A., Williams, R.J., Martinez, N.D., 2002b. Food-web structure and network theory: the role of connectance and size. Proc. Natl. Acad. Sci. U.S.A. 99, 12917–12922.

Dunshea, G., 2009. DNA-based diet analysis for any predator. PLoS One 4, e5252.

Eklöf, A., Jacob, U., Kopp, J., Bosch, J., Castro-Urgal, R., Chacoff, N.P., Dalsgaard, B., de Sassi, C., Galetti, M., Guimarães, P.R., Lomáscolo, S.B., Martín González, A.M., Pizo, M.A., Rader, R., Rodrigo, A., Tylianakis, J.M., Vázquez, D.P., Allesina, S., 2013. The dimensionality of ecological networks. Ecol. Lett. 16, 577–583.

Elbrecht, V., Leese, F., 2015. Can DNA-based ecosystem assessments quantify species abundance? Testing primer bias and biomass—sequence relationships with an innovative metabarcoding protocol. PLoS One 10, e0130324.

Elias, M., Fontaine, C., van Veen, F.J.F., 2013. Evolutionary history and ecological processes shape a local multilevel antagonistic network. Curr. Biol. 23, 1355–1359.

Elton, C.S., 1927. Animal Ecology. Sidgwick and Jackson. London.

Estes, J.A., Tinker, M.T., Williams, T.M., Doak, D.F., 1998. Killer Whale predation on Sea Otters linking oceanic and nearshore ecosystems. Science 282, 473–476.

Evans, D.M., Pocock, M.J.O., Memmott, J., 2013. The robustness of a network of ecological networks. Ecol. Lett. 16, 844–852. https://doi.org/10.1111/ele.12117.

Evans, D.M., Kitson, J.J.N., Lunt, D.H., Straw, N.A., Pocock, M.J.O., 2016. Merging DNA metabarcoding and ecological network analysis to understand and build resilient terrestrial ecosystems. Funct. Ecol. 30, 1904–1916. https://doi.org/10.1111/1365-2435.12659.

Evans, D.M., Gilbert, J.D.J., Port, G.R., 2017. Everything is connected: network thinking in entomology. Ecol. Entomol. 42, 1–3.

Faust, K., Raes, J., 2012. Microbial interactions: from networks to models. Nat. Rev. Microbiol. 10, 538–550. https://doi.org/10.1038/nrmicro2832.

Faust, K., Lahti, L., Gonze, D., de Vos, W.M., Raes, J., 2015a. Metagenomics meets time series analysis: unraveling microbial community dynamics. Curr. Opin. Microbiol. 25, 56–66. https://doi.org/10.1016/j.mib.2015.04.004.

Faust, K., Lima-Mendez, G., Lerat, J.S., Sathirapongsasuti, J.F., Knight, R., Huttenhower, C., Lenaerts, T., Raes, J., 2015b. Cross-biome comparison of microbial association networks. Front. Microbiol. 6, 1–13. https://doi.org/10.3389/fmicb.2015.01200.

Fayle, T.M., Scholtz, O., Dumbrell, A.J., Russell, S., Segar, S.T., Eggleton, P., 2015. Detection of mitochondrial COII DNA sequences in ant guts as a method for assessing termite predation by ants. PLoS One 10, e0122533.

Fischer, M., Strauch, B., Renard, B.Y., 2017. Abundance estimation and differential testing on strain level in metagenomics data. Bioinformatics 33, i124–i132.

Fitter, A.H., Gilligan, C.A., Hollingworth, K., Kleczowski, A., Twyman, R.M., Pitchford, J.W., The Members of the NECR Soil Biodiversity Programme, 2005. Biodiversity and ecosystem function in soil. Funct. Ecol. 19, 369–377.

Fontaine, C., Guimaraes Jr., P.R., Kéfi, S., Loeuille, N., Memmott, J., van der Putten, W.H., van Veen, F.J.F., Thébault, E., 2011. The ecological and evolutionary implications of merging different types of networks. Ecol. Lett. 14, 1170–1181.

Forest, F., Grenyer, R., Rouget, M., Davies, T.J., Cowling, R.M., Faith, D.P., Balmford, A., Manning, J.C., Proches, S., van der Bank, M., Reeves, G., Hedderson, T.A.J., Savolainen, V., 2007. Preserving the evolutionary potential of floras in biodiversity hotspots. Nature 445, 757–760.

Fortuna, M.A., Stouffer, D.B., Olesen, J.M., Jordano, P., Mouillot, D., Krasnov, B.R., Poulin, R., Bascompte, J., 2010. Nestedness versus modularity in ecological networks: two sides of the same coin? J. Anim. Ecol. 79, 811–817.

Friedman, J., Alm, E.J., 2012. Inferring correlation networks from genomic survey data. PLoS Comput. Biol. 8, e1002687. https://doi.org/10.1371/journal.pcbi.1002687.

Gaba, S., Chauvel, B., Dessaint, F., Bretagnolle, V., Petit, S., 2010. Weed species richness in winter wheat increases with landscape heterogeneity. Agric. Ecosyst. Environ. 138, 318–323.

Galimberti, A., De Mattia, F., Bruni, I., Scaccabarozzi, D., Sandionigi, A., Barbuto, M., Casiraghi, M., Labra, M., 2014. A DNA barcoding approach to characterize pollen collected by honeybees. PLoS One 9, e109363.

Gariepy, T.D., Kuhlmann, U., Gillott, C., Erlandson, M., 2008. A largescale comparison of conventional and molecular methods for the evaluation of host–parasitoid associations in non-target risk-assessment studies. J. Appl. Ecol. 15, 481–495.

Genrich, C.M., Mello, M.A.R., Silveira, F.A.O., Bronstein, J.L., Paglia, A.P., 2017. Duality of interaction outcomes in a plant–frugivore multilayer network. Oikos 126, 361–368. https://doi.org/10.1111/oik.03825.

Gerhold, P., Cahill Jr, J.F., Winter, M., Bartish, I.G., Prinzing, A., 2015. Phylogentic patterns are not proxies of community assembly mechanisms (they are far better). Funct. Ecol. 29, 600–614.

Gibson, R.H., Knott, B., Eberlein, T., Memmott, J., 2011. Sampling method influences the structure of plant–pollinator networks. Oikos 120, 822–831.

Gilpin, M.E., 1975. Stability of feasible predator-prey systems. Nature 254, 137–139.

Gnirke, A., Melnikov, A., Maguire, J., Rogov, P., LeProust, E.M., Brockman, W., Fennell, T., Giannoukos, G., Fisher, S., Russ, C., Gabriel, S., Jaffe, D.B., Lander, E.S., Nusbaum, C., 2009. Solution hybrid selection with ultra-long oligonucleotides for massively parallel targeted sequencing. Nat. Biotechnol. 27, 182–189.

Godfray, H.C.J., Beddington, J.R., Crute, I.R., Haddad, L., Lawrence, D., Muir, J.F., Pretty, J., Robinson, S., Thomas, S.M., Toulmin, C., 2010. Food security: the challenge of feeding 9 billion people. Science 327, 812–818.

Goodwin, S., McPherson, J.D., McCombie, W.R., 2016. Coming of age: ten years of next-generation sequencing technologies. Nat. Rev. Genet. 17, 333–351.

Gotelli, N.J., Chao, A., 2013. Measuring and estimating species richness, species diversity and biotic similarity from sampling data. In: Levin, S.A. (Ed.), second ed. In: Encyclopedio of Biodiversity, vol. 5. Academic Press, Waltham, MA, pp. 195–211.

Gotelli, N.J., Colwell, R.K., 2011. Estimating species richness. In: Magurran, A.E., McGill, B.J. (Eds.), Biological Diversity: Frontiers in Measurement and Assessment. Oxford University Press, Oxford, pp. 39–54.

Gounand, I., Mouquet, N., Canard, E., Guichard, F., Hauzy, C., Gravel, D., 2014. The paradox of enrichment in metaecosystems. Am. Nat. 184, 752–763.

Gravel, D., Guichard, F., Loreau, M., Mouquet, N., 2010. Source and sink dynamics in meta-ecosystems. Ecology 91, 2172–2184.

Gravel, D., Massol, F., Leibold, M.A., 2016. Stability and complexity in model meta-ecosystems. Nat. Commun. 7, 12457.

Gray, C., Baird, D.J., Baumgartner, S., Jacob, U., Jenkins, G.B., O'Gorman, E.J., Lu, X., Ma, A., Pocock, M.J.O., Schuwirth, N., Thompson, M., Woodward, G., 2014. FORUM: ecological networks: the missing links in biomonitoring science. J. Appl. Ecol. 51, 1444–1449. https://doi.org/10.1111/1365-2664.12300.

Hadfield, J.D., Krasnov, B.R., Poulin, R., Nakagawa, S., 2014. A tale of two phylogenies: comparative analyses of ecological interactions. Am. Nat. 183, 174–187.

Haegeman, B., Loreau, M., 2015. A graphical-mechanistic approach to spatial resource competition. Am. Nat. 185, E1–E13.

Hairston, N.G., Smith, F.E., Slobodkin, L.B., 1960. Community structure, population control, and competition. Am. Nat. 44, 421–425.

Hampton, S.E., Strasser, C.A., Tewksbury, J.J., Gram, W.K., Budden, A.E., Batcheller, A.L., Duke, C.S., Porter, J.H., 2013. Big data and the future of ecology. Front. Ecol. Environ. 11, 156–162. https://doi.org/10.1890/120103.

Hawksworth, D.L., 2001. The magnitude of fungal diversity: the 1.5 million species estimate revisited. Mycol. Res. 105, 1422–1432.

Hebert, P.D.N., Cywinska, A., Ball, S.L., deWaard, J.R., 2003. Biological identifications through DNA barcodes. Proc. R. Soc. Lond. Biol. 270, 313–321.

Hebert, P.D.N., Penton, E.H., Burns, J.M., Janzen, D.H., Hallwachs, W., 2004. Ten species in one: DNA barcoding reveals cryptic species in the neotropical skipper butterfly Astraptes fulgerator. Proc. Natl. Acad. Sci. U.S.A. 101, 14812–14817.

Hibbett, D.S., Ohman, A., Kirk, P.M., 2009. Fungal ecology catches fire. New Phytol. 184, 279–282.

Hiltunen, T., Hairston, N.G., Hooker, G., Jones, L.E., Ellner, S.P., 2014. A newly discovered role of evolution in previously published consumer-resource dynamics. Ecol. Lett. 17, 915–923. https://doi.org/10.1111/ele.12291.

Holt, R.D., 1997. From metapopulation dynamics to community structure: some consequences of spatial heterogeneity. In: Hanski, I.A., Gilpin, M.E. (Eds.), Metapopulation Biology: Ecology, Genetics, and Evolution. Academic Press, San Diego, pp. 149–164.

Holyoak, M., Leibold, M.A., Holt, R.D., 2005. Metacommunities: Spatial Dynamics and Ecological Communities. University of Chicago Press, Chicago and London.

Hudson, P.J., Dobson, A.P., Lafferty, K.D., 2006. Is a healthy ecosystem one that is rich in parasites? Trends Ecol. Evol. 21, 381–385.

Ings, T.C., Montoya, J.M., Bascompte, J., Blüthgen, N., Brown, L., Dormann, C.F., Edwards, F., Figueroa, D., Jacob, U., Jones, J.I., Lauridsen, R.B., Ledger, M.E., Lewis, H.M., Olesen, J.M., van Veen, F.J.F., Warren, P.H., Woodward, G., 2009. Review: ecological networks—beyond food webs. J. Anim. Ecol. 78, 253–269.

Ives, A.R., Godfray, H.C.J., 2006. Phylogenetic analysis of trophic associations. Am. Nat. 168, E1–E14.

Jacquet, C., Moritz, C., Morissette, L., Legagneux, P., Massol, F., Archambault, P., Gravel, D., 2016. No complexity-stability relationship in empirical ecosystems. Nat. Commun. 7, 12573.

Jakuschkin, B., Fievet, V., Schwaller, L., Ouadah, S., Robin, S., Robin, C., Vacher, C., 2016. Deciphering the pathobiome: intra-and inter-kingdom interactions involving the pathogen Erysiphe alphitoides. Microb. Ecol. 72, 870–880. https://doi.org/10.1007/s00248-016-0777-x.

Ji, Y., Ashton, L., Pedley, S.M., Edwards, D.P., Tang, Y., Nakamura, A., Ktiching, R., Dolman, P.M., Woodcock, P., Edwards, F.A., Larsenn, T.H., Hsu, W.W., Benedick, S., Hamer, K.C., Wilcove, D.S., Bruce, C., Wang, X., Levi, T., Lott, M.,

Emerson, B.C., Yu, D.W., 2013. Reliable, verifiable and efficient monitoring of biodiversity via metabarcoding. Ecol. Lett. 16, 1245–1257. https://doi.org/10.1111/ele.12162.

Jordano, P., 1987. Patterns of mutualistic interactions in pollination and seed dispersal: connectance, dependence asymmetries, and coevolution. Am. Nat. 129, 657–677.

Jordano, P., 2016. Sampling networks of ecological interactions. Funct. Ecol. 30, 1883–1893. https://doi.org/10.1111/1365-2435.12763.

Jordano, P., Bascompte, J., Olesen, J.M.M., 2003. Invariant properties in coevolutionary networks of plant-animal interactions. Ecol. Lett. 6, 69–81.

Kaartinen, R., Stone, G., Hearn, J., Lohse, K., Roslin, T., 2010. Revealing secret liaisons: DNA barcoding changes our understanding of food webs. Ecol. Entomol. 35, 623–638.

Kaiser-Bunbury, C.N., Blüthgen, N., 2015. Integrating network ecology with applied conservation: a synthesis and guide to implementation. AoB Plants 7, plv076, https://doi.org/10.1093/aobpla/plv076.

Kaiser-Bunbury, C.N., Mougal, J., Whittington, A.E., Valentin, T., Gabriel, R., Olesen, J.M., Blüthgen, N., 2017. Ecosystem restoration strengthens pollination network resilience and function. Nature 542, 223–227. https://doi.org/10.1038/nature21071.

Kaiser-Bunbury, C.N., Muff, S., Memmott, J., Muller, C.B., Caflisch, A., 2010. The robustness of pollination networks to the loss of species and interactions: a quantitative approach incorporating pollinator behaviour. Ecol. Lett. 13, 442–452.

Kamenova, S., Bartley, T.J., Bohan, D.A., Boutain, J.R., Colautti, R.I., Domaizon, I., Fontaine, C., Lemainque, A., Le Viol, I., Mollot, G., Perga, M.E., Ravigné, V., Massol, F., 2017. Invasions toolkit: current methods for tracking the spread and impact of invasive species. In: Bohan, D.A., Dumbrell, A.J., Massol, F. (Eds.), Advances in Ecological Research. Academic Press, pp. 85–182.

Karimi, B., Maron, P.A., Chemidlin-Prevost Boure, N., Bernard, N., Gilbert, D., Ranjard, L., 2017. Microbial diversity and ecological networks as indicators of environmental quality. Environ. Chem. Lett. 15, 265–281. https://doi.org/10.1007/s10311-017-0614-6.

Kéfi, S., Berlow, E.L., Wieters, E.A., Navarrete, S.A., Petchey, O.L., Wood, S.A., Boit, A., Joppa, L.N., Lafferty, K.D., Williams, R.J., Martinez, N.D., Menge, B.A., Blanchette, C.A., Iles, A.C., Brose, U., 2012. More than a meal integrating non-feeding interactions into food webs. Ecol. Lett. 15, 291–300.

Kéfi, S., Miele, V., Wieters, E.A., Navarrete, S.A., Berlow, E.L., 2016. How structured is the entangled bank? The surprisingly simple organization of multiplex ecological networks leads to increased persistence and resilience. PLoS Biol. 14, 1–21. https://doi.org/10.1371/journal.pbio.1002527.

Kekkonen, M., Hebert, P.D.N., 2014. DNA barcode-based delineation of putative species: efficient start for taxonomic workflows. Mol. Ecol. Resour. 14, 706–715.

King, R.A., Moreno-Ripoll, R., Agustí, N., Shayler, S.P., Bell, J.R., Bohan, D.A., Symondson, W.O.C., 2011. Multiplex reactions for the molecular detection of predation on pest and nonpest invertebrates in agroecosystems. Mol. Ecol. Resour. 11, 370–373.

Kitson, J.J.N., Warren, B.H., Florens, F.B.V., Baider, C., Strasverg, D., Emerson, B.C., 2013. Molecular characterization of trophic ecology within an island radiation of insect herbivores (Curculionidae: Entiminae: Cratopus). Mol. Ecol. 22, 5441–5455.

Kitson, J.J.N., Hahn, C., Sands, R.J., Straw, N.A., Evans, D.M., Lunt, D.H., 2016. Nested metabarcode tagging: a robust tool for studying species interactions in ecology and evolution. Biorxiv. bioRxiv, 035071. https://doi.org/10.1101/035071.

Kurtz, Z.D., Mueller, C.L., Miraldi, E.R., Littman, D.R., Blaser, M.J., Bonneau, R.A., 2015. Sparse and compositionally robust inference of microbial ecological networks. PLoS Comput. Biol. 11 (5), e1004226.

Lafferty, K.D., Allesina, S., Arim, M., Briggs, C.J., De Leo, G., Dobson, A.P., Dunne, J.A., Johnson, P.T.J., Kuris, A.M., Marcogliese, D.J., Martinez, N.D., Memmott, J., Marquet, P.A., McLaughlin, J.P., Mordecai, E.A., Pascual, M., Poulin, R., Thieltges, D.W., 2008. Parasites in food webs: the ultimate missing links. Ecol. Lett. 11, 533–546. https://doi.org/10.1111/j.1461-0248.2008.01174.x.

Langille, M.G.I., Zaneveld, J., Caporaso, J.G., McDonald, D., Knights, D., Reyes, J.A., Clemente, J.C., Burkepile, D.E., Vega Thurber, R.L., Knight, R., Beiko, R.G., Huttenhower, C., 2013. Predictive functional profiling of microbial communities using 16S rRNA marker gene sequences. Nat. Biotechnol. 31, 814–821.

Layeghifard, M., Hwang, D.M., Guttman, D.S., 2017. Disentangling interactions in the microbiome: a network perspective. Trends Microbiol. 25, 217–228. https://doi.org/10.1016/j.tim.2016.11.008.

Leibold, M.A., Holyoak, M., Mouquet, N., Amarasekare, P., Chase, J.M., Hoopes, M.F., Holt, R.D., Shurin, J.B., Law, R., Tilman, D., Loreau, M., Gonzalez, A., 2004. The metacommunity concept: a framework for multi-scale community ecology. Ecol. Lett. 7, 601–613.

Leinonen, R., Akhtar, R., Birney, E., Bower, L., Cerdeno-Tárraga, A., Cheng, Y., Cleland, I., Faruque, N., Goodgame, N., Gibson, R., Hoad, G., Jang, M., Pakseresht, N., Plaister, S., Radhakrishnan, R., Reddy, K., Sobhany, S., Hoopsen, P.T., Vaughan, R., Zalunin, V., Cochrane, G., 2011. The european nucleotide archive. Nucleic Acids Res. 39, D28–D31.

Leray, M., Knowlton, N., 2017. Random sampling causes the low reproducibility of rare eukaryotic OTUs in Illumina COI metabarcoding. PeerJ 5, e3006.

Leray, M., Yang, J.Y., Meyer, C.P., Mills, S.C., Agudelo, N., Ranwez, V., Boehm, J.T., Machida, R.J., 2013. A new versatile primer set targeting a short fragment of the mitochondrial COI region for metabarcoding metazoan diversity: application for characterizing coral reef fish gut contents. Front. Zool. 10, 34.

Lewinsohn, T.M., Roslin, T., 2008. Four ways towards tropical herbivore megadiversity. Ecol. Lett. 11, 398–416. https://doi.org/10.1111/j.1461-0248.2008.01155.x.

Lewinsohn, T.M., Prado, P.I., Jordano, P., Bascompte, J., Olesen, J.M., 2006. Structure in plant-animal interaction assemblages. Oikos 113, 174–184.

Lewis, H.M., Law, R., 2007. Effect of dynamics on ecological networks. J. Theor. Biol. 247, 64–76.

Li, C., Lim, K.M.K., Chng, K.R., Nagarajan, N., 2016. Predicting microbial interactions through computational approaches. Methods 102, 12–19. https://doi.org/10.1016/j.ymeth.2016.02.019.

Lidicker, W.Z.A., 1979. Clarification of interactions in ecological systems. BioScience 29, 475–477.

Lima-Mendez, G., Faust, K., Henry, N., Decelle, J., Colin, S., Carcillo, F., Chaffron, S., Ignacio-Espinosa, J.C., Roux, S., Vincent, F., Bittner, L., Darzi, Y., Wang, J., Audic, S., Berline, L., Bontempi, G., Cabello, A.M., Coppola, L., Cornejo-Castillo, F.M., d'Ovidio, F., De Meester, L., Ferrera, I., Garet-Delmas, M.-J., Guidi, L., Lara, E., Pesant, S., Royo-Llonch, M., Salazar, G., Sánchez, P., Sebastian, M., Souffreau, C., Dimier, C., Picheral, M., Searson, S., Kandels-Lewis, S., Gorsky, G., Not, F., Ogata, H., Speich, S., Stemmann, L., Weissenbach, J., Wincker, P., Acinas, S.G., Sunagawa, S., Bork, P., Sullivan, M.B., Karsenti, E., Bowler, C., de Vargas, C., Raes, J., 2015. Determinants of community structure in the global plankton interactome. Science 348, 1262073. https://doi.org/10.1126/science.1262073.

Lindeman, R.L., 1942. The trophic-dynamic aspect of ecology. Ecology 23, 399–417.

Liu, S.L., Li, Y., Lu, J., Su, X., Tang, M., Zhang, R., Zhou, L., Zhou, C., Yang, Q., Ji, Y., Yu, D.W., Zhou, X., 2013. SOAPBarcode: revealing arthropod biodiversity through assembly of Illumina shotgun sequences of PCR amplicons. Meth. Ecol. Evol. 4, 1142–1150.

Loreau, M., Holt, R.D., 2004. Spatial flows and the regulation of ecosystems. Am. Nat. 163, 606–615.

Loreau, M., Mouquet, N., Holt, R.D., 2003. Meta-ecosystems: a theoretical framework for a spatial ecosystem ecology. Ecol. Lett. 6, 673–679.

Ma, A., Bohan, D., Canard, E., Derocles, S., Gray, C., Kratina, K., Lu, X., Macfadyen, S., Romero, G., 2017. A replicated network approach to "Big Data" in ecology. Advances in Ecological Research.

MacArthur, R., 1955. Fluctuations of animal populations and a measure of community stability. Ecology 36, 533–536.

Macfadyen, S., Gibson, R., Polaszek, A., Morris, R.J., Craze, P.G., Planqué, R., Symondson, W.O.C., Memmott, J., 2009. Do differences in food web structure between organic and conventional farms affect the ecosystem service of pest control? Ecol. Lett. 12, 229–238. https://doi.org/10.1111/j.1461-0248.2008.01279.x.

Macher, J.N., Zizka, V., Weigand, A.M., Leese, F., 2017. A simple centrifugation protocol increases mitochondrial DNA yield 140-fold and facilitates mitogenomic studies. BioRxiv. 106583.

Marleau, J.N., Guichard, F., Loreau, M., 2015. Emergence of nutrient co-limitation through movement in stoichiometric meta-ecosystems. Ecol. Lett. 18, 1163–1173.

Martinson, V.G., Douglas, A.E., Jaenike, J., 2017. Community structure of the gut microbiota in sympatric species of wild Drosophila. Ecol. Lett. 20, 629–639. https://doi.org/10.1111/ele.12761.

Massol, F., Petit, S., 2013. Interaction networks in agricultural landscape mosaics. Adv. Ecol. Res. 49, 291–338.

Massol, F., Gravel, D., Mouquet, N., Cadotte, M.W., Fukami, T., Leibold, M.A., 2011. Linking ecosystem and community dynamics through spatial ecology. Ecol. Lett. 14, 313–323.

May, R.M., 1972. Will a large complex system be stable? Nature 238, 413–414.

May, R.M., 1973a. Qualitative stability in model ecosystems. Ecology 54, 638–641.

May, R.M., 1973b. Stability and Complexity in Model Ecosystems. Princeton University Press, Princeton.

Mayfield, M.M., Levine, J.M., 2010. Opposing effects of competitive exclusion on the phylogenetic structure of communities. Ecol. Lett. 13, 1085–1093.

Médoc, V., Firmat, C., Sheath, D.J., Pegg, J., Andreou, D., Britton, J.R., 2017. Parasites and biological invasions: predicting ecological alterations at levels from individual hosts to whole networks. Adv. Ecol. Res. 57, 1–54.

Melián, C.J., Bascompte, J., Jordano, P., Krivan, V., Melian, C.J., 2009. Diversity in a complex ecological network with two interaction types. Oikos 118, 122–130. https://doi.org/10.1111/j.1600-0706.2008.16751.x.

Memmott, J., Waser, N.M., Price, M.V., 2004. Tolerance of pollination networks to species extinctions. Proc. R. Soc. Lond. Ser. B Biol. Sci. 271, 2605–2611.

Montoya, J.M., Pimm, S.L., Solé, R.V., 2006. Ecological networks and their fragility. Nature 442, 259–264.

Morriën, E., Hannula, S.E., Snoek, L.B., Helmsing, N.R., Zweers, H., de Hollander, M., Soto, R.L., Bouffaud, M.-L., Buée, M., Dimmers, W., Duyts, H., Geisen, S., Girlanda, M., Griffiths, R.I., Jørgensen, H.-B., Jensen, J., Plassart, P., Redecker, D., Schmelz, R.M., Schmidt, O., Thomson, B.C., Tisserant, E., Uroz, S., Winding, A., Bailey, M.J., Bonkowski, M., Faber, J.H., Martin, F., Lemanceau, P., de Boer, W., van Veen, J.A., van der Putten, W.H., 2017. Soil networks become more connected and take up more carbon as nature restoration progresses. Nat. Commun. 8, 14349.

Morris, R.J., Lewis, O.T., Godfray, H.C.J., 2004. Experimental evidence for apparent competition in a tropical forest food web. Nature 428, 310–313. https://doi.org/10.1038/nature02394.

Morris, R.J., Sinclair, F.H., Burwell, C.J., 2015. Food web structure changes with elevation but not rainforest stratum. Ecography 38, 792–802. https://doi.org/10.1111/ecog.01078.

Mougi, A., 2016. The roles of amensalistic and commensalistic interactions in large ecological network stability. Sci. Rep. 6, 29929.

Mougi, A., Kondoh, M., 2016. Food-web complexity, meta-community complexity and community stability. Sci. Rep. 6 (24478).

Mouquet, N., Loreau, M., 2002. Coexistence in metacommunities: the regional similarity hypothesis. Am. Nat. 159, 420–426.

Mouquet, N., Miller, T.E., Daufresne, T., Kneitel, J.M., 2006. Consequences of varying regional heterogeneity in source-sink metacommunities: a mechanistic model. Oikos 113, 481–488.

Mouquet, N., Gravel, D., Massol, F., Calcagno, V., 2013. Extending the concept of keystone species to communities and ecosystems. Ecol. Lett. 16, 1–8.

Muggleton, S., 1991. Inductive logic programming. New Generat. Comput. 8, 295–318.

Muggleton, S.H., Lin, D., Pahlavi, N., Tamaddoni-Nezhad, A., 2014. Meta-interpretive learning: application to grammatical inference. Mach. Learn. 94, 25–49.

Muggleton, S.H., Lin, D., Tamaddoni-Nezhad, A., 2015. Meta-interpretive learning of higher-order dyadic datalog: predicate invention revisited. Mach. Learn. 100, 49–73.

Mulder, C., Wouterse, M., Raubuch, M., Roelofs, W., Rutgers, M., 2006. Can transgenic maize affect soil microbial communities? PLoS Comput. Biol. 2, e128.

Mulder, C., Ahrestani, F.S., Bahn, M., Bohan, D.A., Bonkowski, M., Griffiths, B.S., Guicharnaud, R.A., Kattge, J., Krogh, P.H., Lavorel, S., Lewis, O.T., Mancinelli, G., Naeem, S., Peñuelas, J., Poorter, H., Reich, P.B., Rossi, L., Rusch, G.M., Sardans, J., Wright, I.J., 2013. Connecting the green and brown worlds: allometric and stoichiometric predictability of above- and below-ground networks. Adv. Ecol. Res. 49, 69–175.

Myers, N., Mittermeier, R.A., Mittermeier, C.G., da Fonseca, G.A.B., Kent, J., 2000. Biodiversity hotspots for conservation priorities. Nature 403, 853–858.

Nakazawa, T., Yamamura, N., 2006. Community structure and stability analysis for intraguild interactions among host, parasitoid, and predator. Popul. Ecol. 48, 139–149.

Neutel, A.M., Heesterbeek, J.A.P., de Ruiter, P.C., 2002. Stability in real food webs: weak links in long loops. Science 296, 1120–1123.

Nguyen, N.H., Song, Z., Bates, S.T., Branco, S., Tedersoo, L., Menke, J., Schilling, J.S., Kennedy, P.G., 2016. FUNGuild: an open annotation tool for parsing fungal community datasets by ecological guild. Fungal Ecol. 20, 241–248.

Nilsson, R.H., Tedersoo, L., Lindahl, B.D., Kjøller, R., Carlsen, T., Quince, C., Aberenkov, K., Pennanen, T., Stenlid, J., Bruns, T., Larsson, K.-H., Kõljalg, U., Kauserud, H., 2011. Towards standardization of the description and publication of next-generation sequencing datasets of fungal communities. New Phytol. 191, 314–318.

Novak, M., Wootton, J.T., Doak, D.F., Emmerson, M., Estes, J.A., Tinker, M.T., 2011. Predicting community responses to perturbations in the face of imperfect knowledge and network complexity. Ecology 92, 836–846.

Oksanen, L., Fretwell, S.D., Arruda, J., Niemelä, P., 1981. Exploitation ecosystems in gradients of primary productivity. Am. Nat. 118, 240–261.

Olesen, J.M., Bascompte, J., Dupont, Y.L., Jordano, P., 2007. The modularity of pollination networks. Proc. Natl. Acad. Sci. U.S.A. 104, 19891–19896.

Ovaskainen, O., Abrego, N., Halme, P., Dunson, D., 2016. Using latent variable models to identify large networks of species-to-species associations at different spatial scales. Meth. Ecol. Evol. 7, 549–555. https://doi.org/10.1111/2041-210X.12501.

Ovaskainen, O., Tikhonov, G., Dunson, D., Grøtan, V., Engen, S., Sæther, B., Abrego, N., 2017a. How are species interactions structured in species-rich communities? A new method for analysing time-series data. Proc. R. Soc. B Biol. Sci. 284, 20170768. https://doi.org/10.1098/rspb.2017.0768.

Ovaskainen, O., Tikhonov, G., Norberg, A., Blanchet, F.G., Duan, L., Dunson, D., Roslin, T., Abrego, N., 2017b. How to make more out of community data? A conceptual framework and its implementation as models and softwares. Ecol. Lett. 20, 561–576.

Paine, R.T., 1966. Food web complexity and species diversity. Am. Nat. 100, 65–75.

Paine, R.T., 1969. A note on trophic complexity and community stability. Am. Nat. 103, 91–93.

Paine, R.T., 1974. Intertidal community structure: experimental studies on the relationship between a dominant competitor and its principal predator. Oecologia 15, 93–120.

Pantel, J.H., Bohan, D., Calcagno, V., David, P., Duyck, P.-F., Kamenova, S., Loeuille, N., Mollot, G., Romanuk, T., Thébault, E., Tixier, P., Massol, F., 2017. 14 questions for invasion in ecological networks. Adv. Ecol. Res. 56, 293–340.

Parravicini, V., Villéger, S., McClanahan, T.R., Arias-González, J.E., Bellwood, D.R., Belmaker, J., Chabanet, P., Floeter, S.R., Friedlander, A.M., Guilhaumon, F., Vigliola, L., Kulbicki, M., Mouillot, D., 2014. Global mismatch between species richness and vulnerability of reef fish assemblages. Ecol. Lett. 17, 1101–1110.

Pearse, I.S., Altermatt, F., 2013. Predicting novel trophic interactions in a non-native world. Ecol. Lett. 16, 1088–1094.

Peay, K.G., Kennedy, P.G., Bruns, T.D., 2008. Fungal community ecology: a hybrid beast with a molecular master. BioScience 58, 799. https://doi.org/10.1641/B580907.

Peralta, G., 2016. Merging evolutionary history into species interaction networks. Funct. Ecol. 30, 1917–1925. https://doi.org/10.1111/1365-2435.12669.

Pillai, P., Gonzalez, A., Loreau, M., 2011. Metacommunity theory explains the emergence of food web complexity. Proc. Nat. Acad. Sci. U.S.A 108, 19293–19298.

Pilosof, S., Fortuna, M.A., Cosson, J.-F., Galan, M., Kittipong, C., Ribas, A., Segal, E., Krasnov, B.R., Morand, S., Bascompte, J., 2014. Host–parasite network structure is associated with community-level immunogenetic diversity. Nat. Commun. 5, 5172.

Pilosof, S., Porter, M.A., Pascual, M., Kéfi, S., 2017. The multilayer nature of ecological networks. Nat. Ecol. Evol. 1, 101. https://doi.org/10.1038/s41559-017-0101.

Pimm, S.L., 1980. Properties of food webs. Ecology 61, 219–225.

Piñol, J., San Andres, V., Clare, E.L., Mir, G., Symondson, W.O.C., 2014. A pragmatic approach to the analysis of diets of generalist predators: the use of next-generation sequencing with no blocking probes. Mol. Ecol. Resour. 14, 18–26. https://doi.org/10.1111/1755-0998.12156.

Pocock, M.J.O., Evans, D.M., Memmott, J., 2012. The robustness and restoration of a network of ecological networks. Science 335, 973–977. https://doi.org/10.1126/science.1214915.

Poelen, J.H., Simons, J.D., Mungal, C.J., 2014. Global biotic interactions: an open infrastructure to share and analyse species-interaction datasets. Eco. Inform. 24, 148–159.

Poisot, T., 2013. An a posteriori measure of network modularity. F1000Res 2, 130. https://doi.org/10.12688/f1000research.2-130.v2.

Poisot, T., Gravel, D., 2014. When is an ecological network complex? Connectance drives degree distribution and emerging network properties. PeerJ 2, e251.

Poisot, T., Canard, E., Mouillot, D., Mouquet, N., Gravel, D., Jordan, F., 2012. The dissimilarity of species interaction networks. Ecol. Lett. 15, 1353–1361. https://doi.org/10.1111/ele.12002.

Poisot, T., Baiser, B., Dunne, J.A., Kéfi, S., Massol, F., Mouquet, N., Romanuk, T.N., Stouffer, D.B., Wood, S.A., Gravel, D., 2016a. Mangal—making ecological network analysis simple. Ecography 39, 384–390.

Poisot, T., Stouffer, D.B., Kéfi, S., 2016b. Describe, understand and predict: why do we need networks in ecology? Funct. Ecol. 30, 1878–1882. https://doi.org/10.1111/1365-2435.12799.

Polis, G.A., 1991. Complex trophic interactions in deserts: an empirical critique of food web theory. Am. Nat. 138, 123–155.

Polis, G.A., Anderson, W.B., Holt, R.D., 1997. Toward an integration of landscape and food web ecology: the dynamics of spatially subsidized food webs. Annu. Rev. Ecol. Syst. 28, 289–316.

Polz, M.F., Cavanaugh, C.M., 1998. Bias in template-to-product ratios in multitemplate PCR. Appl. Environ. Microbiol. 64, 3724–3730.

Pornon, A., Escaravage, N., Burrus, M., Holota, H., Khimoun, A., Mariette, J., Pellizzari, C., Iribar, A., Etienne, R., Taberlet, P., Vidal, M., Winterton, P., Zinger, L., Andalo, C., 2016. Using metabarcoding to reveal and quantify plant-pollinator interactions. Sci. Rep. 6, 27282. https://doi.org/10.1038/srep27282.

Poudel, R., Jumpponen, A., Schlatter, D.C., Paulitz, T.C., McSpadden Gardener, B.B., Kinkel, L.L., Garrett, K.A., 2016. Microbiome networks: a systems framework for identifying candidate microbial assemblages for disease management. Phytopathology 106, 1083–1096. https://doi.org/10.1094/PHYTO-02-16-0058-FI.

Puillandre, N., Lambert, A., Brouillet, S., Achaz, G., 2012. ABGD, Automatic Barcode Gap Discovery for primary species delimitation. Mol. Ecol. 21, 1864–1877.

Quintessence Consortium, 2016. Networking our way to better ecosystem service provision. Trends Ecol. Evol. 31, 105–115. https://doi.org/10.1016/j.tree.2015.12.003.

R Development Core Team, 2017. R: A Language and Environment for Statistical Computing. R Foundation for Statistical Computing, Vienna, Austria.

Rafferty, N.E., Ives, A.R., 2013. Phylogenetic trait-based analyses of ecological networks. Ecology 94, 2321–2333.

Ratnasingham, S., Hebert, P.D.N., 2007. BOLD: the barcode of life data system. Mol. Ecol. Notes 7, 355–364.

Raxworthy, C.J., Martinez-Meyer, E., Horning, N., Nussbaum, R.A., Schneider, G.E., Ortega-Huerta, M.A., Peterson, A.T., 2003. Predicting distributions of known and unknown reptile species in Madagascar. Nature 426, 837–841.

Razgour, O., Clare, E.L., Zeale, M.R.K., Hanmer, J., Baerholm Schnell, I., Rasmussen, M., Gilbert, T.P., Jones, G., 2011. High-throughput sequencing offers insight into mechanisms of resource partitioning in cryptic bat species. Ecol. Evol. 1, 556–570.

Rezende, E.L., Lavabre, J.E., Guimarães, P.R., Jordano, P., Bascompte, J., 2007. Nonrandom coextinctions in phylogenetically structured mutualistic networks. Nature 448, 925–928. https://doi.org/10.1038/nature05956.

Rivera-Hutinel, A., Bustamante, R.O., Marín, V.H., Medel, R., 2012. Effects of sampling completeness on the structure of plant-pollinator networks. Ecology 93, 1593–1603.

Rodríguez-Echeverría, S., Traveset, A., 2015. Putative linkages between below- and above-ground mutualisms during alien plant invasions. AoB Plants 7, plv062.

Rohr, R.P., Naisbit, R.E., Mazza, C., Bersier, L.-F., 2016. Matching–centrality decomposition and the forecasting of new links in networks. Proc. R. Soc. B 283, 20152702.

Rooney, N., McCann, K., Gellner, G., Moore, J.C., 2006. Structural asymmetry and the stability of diverse food webs. Nature 442, 265–269.

Rougerie, R., Smith, M.A., Fernandez-Triana, J.F., Lopez-Vaamonde, C.L., Ratnasingham, S., Hebert, P.D.N., 2011. Molecular analysis of parasitoid linkages (MAPL): gut contents of adult parasitoid wasps reveal larval host. Mol. Ecol. 20, 179–186.

Sander, E.L., Wootton, J.T., Allesina, S., 2017. Ecological network inference from long-term presence-absence data. Sci. Rep. 7, 7154. https://doi.org/10.1038/s41598-017-07009-x.

Säterberg, T., Sellman, S., Ebenman, B., 2013. High frequency of functional extinctions in ecological networks. Nature 499, 468–471. https://doi.org/10.1038/nature12277.

Schiebelhut, L.M., Abboud, S.S., Gómez Daglio, L.E., Swift, H.F., Dawson, M.N., 2017. A comparison of DNA extraction methods for high-throughput DNA analyses. Mol. Ecol. Resour. 17, 721–729.

Schleuning, M., Fründ, J., Schweiger, O., Welk, E., Albrecht, J., Albrecht, M., Beil, M., Benadi, G., Blüthgen, N., Bruelheide, H., Böhning-Gaese, K., Dehling, D.M., Dormann, C.F., Exeler, N., Farwig, N., Harpke, A., Hickler, T., Kratochwil, A., Kuhlmann, M., Kühn, I., Michez, D., Mudri-Stojnić, S., Plein, M., Rasmont, P., Schwabe, A., Settele, J., Vujić, A., Weiner, C.N., Wiemers, M., Hof, C., 2016. Ecological networks are more sensitive to plant than to animal extinction under climate change. Nat. Commun. 7, 13965. https://doi.org/10.1038/ncomms13965.

Schneider, F.D., Brose, U., Rall, B.C., Guill, C., 2016. Animal diversity and ecosystem functioning in dynamic food webs. Nat. Commun. 7, 12718.

Schoener, T.W., 1989. Food webs from the small to the large. Ecology 70, 1559–1589.

Shang, Y., Sikorski, J., Bonkowski, M., Fiore-Donno, A.-M., Kandeler, E., Marhan, S., Boeddinghaus, R.S., Solly, E.F., Schrumpf, M., Schöning, I., Wubet, T., Buscot, F., Overmann, J., 2017. Inferring interactions in complex microbial communities from nucleotide sequence data and environmental parameters. PLoS One 12, e0173765. https://doi.org/10.1371/journal.pone.0173765.

Shokralla, S., Porter, T.M., Gibson, J.F., Dobosz, R., Janzen, D.H., Hallwachs, W., Golding, G.B., Hajibabaei, M., 2015. Massively parallel multiplex DNA sequencing for specimen identification using an Illumina MiSeq platform. Sci. Rep. 5, 9687.

Sickel, W., Ankenbrand, M., Grimmer, G., Holzschuh, A., Hartel, S., Lanzen, J., Steffan-Dewenter, I., Keller, A., 2015. Increased efficiency in identifying mixed pollen samples by meta-barcoding with a dual-indexing approach. BMC Ecol. 15, 20.

Smith, M.A., Woodley, N.E., Janzen, D.H., Hallwachs, W., Hebert, P.D.N., 2006. DNA barcodes reveal cryptic host specificity within the presumed polyphagous members of a genus of parasitoid flies (Diptera: Tachinidae). Proc. Natl. Acad. Sci. U.S.A. 103, 3657–3662.

Smith, M.A., Woodley, N.E., Janzen, D.H., Hallwachs, W., Hebert, P.D.N., 2007. DNA barcodes affirm that 16 species of apparently generalist tropical parasitoid flies (Diptera, Tachinidae) are not all generalists. Proc. Natl. Acad. Sci. U.S.A. 104, 4967–4972.

Smith, M.A., Rodriguez, J.J., Whitfield, J.B., Deans, A.R., Janzen, D.H., Hallwachs, W., Hebert, P.D.N., 2008. Extreme diversity of tropical parasitoid wasps exposed by iterative integration of natural history, DNA barcoding, morphology, and collections. Proc. Natl. Acad. Sci. U.S.A. 105, 12359–12364.

Solow, A.R., Beet, A.R., 1998. On lumping species in food webs. Ecology 79, 2013–2018.

Sommeria-Klein, G., Zinger, L., Taberlet, P., Coissac, E., Chave, J., 2016. Inferring neutral biodiversity parameters using environmental DNA datasets. Sci. Rep. 6, 35644.

Sonet, G., Jordaens, K., Braet, Y., Bourguignon, L., Dupont, E., Backeljau, T., Desmyter, S., 2013. Utility of GenBank and the Barcode of Life Data Systems (BOLD) for the identification of forensically important Diptera from Belgium and France. ZooKeys 365, 307–328.

Srinivasan, U.T., Dunne, J.A., Harte, J., Martinez, N.D., 2007. Response of complex food webs to realistic extinction sequences. Ecology 88, 671–682.

Stamatakis, A., 2014. RAxML version 8: a tool for phylogenetic analysis and post-analysis of large phylogenies. Bioinformatics 30, 1312–1313.

Staniczenko, P.P.A., Lewis, O.T., Jones, N.S., Reed-Tsochas, F., 2010. Structural dynamics and robustness of food webs. Ecol. Lett. 13, 891–899.

Stenseth, N.C., 1985. The structure of food webs predicted from optimal food selection models: an alternative to Pimm's stability hypothesis. Oikos 44, 361–364.

Stouffer, D.B., Camacho, J., Jiang, W., Amaral, L.A.N., 2007. Evidence for the existence of a robust pattern of prey selection in food webs. Proc. R. Soc. Lond. B Biol. Sci. 274, 1931–1940.

Sugihara, G., May, R., Ye, H., Hsieh, C.-H., Deyle, E., Fogarty, M., Munch, S., 2012. Detecting causality in complex ecosystems. Science 338, 496–500. https://doi.org/10.1126/science.1227079.

Symondson, W.O.C., Harwood, J.D., 2014. Special issue on molecular detection of trophic interactions: unpicking the tangled bank. Mol. Ecol. 23, 3601–3604.

Szitenberg, A., John, M., Blaxter, M.L., Lunt, D.H., 2015. ReproPhylo: an environment for reproducible phylogenomics. PLoS Comput. Biol. 11 (9), e1004447.

Tamaddoni-Nezhad, A., Chaleil, R., Kakas, A., Muggleton, S., 2006. Application of abductive ILP to learning metabolic network inhibition from temporal data. Mach. Learn. 64, 209–230.

Tamaddoni-Nezhad, A., Milani, G.A., Raybould, A., Muggleton, S., Bohan, D.A., 2013. Construction and validation of food webs using logic-based machine learning and text mining. Adv. Ecol. Res. 49, 225–289.

Tamaddoni-Nezhad, A., Bohan, D.A., Raybould, A., Muggleton, S., 2015. Towards machine learning of predictive models from ecological data. Proceedings of the 24th International Conference on Inductive Logic Programming Springer-Verlag, pp. 154–167.

Tang, M., Hardman, C.J., Ji, Y., Meng, G., Liu, S., Tan, M., Yang, S., Moss, E.D., Wang, J., Yang, C., Bruce, C., Nevard, T., Potts, S.G., Zhou, X., Yu, D.W., 2015. High-throughput monitoring of wild bee diversity and abundance via mitogenomics. Methods Ecol. Evol. 6, 1034–1043.

Tang, S., Pawar, S., Allesina, S., 2014. Correlation between interaction strengths drives stability in large ecological networks. Ecol. Lett. 17, 1094–1100. https://doi.org/10.1111/ele.12312.

Thébault, E., Fontaine, C., 2010. Stability of ecological communities and the architecture of mutualistic and trophic networks. Science 329, 853–856. https://doi.org/10.1126/science.1188321.

Thompson, P.L., Gonzalez, A., 2017. Dispersal governs the reorganization of ecological networks under environmental change. Nat. Ecol. Evol 1. Article number: 0162 (2017). https://doi.org/10.1038/s41559-017-0162.

Thompson, R.M., Brose, U., Dunne, J.A., Hall Jr., R.O., Hladyz, S., Kitching, R.L., Martinez, N.D., Rantala, H., Romanuk, T.N., Stouffer, D.B., Tylianakis, J.M., 2012. Food webs: reconciling the structure and function of biodiversity. Trends Ecol. Evol. 27, 689–697. https://doi.org/10.1016/j.tree.2012.08.005.

Thompson, M.S.A., Bankier, C., Bell, T., Dumbrell, A.J., Gray, C., Ledger, M.E., Lehmann, K., McKew, B.A., Sayer, C.D., Shelley, F., Trimmer, M., Warren, S.L., Woodward, G., 2016. Gene-to-ecosystem impacts of a catastrophic pesticide spill: testing a multilevel bioassessment approach in a river ecosystem. Freshw. Biol. 61, 2037–2050.

Thomsen, P.F., Willerslev, E., 2015. Environmental DNA—an emerging tool in conservation for monitoring past and present biodiversity. Biol. Conserv. 183, 4–18.

Tiede, J., Wemheuer, B., Traugott, M., Daniel, R., Tscharntke, T., Ebeling, A., Scherber, C., 2016. Trophic and non-trophic interactions in a biodiversity experiment assessed by next-generation sequencing. PLoS One 11 (2), e0148781.

Tilman, D., 1982. Resource Competition and Community Structure. Princeton University Press, Princeton.

Toju, H., Sato, H., Yamamoto, S., Kadowaki, K., Tanabe, A.S., Yazawa, S., Nishimura, O., Agata, K., 2013. How are plant and fungal communities linked to each other in below-ground ecosystems? A massively parallel pyrosequencing analysis of the association specificity of root-associated fungi and their host plants. Ecol. Evol. 3, 3112–3124. https://doi.org/10.1002/ece3.706.

Toju, H., Guimaraes, P.R., Olesen, J.M., Thompson, J.N., 2014. Assembly of complex plant-fungus networks. Nat. Commun. 5, 5273. https://doi.org/10.1038/ncomms6273.

Traugott, M., Bell, J.R., Broad, G.R., Powell, W., van Veen, F.J.F., Vollhardt, I.M., Symondson, W.O.C., 2008. Endoparasitism in cereal aphids: molecular analysis of a whole parasitoid community. Mol. Ecol. 17, 3928–3938.

Tylianakis, J.M., Tscharntke, T., Lewis, O.T., 2007. Habitat modification alters the structure of tropical host-parasitoid food webs. Nature 445, 202–205. https://doi.org/10.1038/nature05429.

Vacher, C., Piou, D., Desprez-Loustau, M.-L., 2008. Architecture of an antagonistic tree/fungus network: the asymmetric influence of past evolutionary history. PLoS One 3, e1740.

Vacher, C., Tamaddoni-Nezhad, A., Kamenova, S., Peyrard, N., Moalic, Y., Sabbadin, R., Schwaller, L., Chiquet, J., Smith, M.A., Vallance, J., Fievet, V., Jakuschkin, B., Bohan, D.A., 2016. Learning ecological networks from next-generation sequencing data. In: Woodward, G., Bohan, D.A. (Eds.), Advances in Ecological Research. In: vol. 54. Academic Press, Oxford, pp. 1–39. https://doi.org/10.1016/bs.aecr.2015.10.004.

Valentini, A., Pompanon, F., Taberlet, P., 2009a. Barcoding for ecologist. Trends Ecol. Evol. 24, 110–117.

Valentini, A., Pompanon, F., Taberlet, P., 2009b. DNA barcoding for ecologists. Trends Ecol. Evol. 24, 110–117.

van der Heijden, M.G.A., Bardgett, R.D., van Straalen, N.M., 2008. The unseen majority: soil microbes as drivers of plant diversity and productivity in terrestrial ecosystems. Ecol. Lett. 11, 296–310. https://doi.org/10.1111/j.1461-0248.2007.01139.x.

van Veen, F.J.F., Morris, R.J., Godfray, H.C.J., 2006. Apparent competition, quantitative food webs, and the structure of phytophagous insect communities. Annu. Rev. Entomol. 51, 187–208.

Vanbergen, A.J., Woodcock, B.A., Heard, M.S., Chapman, D.S., 2017. Network size, structure and mutualism dependence affect the propensity for plant-pollinator extinction cascades. Funct. Ecol. 31, 1285–1293. https://doi.org/10.1111/1365-2435.12823.

Vesty, A., Biswas, K., Taylor, M.W., Gear, K., Douglas, R.G., 2017. Evaluating the impact of DNA extraction method on the representation of human oral bacterial and fungal communities. PLoS One 12, e0169877.

Wang, S., Loreau, M., Arnoldi, J.-F., Fang, J., Rahman, K.A., Tao, S., de Mazancourt, C., 2017. An invariability-area relationship sheds new light on the spatial scaling of ecological stability. Nat. Commun. 8, 15211.

Webb, C.O., Ackerly, D.D., McPeek, M.A., Donoghue, M.J., 2002. Phylogenies and community ecology. Annu. Rev. Ecol. Evol. Syst. 33, 475–505.

Weiss, S., Van Treuren, W., Lozupone, C., Faust, K., Friedman, J., Deng, Y., Xia, L.C., Xu, Z.Z., Ursell, L., Alm, E.J., Birmingham, A., Cram, J.A., Fuhrman, J.A., Raes, J., Sun, F., Zhou, J., Knight, R., 2016. Correlation detection strategies in microbial data sets vary widely in sensitivity and precision. ISME J. 10, 1669–1681. https://doi.org/10.1038/ismej.2015.235.

Wells, J.D., Sperling, F.A., 2000. A DNA-based approach to the identification of insect species used for postmortem interval estimation and partial sequencing of the cytochrome oxydase b subunit gene I: a tool for the identification of European species of blow flies for postmortem interval. J. Forensic Sci. 45, 1358–1359.

Wells, J.D., Stevens, J.R., 2008. Application of DNA-based methods in forensic entomology. Annu. Rev. Entomol. 53, 103–120.

Whitman, W.B., Coleman, D.C., Wiebe, W.J., 1998. Prokaryotes: the unseen majority. PNAS 95, 6578–6583.

Williams, R.J., Martinez, N.D., 2000. Simple rules yield complex food webs. Nature 404, 180–183.

Williams, R.J., Martinez, N.D., 2004. Limits to trophic levels and omnivory in complex food webs: theory and data. Am. Nat. 163, 458–468.

Wirta, H.K., Hebert, P.D.N., Kaartinen, R., Prosser, S.W., Varkonyi, G., Roslin, T., 2014. Complementary molecular information changes our perception of food web structure. Proc. Natl. Acad. Sci. U.S.A. 111, 1885–1890.

Woodward, G., Brown, L.E., Edwards, F.K., Hudson, L.N., Milner, A.M., Reuman, D.C., Ledger, M.E., 2012. Climate change impacts in multispecies systems: drought alters food web size structure in a field experiment. Phil. Trans. R. Soc. B 367, 2990–2997.

Ye, H., Sugihara, G., 2016. Information leverage in interconnected ecosystems: overcoming the curse of dimensionality. Science 353, 922–925. https://doi.org/10.1126/science.aag0863.

Yodzis, P., 1998. Local trophodynamics and the interaction of marine mammals and fisheries in the Benguela ecosystem. J. Anim. Ecol. 67, 635–658.

Yu, D.W., Ji, Y., Emerson, B.C., Wang, X., Ye, C., Ding, Z., 2012. Biodiversity soup: metabarcoding of arthropods for rapid biodiversity assessment and biomonitoring. Meth. Ecol. Evol 3, 613–623.

Zeale, M.R.K., Butlin, R.K., Barker, G.L.A., Lees, D.C., Jones, G., 2011. Taxon-specific PCR for DNA barcoding arthropod prey in bat faeces. Mol. Ecol. Resour. 11, 236–244.

Zhang, J., Kapli, P., Stamatakis, A., 2013. A general species delimitation method with applications to phylogenetic placements. Bioinformatics 29, 2869–2876. https://doi.org/10.1093/bioinformatics/btt499.

Zorita, E., Cuscó, P., Filion, G.J., 2015. Starcode: sequence clustering based on all-pairs search. Bioinformatics 31, 1913–1919.

FURTHER READING

Gotelli, N.J., 2004. A taxonomic wish-list for community ecology. Phil. Trans. R. Soc. Lond. B. 359, 585–597. https://doi.org/10.1098/rstb.2003.1443.

Gotelli, N.J., Colwell, R.K., 2001. Quantifying biodiversity: procedures and pitfalls in the measurement and comparison of species richness. Ecol. Lett. 4, 379–391.

CHAPTER TWO

Why We Need Sustainable Networks Bridging Countries, Disciplines, Cultures and Generations for Aquatic Biomonitoring 2.0: A Perspective Derived From the *DNAqua-Net* COST Action

Florian Leese[*,†,1], Agnès Bouchez[‡], Kessy Abarenkov[§],
Florian Altermatt[¶,‖], Ángel Borja[#], Kat Bruce[**], Torbjørn Ekrem[††],
Fedor Čiampor Jr.[‡‡], Zuzana Čiamporová-Zaťovičová[‡‡],
Filipe O. Costa[§§], Sofia Duarte[§§], Vasco Elbrecht[*,¶¶],
Diego Fontaneto[‖‖], Alain Franc[##], Matthias F. Geiger[***],
Daniel Hering[†,†††], Maria Kahlert[‡‡‡], Belma Kalamujić Stroil[§§§],
Martyn Kelly[¶¶¶], Emre Keskin[‖‖‖], Igor Liska[###],
Patricia Mergen[****,††††], Kristian Meissner[‡‡‡‡], Jan Pawlowski[§§§§],
Lyubomir Penev[¶¶¶¶], Yorick Reyjol[‖‖‖‖], Ana Rotter[####],
Dirk Steinke[¶¶,*****], Bas van der Wal[†††††], Simon Vitecek[‡‡‡‡‡,§§§§§],
Jonas Zimmermann[¶¶¶¶¶], Alexander M. Weigand[*,†,‖‖‖‖‖]

*Aquatic Ecosystem Research, University of Duisburg-Essen, Essen, Germany
†Center of Water and Environmental Research (ZWU), University of Duisburg-Essen, Essen, Germany
‡INRA UMR CARRTEL, Thonon-les-bains, France
§University of Tartu, Tartu, Estonia
¶Eawag, Dübendorf, Switzerland
‖University of Zurich, Zürich, Switzerland
#AZTI, Pasaia, Spain
**NatureMetrics, CABI Site, Surrey, United Kingdom
††Norwegian University of Science and Technology, Trondheim, Norway
‡‡Zoology Lab, Plant Science and Biodiversity Center, Slovak Academy of Sciences, Bratislava, Slovakia
§§Centre of Molecular and Environmental Biology (CBMA), University of Minho, Braga, Portugal
¶¶Centre for Biodiversity Genomics, University of Guelph, Guelph, ON, Canada
‖‖National Research Council of Italy, Institute of Ecosystem Study, Verbania Pallanza, Italy
##BIOGECO, INRA, Univ. Bordeaux, Cestas, and Pleiade Team, INRIA Sud-Ouest, Talence, France
***Zoologisches Forschungsmuseum Alexander Koenig, Leibniz Institute for Animal Biodiversity, Bonn, Germany
†††Aquatic Ecology, University of Duisburg-Essen, Essen, Germany
‡‡‡Swedish University of Agricultural Sciences, Uppsala, Sweden
§§§University of Sarajevo—Institute for Genetic Engineering and Biotechnology, Sarajevo, Bosnia and Herzegovina
¶¶¶Bowburn Consultancy, Durham, United Kingdom

Advances in Ecological Research, Volume 58
ISSN 0065-2504
https://doi.org/10.1016/bs.aecr.2018.01.001

||||||Evolutionary Genetics Laboratory (eGL), Ankara University Agricultural Faculty, Ankara, Turkey
###ICPDR Permanent Secretariat, Vienna International Centre, Vienna, Austria
****Botanic Garden Meise, Meise, Belgium
††††Royal Museum for Central Africa, Tervuren, Belgium
‡‡‡‡Finnish Environment Institute, General Director's Office, Jyväskylä, Finland
§§§§University of Geneva, Geneva, Switzerland
¶¶¶¶Pensoft Publishers, Sofia, Bulgaria
||||||||AFB, The French Agency for Biodiversity, Direction de la Recherche, Vincennes, France
####National Institute of Biology, Ljubljana, Slovenia
*****University of Guelph, Guelph, ON, Canada
†††††STOWA, Stichting Toegepast Onderzoek Waterbeheer, Amersfoort, The Netherlands
‡‡‡‡‡University of Vienna, Vienna, Austria
§§§§§Senckenberg Research Institute and Natural History Museum, Frankfurt am Main, Germany
¶¶¶¶¶Botanic Garden and Botanical Museum, Freie Universität Berlin, Berlin, Germany
||||||||||Musée National d'Histoire Naturelle de Luxembourg, Luxembourg, Luxembourg
[1]Corresponding author: e-mail address: florian.leese@uni-due.de

Contents

Abstract

Aquatic biomonitoring has become an essential task in Europe and many other regions as a consequence of strong anthropogenic pressures affecting the health of lakes, rivers, oceans and groundwater. A typical assessment of the environmental quality status, such as it is required by European but also North American and other legislation, relies on matching the composition of assemblages of organisms identified using morphological criteria present in aquatic ecosystems to those expected in the absence of anthropogenic pressures. Through decade-long and difficult intercalibration exercises among networks of regulators and scientists in European countries, a pragmatic biomonitoring approach was developed and adopted, which now produces invaluable information. Nonetheless, this approach is based on several hundred different protocols, making it susceptible to issues with comparability, scale and resolution. Furthermore, data acquisition is often slow due to a lack of taxonomic experts for many taxa and regions and time-consuming morphological identification of organisms. High-throughput genetic screening methods such as (e)DNA metabarcoding have been proposed as a possible solution to these shortcomings. Such "next-generation biomonitoring", also termed "biomonitoring 2.0", has many advantages over the traditional approach in terms of speed, comparability and costs. It also creates the potential to include new bioindicators and thereby further improves the assessment of aquatic ecosystem health. However, several major conceptual and technological challenges still hinder its implementation into legal and regulatory frameworks. Academic scientists sometimes tend to overlook legal or socioeconomic constraints, which regulators have to consider on a regular basis. Moreover, quantification of species abundance or biomass remains a significant bottleneck to releasing the full potential of these approaches. Here, we highlight the main challenges for next-generation aquatic biomonitoring and outline principles and good practices to address these with an emphasis on bridging traditional disciplinary boundaries between academics, regulators, stakeholders and industry.

1. STATE AND FATE OF AQUATIC ECOSYSTEMS

The decline of biodiversity and ecosystem functioning in aquatic systems has direct effects on human well-being through alteration of ecosystem services (Millennium Ecosystem Assessment, 2005; Mulder et al., 2015; Vörösmarty et al., 2010; WWF, 2016). It is caused by many human activities, in particular pollution, habitat degradation, flow modification, overexploitation and the spread of invasive species as well as the direct and indirect effects of climate change (Dudgeon et al., 2006). To counteract the degradation of aquatic ecosystems, various pieces of legislation and associated mitigation and restoration strategies have been put into place, ranging from global initiatives such as the Convention on Biological Diversity (CBD, 1992), through European Union directives such as the Water Framework Directive (WFD, Directive 2000/60/EC), the Marine Strategy Framework Directive (MSFD, Directive 2008/56/EC), the Groundwater Directive (GWD, Directive 2006/118/EC) or the Habitats Directive (Directive 92/43/EEC), to a multitude of national, regional (including transboundary) and local programmes. All these require data on the condition of aquatic ecosystems. In most cases, these data are not restricted to environmental variables but also include biological data to measure ecosystem integrity and health for rivers, lakes, wetlands, estuaries and oceans (Birk et al., 2012; Borja et al., 2010, 2016; Hawkins et al., 2000). In a legal context, such assessments are expected to be performed in a recurrent and standardised fashion across those states that ratified the legislation, so that baselines can be established and responses to environmental change are consistent among participants. The biomonitoring programmes for aquatic ecosystems range from simple recording of presence (or abundance) of individual species to sophisticated multilayer programmes involving multiple organismal groups, often representing the result of many years of intense dialogue and regular adjustments. But now, novel genetic tools are emerging, which could revolutionise how we assess and document our environment and communicate the results (Bohmann et al., 2014; Bourlat et al., 2013; Deiner et al., 2016; Hajibabaei et al., 2011; Kermarrec et al., 2013; Taberlet et al., 2012; Valentini et al., 2016; Vasselon et al., 2017a; Woodward et al., 2013). Sometimes termed "biomonitoring 2.0" (Baird and Hajibabaei, 2012; Woodward et al., 2013) or "next-generation monitoring" (Bohan et al., 2017; Valentini et al., 2016), these approaches are increasingly applied in academic studies (Fig. 1). However, outside academia, they have yet to play a significant role in local to large-scale formal bioassessment programmes. The only exception are a few initiatives working on the detection of single rare/endangered or

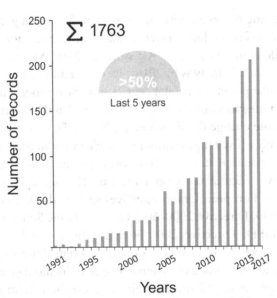

Fig. 1 Growing number of almost exclusively academic genetic biomonitoring studies as revealed by a web of science analysis. The search string "(("*monitoring" OR "*assessment") AND ("*water" OR "marine" OR "aquatic" OR "lake" OR "stream" OR "ocean" OR "river" OR "catchment") AND ("molecular" OR "genetic" OR "DNA") AND ("identif*" OR "characteri*" OR "*barcod*") AND ("PCR" OR "NGS" OR "HTS" OR "sequenc*"))" was applied on 02.01.2018. *http://dnaqua.net.*

invasive alien species, such as the Great Crested Newt in the United Kingdom (Biggs et al., 2015), the Asian carp species in the Great Lakes (Jerde et al., 2013) and programmes aimed at detecting invasive species in Switzerland (Mächler et al., 2014). The higher organisational levels of communities, food webs and ecosystems have been largely ignored.

We argue that the current situation is not only the result of technical constraints but also largely due to the consequence of a substantial information gap between academic and regulatory communities, in part driven by the fear of difficult, complex and costly adaptations of well-established monitoring programmes (Friberg et al., 2011). The tremendous progress made in the past few years has largely gone unnoticed among a large user group due to a communication gap between disjunct communities and as a consequence of deeply rooted national traditions (Kelly et al., 2015). In fact, rigorous DNA-based biomonitoring studies dealing with real monitoring data and questions of intercalibration are still lacking: sustainable interdisciplinary networks are clearly needed to identify and establish state-of-the-art biomonitoring strategies, and to cover understudied yet important ecosystems such as groundwater.

In this review, we (i) provide a brief introduction to the history of aquatic biomonitoring programmes, (ii) highlight the main impacts of novel genetic methods on aquatic ecosystem assessments, (iii) describe the key technological and conceptual challenges preventing their use in biomonitoring programmes to date and (iv) introduce the recently established *DNAqua-Net* COST Action (European Cooperation in Science and Technology), which aims to bridge the gap between science and application in using DNA-based methods in biomonitoring.

2. ADVANCEMENT OF AQUATIC BIOMONITORING WITH A FOCUS ON EUROPE

Aquatic biomonitoring can be dated back to at least the mid-19th century. Between 1870 and 1930, marine and lake biological stations proliferated across Western Europe and North America (Bont, 2015; Egerton, 2014), forming an important, yet largely academic network to study marine and freshwater biodiversity over time. However, these networks were typically restricted to one or a few ecosystems per country (Bont, 2015). Furthermore, the approach to assess and study ecosystems varied greatly due to the lack of regular interaction among researchers and practitioners in the predigital information age. Aquatic biomonitoring in its strict sense of identifying changes in response to human pressures over time only started in earnest in the early 20th century, when we learned that organic pollution of water bodies negatively impacted inland water ecosystems and thereby affected human well-being (Butcher, 1946). One of the first bio-assessment protocols to be applied was the Saprobic System for running waters, an approach that dates back to the mid-19th century (widely Cohn, 1853), although its use for rivers started in the early 20th century (Kolkwitz and Marsson, 1902, 1908, 1909). More sophisticated measures to quantify the condition of aquatic ecosystems were developed over the following decades specifically to assess the degree of organic pollution by using benthic invertebrates, as well as introducing bacteria, ciliates and diatoms as new indicator groups (Sládeček, 1965). Unfortunately, and partly due to difficulties in peer–network communication, many of these methods remained country-specific and there were few examples of larger scale integration.

During the late 1960s, pollution in the United States became such a problem that the Clean Water Act (CWA; United States, 1972), a milestone of environmental legislation, was passed in 1972. The CWA requires each US state to reach clearly defined water quality standards with respect to chemical, physical and biological factors. Although there was a federal law to control aquatic pollution, the CWA of 1972 can be regarded as a starting point for worldwide national and international environmental legislation.

Marine monitoring in a stricter sense started later than in freshwater ecosystems, i.e. during the 1970s, when environmental degradation in marine ecosystems became more apparent in a number of developed countries. Around that time, the theoretical basis of the response of marine communities to human disturbance started to be better understood (Pearson and Rosenberg, 1978): in the 1980s and 1990s, researchers used biomonitoring data to develop well-defined numerical methods with the goal to detect and reflect stress in those communities (Gray and Elliott, 2009). Their work resulted in methods and indicators that went beyond the application of classical saproby-based indices (see Karr, 1991 for freshwater ecosystems) and included primary (i.e. abundance, richness and biomass) and derived structural variables (i.e. diversity indices, abundance and biomass ratios and evenness indices) (e.g. Diaz et al., 2004; Gray and Elliott, 2009). This also included a conceptual shift from "bioindication," i.e., measuring the magnitude of pollution using specific indicator taxa—to a more systemic approach assessing ecosystem integrity.

In Europe, the WFD and some of its more local precursors built on these ideas as it pushed for the development of standardised protocols and the use of multiple organismal groups to assess the ecological status of different biological elements. The WFD encouraged regulators to move away from measuring and assessing the magnitude of individual stressors and looked towards developing more integrated assessments of aquatic ecosystems by identifying the disparity between the observed and the expected status of a water body. At the core of the WFD-compliant assessment methods lies the definition of type-specific reference condition, against which the observed community is compared. This can be done either by comparing the values of biotic metrics or alternatively through prediction systems (Moss et al., 1987). Furthermore, the WFD stipulates which organism groups are to be monitored for the individual aquatic ecosystem types: i.e. phytoplankton, attached plants and algae, benthic invertebrates and fish. At the end of the 1990s, and especially after 2000, a plethora of methods, also metaphorically described as an "adaptive radiation of assessment methods" (Kelly et al., 2015), were published and implemented, often piecemeal, in European countries. Today, these add up to more than 300 national methods for the WFD alone (Birk et al., 2012). In reality, however, a larger subset of those are adaptations of more generally applicable methods, which individual European countries favoured rather than standardised Europe-wide methods, accounting for particular taxa, stress combinations and river or lake types abundant and relevant to their respective countries. Still, differences between countries with similar water bodies may be regarded to some extent as a reflection of the specific national traditions, rather than the sole result of scientific evidence per se (Kelly et al., 2015). In essence, all these methods have similar targets: (i) to quantify the environmental state and difference from reference conditions, and (ii) to identify the causes of degradation as important background information from which appropriate management measures can be derived.

While the methods differ between aquatic ecosystem types, countries and organismal groups ("biological quality elements", BQEs), there are broad similarities in the general steps of the assessment workflow and all approaches rely on traditional morphological taxonomy. The typical six steps of bioassessment programs are shown in Fig. 2. Step 1: Samples are obtained from aquatic ecosystems using taxon-specific gear and methods (e.g. kick-nets, plankton-samplers, electrofishing). Step 2: These samples are processed in the field or lab and in Step 3 identified to the required

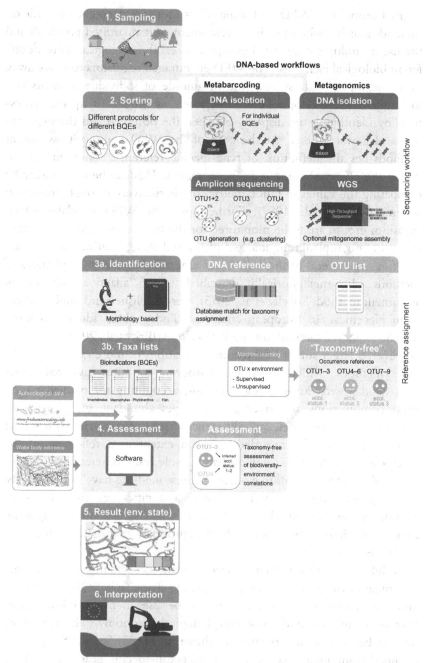

Fig. 2 Overview of the typical workflow used in aquatic bioassessment (Steps 1–6, *left*; *boxes with green headers*): (1) samples are collected in the field, usually independently for each "biological quality element" (BQE), (2) sorted or subsampled in the lab, (3)

or possible taxonomic level, which can be species (usually for fish, diatoms, macrophytes and some cases for macroinvertebrates) or genus to family level. The generated taxa lists are used in Step 4 for bioassessment, typically calculating biotic indices. Here, the taxa lists are linked to information on their ecological preferences using either taxonomic or functional (i.e. bioecological traits) information (e.g. Schmidt-Kloiber and Hering, 2015) to calculate metrics. Those can either be used individually or in combination (e.g. into multimetric indices), or via prediction systems that just compare taxa lists to reference conditions without calculating indices (Apothéloz-Perret-Gentil et al., 2017; Cordier et al., 2017). The results are used in Step 5 to identify the ecological/environmental status (Borja et al., 2012) by calculating the statistical distance of the observed values from values expected in reference conditions (i.e. the "high" status). While Steps 1–4 vary among countries, intercalibration procedures guaranteed the comparability of the ecological status classes used in different countries. Finally, in Step 6, the assessment results are used to inform management decisions, in particular for water bodies not meeting the desired "good ecological status".

With respect to the classical workflow shown (Fig. 2), we argue that next-generation aquatic biomonitoring based on genetic data could substantially improve several of the six steps in bioassessment and contribute to a better management of aquatic ecosystems.

---

identified using morphological traits of the BQEs to produce taxa lists. (4) The actual assessment is performed using the taxa lists obtained and by comparing them to expected communities for the water body type studied (horizontal lines), according to "reference" conditions. (5) The assessment leads to an assignment of an ecological quality class. (6) Assessment results are interpreted. In case of a greater than permitted mismatch with reference conditions (e.g. less than "good" ecological quality status for the WFD) appropriate measures have to be taken in order to improve ecological status. Genetic data can be used in the process in different ways using either amplicon (metabarcoding, *middle column*) or whole-genome shotgun sequencing (WGS) metagenomics data (*right column*). The data produced by both DNA-based approaches can be used to query obtained Operational Taxonomic Units (OTUs) against reference databases in order to produce taxa lists and continue with the traditional assessment (without abundance data). However, it is also possible to avoid the assignment of OTUs to Linnaean taxonomy but assign OTUs independent of their classification to environmental conditions either using training data (supervised machine learning) or through environmental correlations (unsupervised machine learning). Through this approach, a much broader set of biota can be used to derive novel indices for environmental assessment. *http://dnaqua.net.*

3. A DNA-BASED NEXT GENERATION OF AQUATIC BIOMONITORING?

3.1 Revolutions in Sequencing Technology Drive Academic Progress

Today, we are witnessing a revolution in the field of genomics. The introduction of new high-throughput sequencing (HTS) platforms allows for the analysis and identification not only of individual specimens but also of whole communities. High-throughput amplicon sequencing, commonly termed DNA metabarcoding (Taberlet et al., 2012), is an example of how technology initially developed for applications in the fields of medicine, forensics and microbiology is now being adapted to advance environmental monitoring (Staats et al., 2016). Today's HTS machines such as Illumina's NovaSeq are capable of producing billions of sequences in a single run, making it possible to analyse hundreds of samples in parallel and to identify hundreds of species per sample each. However, such an explosion challenges current analytical protocols due to the increase of computation load, and suggests that a move towards high performance computing (HPC) could be relevant. The number of academic studies that utilise high-throughput genetic bioassessment is increasing at an accelerating rate (Fig. 1). Perhaps unsurprisingly, given the different logistical challenges involved in biomonitoring, the first attempt to extend this academic work into applied biomonitoring happened in Canada: The Canadian Aquatic Biomonitoring Network (CABIN) has to monitor a huge number and a vast area of water bodies covering over 9 million km^2, with considerably fewer human resources and more limited taxonomic baseline data than is the case for most European countries. Consequently, in order to ensure sustainable monitoring of Canada's aquatic ecosystems, several different approaches using genetic data have been suggested and are now in regular use (Baird and Hajibabaei, 2012; Gibson et al., 2015; Hajibabaei et al., 2011; Hajibabaei et al., 2016) to improve speed, resolution, costs and comparability in cooperation between CABIN and WWF (World Wildlife Fund).

3.2 Metabarcoding and Other Genetic Approaches for Bioassessment

The methods proposed represent an extension to the classical DNA barcoding approach (Hebert et al., 2003), moving from identifying single specimens, where the question is "what species is this?", to describing community composition of mixed taxon samples, where the question is "what species are there?". These mixed samples can take various forms including bulk invertebrate or biofilm samples ("biodiversity soup" sensu Yu et al., 2012) and mixed-species DNA extracted from water or sediment (=environmental or "eDNA"). In short, eDNA or DNA of bulk samples is extracted and subjected to one of a number of different molecular approaches (see Fig. 3). DNA metabarcoding (Taberlet et al., 2012) probably represents the most widely tested and validated approach to date for processing mixed taxon samples. Over the last decade, metabarcoding and metagenomic studies have led to an enormous increase in the amount of available genetic data for organisms, communities and habitats (e.g. Bik et al., 2012; de Vargas et al., 2015; Elbrecht and Leese, 2017; Radom

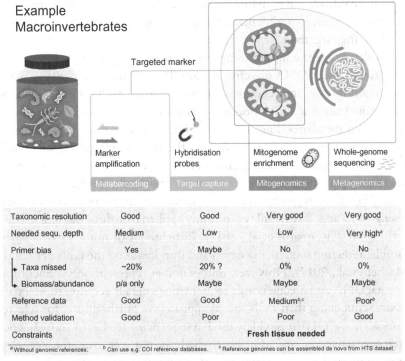

	Metabarcoding	Target capture	Mitogenomics	Metagenomics
Taxonomic resolution	Good	Good	Very good	Very good
Needed sequ. depth	Medium	Low	Low	Very high[a]
Primer bias	Yes	Maybe	No	No
↳ Taxa missed	~20%	20% ?	0%	0%
↳ Biomass/abundance	p/a only	Maybe	Maybe	Maybe
Reference data	Good	Good	Medium[b,c]	Poor[b]
Method validation	Good	Poor	Poor	Good
Constraints			Fresh tissue needed	

[a] Without genomic references.   [b] Can use e.g. COI reference databases.   [c] Reference genomes can be assembled de novo from HTS dataset.

Fig. 3 The potential of different DNA-based approaches is currently being assessed for application in biomonitoring programmes, as shown here for benthic macroinvertebrates. All methods have different advantages and biases. To date, metabarcoding is a well validated method and has good reference databases. This makes this approach advantageous for next-generation biomonitoring in the near future, whereas the other approaches need further validation or other improvements but could be even more promising in the long run. *http://dnaqua.net, modified after Elbrecht, V., April 2017. Development of DNA metabarcoding methods for stream ecosystem assessment. PhD Thesis, University of Duisburg-Essen.*

et al., 2012; Thomsen et al., 2012; Vasselon et al., 2017a) with most studies comparing the performance of morphological and metabarcoding methods for monitoring purposes (Couto et al., 2016; Elbrecht et al., 2017b; Kermarrec et al., 2013; Lejzerowicz et al., 2015; Stoeckle et al., 2017; Vasselon et al., 2017a; Vivien et al., 2015; Zimmermann et al., 2015). Available data suggest that in principle morphological- and metabarcoding-based assessment systems are compatible and can even use the very same sampling protocols. Community metabarcoding relies on PCR amplification using "universal", i.e., nonspecific primers and thus induces amplification bias when the primers inevitably match some taxa better than others (Elbrecht and Leese, 2015; Krehenwinkel et al., 2017). The use of hybridisation probes could potentially counteract this bias by allowing the targeted capture of barcoding genes (Dowle et al., 2016; Shokralla et al., 2016), although biases may still be introduced in hybridisation when too few probes for certain taxa are available in the experimental setup (i.e. being saturated) or are not a good match to the targeted taxa.

An alternative method to using specific probes or primers is to sequence the extracted bulk DNA directly, without PCR (Zhou et al., 2013). These metagenomics or mitochondrial metagenomics techniques (see, e.g., Crampton-Platt et al., 2016) omit the majority of problems associated with PCR-based metabarcoding, i.e., the ability to perform direct quantitative measures for certain taxa by hampering the relationship of sequence reads and taxon biomass/abundance, and even the loss of some taxonomic groups due to incompatible primer binding sites (Pinol et al., 2015). Such a metagenomic sequencing technique is well established for bacterial communities (Tseng and Tang, 2014) and was recently applied to arthropods (e.g. Arribas et al., 2016; Cicconardi et al., 2017). Enrichment of mitochondria is also possible, reducing sequencing depth and thus lowering the costs per sample (Macher et al., 2017). However, mitogenomes represent only a fraction of the total DNA and metagenomic sequencing generally requires substantially deeper sequencing than PCR-based approaches. Furthermore, for selected genes such as COI that is widely used in sequencing studies of animals, good reference databases are available, which is not yet true for complete mitogenomes of aquatic organisms. At this point, the use of metabarcoding with carefully designed and ecosystem-specific target-gene primer sets is advisable: in vitro and in silico tests should be performed to validate the applicability of the primer pair in the desired ecosystem (e.g. marine coastal habitat, high alpine lake, lowland streams, etc.) and taxonomic context (different ecosystems typically host different taxa) and to allow robust interpretations of genetic biomonitoring results. The newer PCR-free techniques have huge potential, but have not been sufficiently validated yet and lack proper reference data for deployment in routine monitoring schemes. Once established, they could, however, be employed in the future to reanalyse and "hindcast" archived DNA samples and provide additional options for the analysis of DNA-based biomonitoring.

Another, more radical alternative to the taxonomy-based (metabarcoding or metagenomics) approach could be the so-called taxonomy-free approach (Apothéloz–Perret–Gentil et al., 2017; Cordier et al., 2017), founding ecological assessments directly on genetic data as a proxy for communities without a Linnaean assignment step (or taxonomy assigned only to higher taxonomic levels such as phyla, orders or families) but simply using machine learning algorithms to link presence of genetic entities with environmental factors (see Fig. 2). Furthermore, extracting information on individual sequence variants (i.e. all nonidentical DNA sequences, omitting the OTU clustering step) from bulk metabarcoding datasets holds great potential for monitoring (Callahan et al., 2017). With this approach, also changes in intraspecific genetic diversity can be monitored (Amir et al., 2017; Callahan et al., 2016; Eren et al., 2013). While these methods have successfully been applied to identify sequence variants in microbial metabarcoding datasets, applying them to macroinvertebrate bulk samples is more challenging because specimen biomasses often vary by several orders of magnitudes and sequence variants of small specimens might go undetected because they are overshadowed by sequences of biomass-rich specimens, especially if sequencing depths is too shallow (but see Elbrecht et al., 2017a).

4. THE GRAND CHALLENGES FOR NEXT-GENERATION AQUATIC BIOMONITORING

Current sample processing and taxonomic identification could be shifted towards the use of modern HTS approaches, in particular DNA metabarcoding. The traditional workflow could remain largely unchanged, with DNA-based tools simply replacing morphological work, which is still rooted in labour-intensive light microscopy (see Fig. 2). Initial tests show similar results for both marine and freshwater biomonitoring (Aylagas et al., 2014; Elbrecht et al., 2017b; Lejzerowicz et al., 2015), suggesting that a rather straightforward implementation into the classical workflow may be achievable. However, a shift from traditional morphology-based assessments to DNA-based methods may be

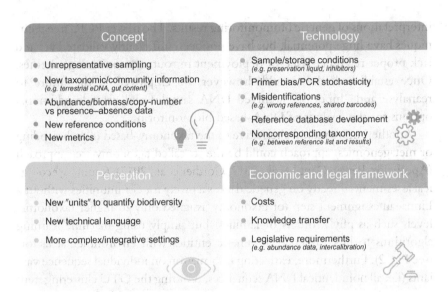

Fig. 4 The main challenges hindering the straightforward application of DNA-based tools in aquatic biomonitoring. See text for further explanation. *http://dnaqua.net.*

complicated by a range of issues related to technology, concept, perception as well as economic and legal frameworks (see Fig. 4). These are discussed in more detail below.

4.1 Genetic Data Cannot Deliver All Information Required by Legislation

The WFD and MSFD explicitly state that in bioassessments "composition and abundance" (Annex V and III, respectively) of BQEs need to be accounted for. Until now, issues such as unequal biomass of multicellular organisms and in particular primer bias greatly hamper the (current) ability of metabarcoding to deliver abundance data for a wide range of indicator taxa. Although several studies have shown that this bias can be reduced with optimised and in silico evaluated primer combinations (Elbrecht and Leese, 2017) and that correction factors can be calculated under certain assumptions (e.g. Krehenwinkel et al., 2017; Thomas et al., 2016), it is clear that metabarcoding cannot deliver absolute abundance data for complex communities of multicellular organisms (Elbrecht et al., 2017b). If primer bias can be reduced or eliminated, e.g., through the use of PCR-free HTS approaches, it may be possible to use mito-read number as a proxy for biomass (Choo et al., 2017). This would be a major breakthrough for bioassessment options but also for novel ecological assessments, as more direct links to food web structure, ecosystem processes and ecosystem services. However, varying biomass within species and varying mitochondrial copy number among cells, developmental stages, organisms and taxa mean that the link with abundance will remain weak, particularly for samples that contain a diverse assemblage of taxa or eDNA. In addition to hampering inference of abundance measures, primer bias can cause important taxa to be missed, particularly where a single set of primers is relied upon for analysis of diverse assemblages. Subsampling options or site-occupancy approaches might be options to quantify abundance. However, in terms of the current bioassessment workflow, it is interesting to note that even after the omission of a number of taxa due to their lack in reference databases, ecological status assessments can be robust (Vasselon et al., 2017b), even if only presence–absence data are used (Aylagas et al., 2014; Elbrecht et al., 2017a,b).

4.2 Adjusting the Reference Conditions

The "good ecological status" of the WFD and the "good environmental status" of the MSFD require comparisons with undisturbed, and in the context of the WFD, type-specific, reference conditions (Borja et al., 2012; Hering et al., 2006). A substantial amount of work in the past two decades has been invested into defining supposed reference conditions for a multitude of water body types (when possible), and exploring how biotic communities in degraded water bodies deviate from them. Current reference conditions of water bodies (both reference taxa lists and reference metric values) have been based on morphological identification methods and are only considered valid if these methods are applied. Recent data indicate that outcomes from DNA-based studies seem to be very compatible with morphology-based assessments, but a shift to DNA-based tools comes with both gains and losses. Metabarcoding typically retrieves taxa overlooked by morphological analyses, particularly cryptic species, endoparasites, ingested organisms or species that are represented by small, unidentifiable juvenile states or sexes. In some cases, the number of taxa is quite high compared to that found in morphological analysis, even when including biological replicates and strict error filtering (e.g. Elbrecht et al., 2017b). Furthermore, genetic analyses can miss several potentially important indicator taxa such as nematodes that fail to amplify for the commonly used primers in the PCR step (see below) and also simply due to the currently incomplete reference databases. Consequently, it will be important that any future inclusion of genetic tools is accompanied by the addition of DNA-based descriptions of biotic communities at reference sites, and that the regular update of reference conditions takes this aspect into account. Otherwise, the finding of many more taxa with DNA-based approaches might bias richness-based indices and report a water body as being systematically (but inaccurately) better than would be the case with a traditional assessment. This point requires careful assessment and planning and forms a central topic of *DNAqua-Net* working groups (WG2 and 5, Fig. 5; see also Leese et al., 2016).

Taxonomic information is an integral part of all current bioassessment protocols. For pragmatic reasons (time/money constraints as well as identification problems) many focus on higher level taxonomy (typically genus or family) because all or most of the members of the respective taxa are considered to indicate similar ecological conditions, following the phylogenetic niche conservatism concept (e.g. Keck et al., 2016). According to this concept, closely related species are assumed to possess similar ecological characteristics and thus should indicate similar environmental conditions. However, species within genera can differ markedly in their ecological preferences and thus species-level resolution data sometimes provide more precise ecological information, which should, in theory, improve ecosystem health assessments (Macher et al., 2016; Schmidt-Kloiber and Nijboer, 2004). In principle, this may be especially relevant for assessments of biota having (partly) arised from sympatric speciation processes where different ecologies (ecological character displacement) rather than distinct distributions promoted diversification, e.g. plankton communities. Most frequently, specimens are identified using morphological criteria, and the Linnaean binomial nomenclature is used to query a database or software to retrieve taxon-specific ecological information. Genetic markers could provide similar or higher than species-level identification and thereby act as a link to reference databases of OTUs in the same manner. Overviews of markers and databases used are given in, e.g., Pawlowski et al. (2012) and Creer et al. (2016). The application of DNA barcoding has been hampered by a lack of reliable and comprehensive reference databases. This has been solved for several taxonomic groups and regions through national and international barcoding campaigns (see above). However, even now, there is no European or even global DNA barcode reference library for aquatic BQEs. A first overview of aquatic BQEs for selected regions was prepared by members of Working Group 1 of *DNAqua-Net* (Table 1; see Section 5). While reference libraries are fairly comprehensive for freshwater and marine fishes

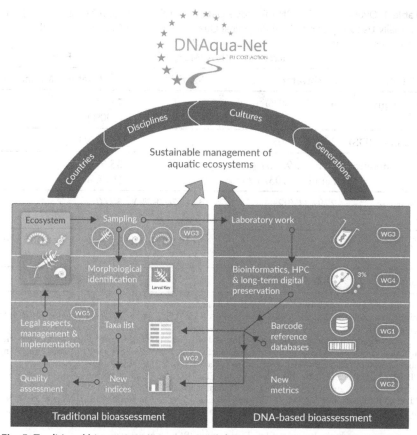

Fig. 5 Traditional biomonitoring in practice (*left; green*) and ways to include DNA-based methods (*right; blue*) into a "next-generation" programme for sustainable water-resource management. The roles of *DNAqua-Net*'s five Working Groups (WG1–5) are indicated. WG1: DNA barcode references; WG2: biotic indices and metrics; WG3: field and lab protocols; WG4: data analysis and storage (*HPC* = high-performance computing); WG5: implementation strategy and legal issues. Further, the international and intersectoral *DNAqua-Net* consortium aims at developing novel genetic tools for bioassessment of aquatic ecosystems in Europe and beyond, thereby bridging countries, disciplines, cultures (see Kelly et al., 2015) and generations. *http://dnaqua.net.*

as well as freshwater benthic macroinvertebrates, for major groups of European marine benthic macroinvertebrates, considerable compilation effort is still needed, as even in major groups such as Annelida, Crustacea and Mollusca (MZB) below 50% of the taxa relevant for routine monitoring possess reference barcodes (Table 1). Reviewing available DNA barcode data revealed that even in groups with relatively good species coverage, we still

Table 1 Overview of COI DNA Barcode Data Available (**Bold**) and Missing (*Italics*) for Formally Used Aquatic Animal Biological Quality Elements (BQEs) for Selected Countries/Regions

| Country | Macrozoobenthos (MZB) | | Fish | |
	Freshwater	Marine	Freshwater	Marine
Germany	**611**/*79* (89%)		**72**/*0* (100%)	
Danube (JDS)	**280**/*72* (80%)		**62**/*4* (94%)	
Luxembourg[a]	Genera: **95**/*61* (61%) Families: **103**/*19* (84%)		**35**/*0* (100%)	
Norway	**229**/*21* (92%)	**311**/*70* (82%)	**37**/*0* (100%)	
Europe		**1119**/*1406* (44%)[b]	**544**/*86* (86%)[c]	**1188**/*306* (80%)[d]
Slovakia	**982**/*883* (53%)		**65**/*4* (94%)	

[a]Reference list for MZB is based on family-, subfamily- and/or genus-level entries only. Coverage is given for family- and genus-level entries based on the checklist function implemented in BOLD (as on 02.01.2018).
[b]Checklist generated from AMBI's species list, including all Annelida, Crustacea and Mollusca species reported for Europe (excluding Barents Sea and Black Sea). All available DNA barcodes, including unreviewed, unpublished data and specimens collected outside European marine regions.
[c]Based on all available DNA barcodes from the FREDIE consortium (www.fredie.eu) including yet unpublished data, including 21 extinct and 44 alien species.
[d]Based on ERMS checklist, only Actinopterygii, Elasmobranchii and Holocephali. All available DNA barcodes, including unreviewed, unpublished data and specimens collected outside European marine regions.
Percent coverage in the Barcode of Life Data Systems (BOLD) in *parenthesis*.

lack baseline information on local genetic variability, i.e., barcodes typically come from a few campaigns conducted in a small subset of the range of a given species and do not cover its genetic diversity. For example, almost all fish species known to occur in Slovakia are represented by DNA barcodes; however, only 4% of them are derived from samples actually collected in Slovakia. The situation is similar for freshwater macroinvertebrates, with only 2%–3% of all referenced species possessing barcodes from Slovakian specimens. Bergsten et al. (2012) showed that geographical coverage of sampling is important in order to correctly apply DNA barcoding and, e.g., not split two geographically distinct population of the same species into two different species. This highlights the need to support local barcoding campaigns in order to ensure better coverage of species ranges but also to link faunistic/floristic databases such as the Fauna Europaea (https://fauna-eu.org) to DNA barcode databases, as otherwise the uncertainty of identification can increase with the growing data available (Bergsten et al., 2012).

For diatoms, the chloroplast marker rbcL is used for barcoding. However, the taxonomic coverage is still very sparse due to the complications associated with isolating and culturing species prior to barcoding (Rimet et al., 2016). For example, 1437 taxa are reported on the German red list (Lange-Bertalot and Steindorf, 1996). Of these, however, only 19.5% are available in a barcode reference library (Zimmermann J., personal communication). For Slovakia, the freshwater diatoms reference list used for WFD implementation includes 853 taxa (species/subspecies). Out of them 156 (18%) possess records in the Barcode of Life Data System (BOLD), but only four species (<0.5%) have a barcode. For Sweden, 1317 taxa are on the official diatom taxa list, for which only 13% and 16% (rbcL and 18S barcode marker, respectively) are represented by reference barcode data (M. Kahlert, personal communication). In France, for the 600 more common diatom species, only 140 (23.3%) have an rbcL barcode (A. Bouchez, personal communication). However, the increasing quantity of HTS sequences obtained from environmental diatom samples may be efficiently used to fuel current databases (Rimet et al., 2018). Rimet et al. (2018) show that it is possible to link hitherto unidentified barcodes from HTS output to Linnaean binomials using phylogenetic tools with high certainty as long as certain criteria are met (e.g. low diversity diatom assemblages and relatively high frequency of target organism in both HTS output and corresponding light microscopical analysis).

While many national barcoding campaigns have contributed substantial amounts of sequence data, coverage remains spatially fragmented, taxonomically incomplete and disconnected from available autecological trait databases. Recent studies (Elbrecht et al., 2017b; Vasselon et al., 2017a) have documented high proportions of taxa that were identified using morphological traits but not found in the genetic analyses of the bulk samples (30% for freshwater macroinvertebrates and 68% for diatoms). This can to some extent be attributed to a lack of available reference barcodes (especially for diatoms), as only those taxa that have a corresponding record in the database can be detected and assigned with genetic methods. Other factors that may contribute to these differences include misidentifications and primer bias. In order to improve both speed and robustness for DNA-based assessment, skilled taxonomists are needed to fill those gaps, since precision of assessment increases with the number of taxa assigned unequivocally

(Haase et al., 2010; Rosser, 2017; Schmidt-Kloiber and Nijboer, 2004; Stein et al., 2014).

Another reason to embed expert taxonomists in the process of developing a next-generation biomonitoring concept is the fact that species names (or species hypotheses), especially among more cryptic taxa, are often dynamic entities under constant revision and therefore subject to change. Many species are still being lumped, split, synonymised or transferred between higher taxonomic levels and many monitoring lists use rather outdated taxonomy as they are not updated regularly. This is the case for WFD assessments using phytoplankton, for instance, where analysts often use identification keys written in the 1970s, which do not follow the current taxonomy of microalgae. Current literature on microalgae taxonomy now integrates DNA data and electron microscopy observations to delimit species boundaries. Such features are of little use in determinations based on light microscopy. Other organism groups have cryptic or nearly cryptic species that are difficult to distinguish using light microscopy (e.g. Jörger and Schrödl, 2013; Lin et al., 2017). Simple comparisons of taxa lists used by managers with DNA reference databases without questioning data validity can lead to gross underestimates of availability and coverage of available DNA barcode references, because the species names listed in the manager's lists might be outdated and not connected to the most current taxonomic classification used in the reference database, or vice versa. For instance, a gap analysis for fish species used for biomonitoring in Luxembourg utilising BOLD's checklist function revealed several taxa with no available DNA barcode reference data (Weigand A.M., unpublished data). However, after updating the taxonomy of the WFD-relevant taxa list and including a synonomy list in BOLD, the coverage increased from 86% to 100%. The opposite is also possible: a straightforward comparison of the marine European fish species against the DNA barcode reference library entries revealed that 39% of species had ambiguous DNA barcode references associated. However, this amount decreased to only 13% after critical revision of the records' incongruences (Oliveira et al., 2016). About 60% of the detected mismatches resulted from syntax errors, synonyms, mislabelling and contamination.

Another conceptual challenge with respect to barcode-based species assessment is that different barcodes do not necessarily represent different species and vice versa, identical barcodes may include more than one species (e.g. Bergsten et al., 2012). Hence, reference data revision and curation as well as the inference of biological trait data are much needed and, ideally, should result from the coordinated efforts of taxonomists, ecologists, genet-icists and database managers (Ekrem et al., 2007). Linking and promoting taxonomic research through the network is one of the central goals of *DNAqua-Net*. Our strategy is to collaborate with the Consortium of Euro-pean Taxonomic Facilities (CETAF, http://www.cetaf.org) together with the international Barcode of Life project (iBOL, http://www.ibol.org) and national barcoding campaigns.

4.3 Stressor–Biodiversity Relationships

Even once we have standardised our baseline reference data, we will need to understand how these deviate from the desired state under environmental stress. At present, we still do not fully understand many of the details in stressor–biodiversity responses, especially when considering biodiversity responses from the level of genes through to community-level and related functional aspects. Stressors often co-occur in natural systems in space and/or time and the response of biota to them is not necessarily simple or linear as they may interact in antagonistic, synergistic and additive ways, generating "ecological surprises" (Hering et al., 2015; Jackson et al., 2016a; Townsend et al., 2008). The next generation of biomonitoring programmes will not necessarily solve these problems, but they could make important steps towards that goal given their ability to generate a far more comprehensive range of OTUs or even individual genotypes and hence response variables that could act in more nuanced ways than is the case with the coarse traditional methods currently in use (Jackson et al., 2016b; Macher et al., 2016). The role of genetic data is key as they (i) can diminish taxonomic uncertainty, (ii) provide information on additional taxa (such as bacterial or protist communities) and ecosystems, not currently included into traditional monitoring programmes, and (iii) deliver information on ecosystem functions, which may be affected more directly by multiple stressors. In times of increasing human impacts on ecosystems and climate change, this task will require large-scale coordinated research projects over the coming decades.

4.4 Technological Progress Hampers Continuity and Demands Standardisation

In Section 3, we highlighted the currently explored methods for DNA-based biomonitoring and it became obvious that some are quite advan-ced (metabarcoding), whereas others are still at the developmental stage (hybridisation techniques, metagenomics). This makes it difficult at the moment to provide clear guidance to regulators on which method to use. The EU's WFD, as a major piece of legislation with direct relevance to *DNAqua-Net*, is due to have a revision of procedures in 2019. Thus, consultations on revisions and amendments as well as decisions on which steps to include in the next monitoring cycle need to happen soon. Therefore, the rather exploratory approaches using target capture and metagenomics seem to be an inappropriate strategy to propose at the moment, compared to the relatively robust species presence–absence estimates provided by DNA metabarcoding. However, what is now exploratory might soon become standard. Therefore, revisions need to accommodate the possibility of new and powerful tools becoming available, despite current limitations. It may thus be more important to consider and develop Quality Assurance/ Quality Control (QA/QC) measures as well as intercalibration with previous approaches for any new DNA-based method to be used in biomonitoring rather than recommending specific tools and protocols. This could include the use of standardised, artificially added DNA of well-defined (yet unknown to the analyst) "mock communities" (i.e. combination of species-specific DNA isolates in the laboratory to design artificial communities for test purpose). The DNA or even complete samples can be routinely processed alongside real samples as internal quality checks, as well as ring tests and proficiency tests as external checks on the performance of laboratories. As a first example of its use in this field, proficiency testing was introduced for eDNA detection of the Great Crested Newt in the United Kingdom in 2017. The choice of methods to implement and evaluate novel DNA-based approaches in biomonitoring is open to debate, and a mutual understanding of their capacities, benefits and pitfalls is important. An understanding of legal frameworks will be necessary to inform these decisions. Moreover, it is also important to consider trade-offs in practicality and cost-effectiveness of the available approaches that will allow DNA-based monitoring to be carried out routinely by nonacademics in a robust and replicable manner. For example, it may not be realistic to propose a method that requires samples to be immediately transported to the laboratory on ice as the majority of field ecologists will not have access to the necessary resources and facilities for this. The question of promoting a possibility of routine processing with QA/QC for data analysis (bioinformatic pipelines) should be addressed as well, taking care of accessibility, versioning and repeatability of the selected bioinformatic filtering.

4.5 Costs and Accessibility: A Janus-Headed Debate?

At this point, European monitoring is in a difficult situation: traditional bio-monitoring approaches are—after more than a decade of testing and negotiations—in place and thus regulators and water managers have little interest in changing these established protocols (Friberg et al., 2011). Implementation of the WFD is already very complex to conduct and stability is desired. Intercalibration procedures have been implemented with some success and so have quality standards for many countries outside the EU (Poikane et al., 2014). As the success of the traditional biomonitoring workflow (Fig. 2) does not rely on sophisticated technological devices but rather on trained personnel, the process chain works in all European countries independently of their technological or economic development. However, in some EU countries with extended monitoring networks, both policy makers and water managers see the possibility that new DNA-based approaches may save money, as they could reduce the costs of labour-intensive species identifications substantially. This holds true in particular for countries in which labour costs are high compared to sequencing costs and where sequencing technologies and labs are readily available. Here it becomes important to offset labour costs by making theoretical and methodological knowledge available to all countries.

We also see a need for new funding streams for research on bio-monitoring data acquisition using both traditional and novel methods in a period of "handshaking" between the new and old technologies. This would allow researchers and practitioners to explore how best to exploit genetic methods without sacrificing backward-compatibility, which is a major bottleneck from a regulatory perspective in the context of WFD and MSFD. However, stakeholders and regulators often view such studies as R&D activities that should not be funded by their budget, making their full engagement with academic programmes challenging. While there are notable exceptions such as the funding of the German Barcode of Life II phase ("Bridging science and application gap", funded by the German Federal Ministry of Education and Research) and the WFD morphological identification vs environmental DNA comparison on the island of Mayotte (where the WFD applies) by the French ONEMA-AFB (Vasselon et al., 2017a), international funding or transnational programmes for such studies in the "borderland" (sensu Kelly et al., 2015) between research and management are of central importance. Many of the challenges we have highlighted are now being addressed by networks of working groups (see Fig. 5) mostly using individual, institutional, national or European research funds, and other research groups elsewhere in the world (Hajibabaei et al., 2016). One of the intended outputs is a streamlined array of good-practice strategies from sampling to assessment, but further international programmes of research funding agencies will need to prioritise cross-method comparisons to take these forward into a legally robust, backward-compatible toolkit ready for future monitoring programmes starting after the end-points of MSFD and WFD in 2020 and 2027, respectively. Genetic tools hold the promise to make aquatic biomonitoring better, cheaper (but see Section 4.5) and more reliable (or, at least, as reliable for a comparable cost), but the support mechanisms to develop them across sectors and countries need to be implemented as soon as possible to ensure early adoption of good-practice approaches.

4.6 The Importance of Transdisciplinary Dialogue

It has never been an easy task to link science, industry, decision making and practice, especially due to different motivations and organisational mechanisms that govern these sectors, but also because of the insular traditions that kept them "independent" from each other (Friberg et al., 2011). As a result, dissemination strategies are most often focussing on single sectors, and rarely cross their borders. This may explain why despite significant progress in academic development and application of molecular tools for bioassessment (see sections above), it appears that there is still limited immediate demand for biomonitoring 2.0 by regulatory agencies and little awareness in the general public. At the same time, and from a policy maker's and water manager's perspective, there is increased acknowledgement that scientific findings and recommendations on practical aspects such as biomonitoring must be communicated and disseminated outside existing research networks, if scientists want these methods to become attractive more rapidly beyond the strictly academic sphere (Mea et al., 2016). Most large-scale research funding now has to show effective dissemination and exploitation strategies for communicating project activities and results to external stakeholders, enterprises and the general public, thereby linking academic and public sectors (Elliott et al., 2017). Despite these tasks being seemingly straightforward and intuitive, as well as a required deliverable from grant funding, they are embedded within formal academic training, so the urgent need for filling these knowledge gaps is still largely unrecognised outside academia.

Communication of new approaches and methods in next-generation biomonitoring needs to primarily target policy makers and water managers at the water basin scale, explaining the efficiency, cost-effectiveness and compatibility of the new approaches. However, the public also needs to be aware of and therefore informed on the potential benefits of these methods for environmental protection and management. There is still a long way to go, as even an established intersectoral communication strategy does not automatically lead to implementation of the new methods, especially since it also needs an appropriately skilled workforce. Globally, it is now an important task to find ways to efficiently inform users and providers of routine biomonitoring of new genetic techniques (even at a basic but unified level), for example, by using Massive Open Online Courses (MOOCs) or sharing tutorials on this topic, which could for instance be disseminated by high-level authorities coordinating biomonitoring implementation, or leading national agencies involved in routine biomonitoring (Lowndes et al., 2017). Fostering an informed start-up environment among the upcoming generation of monitoring specialists may better link to industry. Here, academia plays a central role as it should provide training via specialised master courses, PhDs and postdocs.

5. THE AIM OF *DNAqua-Net*

All of these challenges and opportunities for developing the next generation of aquatic biomonitoring provided the impetus for the EU COST Action *DNAqua-Net* (CA15219), established in November 2016 under the COST (Co-Operation in Science and Technology) programme. The principal goal of the network that runs until October 2020 is to connect the relevant actors working on aquatic biomonitoring and DNA-based tool development to develop concepts and a roadmap for the application of new genetic tools in aquatic bioassessment programmes in Europe and beyond (Leese et al., 2016; see http://DNAqua.Net). The international consortium consists of 39 countries of the expanded European Union (i.e. EU member states, near neighbour countries and Israel) and eight international partner countries (i.e. Australia, Brazil, Canada, Iran, New Zealand, Russia, United States of America and Uruguay), represented by ~370 participants one year after the network launch. *DNAqua-Net*'s five working groups (WGs in Fig. 4) are targeting the further development and curation of aquatic DNA barcode reference libraries (WG1), the adjustment and development of (ecogenomic) biotic indices and metrics (WG2), the optimisation, development and evaluation of field and lab protocols for DNA-based biomonitoring (WG3), the storage and analysis of "big biodiversity data" (WG4) and the legal implementation and other relevant legislative issues (WG5). The main idea is to connect people and expertise and not to fund primary research. Furthermore, there is a strong focus on the integration of stakeholders from the outset and the promotion of early-career investigators to develop the human resource capacity needed for the next generation for researchers and practitioners. Since August 2017, ~10% of all participants are active stakeholders and 35% of the management committee are junior researchers. Moreover, several small and medium enterprises represented in the consortium, or people working in private structures for the collection of the biodiversity data further foster knowledge transfer from academia into practice, e.g., by training schools that are organised (e.g. 2018 on DNA-based biomonitoring for the fourth Joint Danube Survey, JDS4). A further step towards inclusion of a broader academic and nonacademic audience is the establishment of specialised forums explicitly addressing regulator needs, such as the open access Metabarcoding and Metagenomics journal (Leese et al., 2017) launched by *DNAqua-Net* members and Pensoft (Ltd.) with specific article formats such as applied studies (e.g. Theissinger et al., 2018), biomonitoring schemes, primer/probe validation for bioassessment (e.g. Vamos et al., 2017) and DNA barcode reference libraries. Moreover, recommendations of *DNAqua-Net* Working Groups will be made public via announcements on the website (http://DNAqua.Net), publications in relevant stakeholder and water manager journals (e.g. Leese et al., 2017), during stakeholder meetings and workshops (e.g. Norman-Network meeting 2017, Joint Danube Survey meeting 2017, Rhine Commission meeting 2017) or by special invitations to targeted conferences (*DNAqua-Net* conference in March 2017, Essen, Germany) or via special sessions (e.g. 10th European Symposium for Freshwater Sciences, Czech Republic and 7th International Barcode of Life Conference, South Africa), to raise awareness for the importance of upgraded bioassessment schemes. In particular, water-resource management stakeholders will profit from information generated in *DNAqua-Net* and will be targeted via suitable channels, e.g., (i) specific journals addressing the audience involved in all sorts of water related issues (e.g. Leese et al., 2017), (ii) workshops and (iii) specialised dissemination materials.

5.1 Networks Among Countries, Across Generations and Disciplines

In some countries not bound by EU legislation, in particular in southeast Europe, budget limitations as well as the lack of support by stakeholders favour temporally restricted studies over standardised and continuous (bio)surveillance. When considering the development of the next generation of genetic biomonitoring tools, an obvious challenge is the unequal access of countries to information and technology. Advanced genomic analyses (e.g. HTS) are still underutilised in developing countries with low R&D spending due to insufficient national funding, limited access to international grants, newest biotechnological infrastructures and training programmes, up-to-date literature and high costs of licensed analysis tools (Helmy et al., 2016). Network programmes such as the COST Action *DNAqua-Net* are of vital importance in building permanent cooperation platforms to help overcome knowledge or framework gaps between countries and to develop harmonised transnational approaches to common biomonitoring issues. The diversity of countries participating in *DNAqua-Net* ranges from those with strong government–stakeholders–academia links to those that are far less research intensive (i.e. inclusiveness target countries; ITCs). This provides added value in form of combining different experiences and practices and gives participants that would otherwise be excluded access to these emerging technologies. While this has little disadvantages to countries with strong research infrastructure, it has significant advantages to the less research–intensive ITCs. Capacity building in the field of biomonitoring in ITCs is further enhanced through short-term scientific missions implemented within COST programmes that offer a possibility to early-career investigators from ITCs to gain practical experience in a given field of action, to increase their visibility in the broader scientific community and to establish international cooperation networks. In the year 2017, 10 such multinational exchange programmes have been successfully conducted during the launch phase of *DNAqua-Net*. *DNAqua-Net* has non-European countries (Canada, USA) as partners in the network and as official "Management Committee observers" included from the very beginning. Other non-European countries have since then joined in, making *DNAqua-Net* a global network to develop DNA-based biomonitoring concepts for aquatic ecosystems.

Integration of the novel DNA-based or other advanced tools into current and future bioassessment workflows requires strong and sustained interdisciplinary dialogue among academics, regulators and industry, especially since the technological advancement is fast. Collaborative networks planning to distil and funnel scientific progress into the current regulatory framework must therefore offer to participate in the implementation process, taking into account potential inertia at the national and regional levels of organisation. In Europe, the creation of such networks is currently fostered by, e.g., large international COST Actions and the Joint Programming Initiative knowledge hubs. In the context of aquatic bioassessment in Europe, facilitation of uptake of new methods requires dialogue with the relevant organisations of the regulatory framework (e.g. ECOSTAT for the WFD and Directorate-General for the Environment, http://ec.europa.eu/dgs/environment/). Thus, interdisciplinary networks such as the COST Action *DNAqua-Net* and also Joint Programming Initiative knowledge hubs link academics with regulators, industry and policy makers but are short-lived in terms of funding and infrastructure (i.e. 4 years for COST Actions and typically 3 years for Joint Programming Initiative knowledge hubs). Another prerequisite for the uptake of novel methods is that these become mature enough to be processed by the European Committee for Standardization (CEN, http://www.cen.eu), either into standards or through the modification of existing standards. Since method integration into European standards is a slow and separate process, it usually extends beyond the lifespan of ongoing scientific networks dedicated to this subject. Therefore, networks can often only lay the foundation for future consideration of mature methods into standards and update into legal frameworks such as the European Directives. However, in the case of *DNAqua-Net*, network members and members of CEN Working Group 2 of Technical Committee 230 on "Biological methods" have agreed upon the need for a new permanent CEN working group that will pursue aquatic DNA-based method standardisation in the future.

6. NEXT-GENERATION BIOMONITORING OPENS NEW DOORS

Aside from the advantages that new methods offer to current biomonitoring programmes, there are additional potentially ground-breaking opportunities: the amount and type of data will allow researchers to address fundamental ecological questions in unprecedented ways. Many of the core questions in ecology are still centred around how many and which species coexist in given ecosystems, how they shape ecosystem processes and what are the spatial and temporal scales of these dynamics. To date most experimental ecological work has focussed on small scales and short time spans, while many of the major questions related to large scales and long time spans have been left unanswered. Such formerly data-restricted research is now being transformed and "big data", as opposed to too little data, are the growing challenge that will face the next generations of ecologists (Keck et al., 2017). Metabarcoding and metagenomics, potentially in combination with remote sensing and new machine learning/"big data" algorithms, offer novel ways of providing new and fundamental insight into ecological dynamics from local to global scales and across organisational levels (Bohan et al., 2017; Bush et al., 2017). For example, DNA-based methods were recently used to show that remarkably similar ecosystem functions can be observed irrespective of high taxonomic turnover in bacterial and archaeal community composition in the "miniature aquatic ecosystems" maintained in bromeliad plants (Louca et al., 2016). Another recent example shows that multitrophic diversity scales highly nonlinearly across space (Schuldt et al., 2015), and yet another shows how gene-to-ecosystem impacts of a pesticide spill can be traced through from individual functional genes to the entire food web (Andújar et al., 2017). One of the most fascinating prospects is the potential use of sequence data to reconstruct complete food webs and species associations (Bohan et al., 2017; Morueta-Holme et al., 2016; Roslin and Majaneva, 2016) and to measure as well as forecast ecosystem processes by linking them to the measured expression of functional genes (Jackson et al., 2016b). An important step towards this goal is to stratify sampling campaigns at orders of magnitudes higher (across space and time) than in the past, as opposed to only aiming to maximise the taxonomic resolution and speed as part of the available sampling and monitoring campaigns. Thereby, an involvement of scientists early on in the planning of applied (e)DNA studies is crucial to maximise these potential synergies, especially as many ecologists are still unaware of the full potential of these tools, and the molecular scientists may likewise not appreciate the bigger questions that can be asked using their approaches. In addition, ecological concepts and models will increasingly need to be adapted in ways that allow the use of nontraditional abundance/biomass estimates (Choo et al., 2017) or even presence/absence data, at least until the issue of molecular based quantification of these parameters has been resolved. It now seems inevitable that ecology will make huge leaps forward with respect to predictability and causality, in ways that are comparable to the rate and scale of advancements made in climate science over recent decades. A central goal of DNAqua-Net and parallel research programmes is to stimulate such a process and connect research and applied communities alike.

ACKNOWLEDGEMENTS

This chapter is based upon work from COST Action *DNAqua-Net* (CA15219), supported by the COST (European Cooperation in Science and Technology) programme. We thank all *DNAqua-Net* members for their input and in particular Donald Baird, Andrew Mahon, Kristy Deiner, Eric Stein, Roger Sweeting and Mehrdad Hajibabaei for helpful discussions. Thanks also to Slavena Peneva (Pensoft Publishers) for bringing the figures into shape. We are extremely thankful to Guy Woodward for meticulously reviewing and critically commenting on this manuscript and thereby improving it substantially in terms of content and style. Furthermore, we thank Sarah Kückmann for administrative help as well as Mafalda Quintas and Rose Cruz Santos of the EU COST office (Brussels) for their permanent support.

REFERENCES

Amir, A., Mcdonald, D., Navas-Molina, J.A., Kopylova, E., Morton, J.T., Zech Xu, Z., Kightley, E.P., Thompson, L.R., Hyde, E.R., Gonzalez, A., Knight, R., 2017. Deblur rapidly resolves single-nucleotide community sequence patterns. mSystems 2, e00191–16.

Andújar, C., Arribas, P., Gray, C., Bruce, C., Woodward, G., Yu, D.W., Vogler, A.P., 2017. Metabarcoding of freshwater invertebrates to detect the effects of a pesticide spill. Mol. Ecol. 1–21. https://doi.org/10.1111/mec.14410.

Apothéloz-Perret-Gentil, L., Cordonier, A., Straub, F., Iseli, J., Esling, P., Pawlowski, J., 2017. Taxonomy-free molecular diatom index for high-throughput eDNA bio-monitoring. Mol. Ecol. Resour. 17, 1231–1242.

Arribas, P., Andujar, C., Hopkins, K., Shepherd, M., Vogler, A.P., 2016. Metabarcoding and mitochondrial metagenomics of endogean arthropods to unveil the mesofauna of the soil. Methods Ecol. Evol. 7, 1071–1081.

Aylagas, E., Borja, A., Rodriguez-Ezpeleta, N., 2014. Environmental status assessment using DNA metabarcoding: towards a genetics based Marine Biotic Index (gAMBI). PLoS One 9, e90529.

Baird, D.J., Hajibabaei, M., 2012. Biomonitoring 2.0: a new paradigm in ecosystem assessment made possible by next-generation DNA sequencing. Mol. Ecol. 21, 2039–2044.

Bergsten, J., Bilton, D.T., Fujisawa, T., Elliott, M., Monaghan, M.T., Balke, M., Hendrich, L., Geijer, J., Herrmann, J., Foster, G.N., Ribera, I., Nilsson, A.N., Barraclough, T.G., Vogler, A.P., 2012. The effect of geographical scale of sampling on DNA barcoding. Syst. Biol. 61, 851–869.

Biggs, J., Ewald, N., Valentini, A., Gaboriaud, C., Dejean, T., Griffiths, R.A., Foster, J., Wilkinson, J.W., Arnell, A., Brotherton, P., Williams, P., Dunn, F., 2015. Using eDNA to develop a national citizen science-based monitoring programme for the great crested newt (Triturus cristatus). Biol. Conserv. 183, 19–28.

Bik, H.M., Porazinska, D.L., Creer, S., Caporaso, J.G., Knight, R., Thomas, W.K., 2012. Sequencing our way towards understanding global eukaryotic biodiversity. Trends Ecol. Evol. 27, 233–243.

Birk, S., Bonne, W., Borja, A., Brucet, S., Courrat, A., Poikane, S., Solimini, A., Van De Bund, W.V., Zampoukas, N., Hering, D., 2012. Three hundred ways to assess Europe's surface waters: an almost complete overview of biological methods to implement the Water Framework Directive. Ecol. Indic. 18, 31–41.

Bohan, D.A., Vacher, C., Tamaddoni-Nezhad, A., Raybould, A., Dumbrell, A.J., Woodward, G., 2017. Next-generation global biomonitoring: large-scale, automated reconstruction of ecological networks. Trends Ecol. Evol. 32, 477–487.

Bohmann, K., Evans, A., Gilbert, M.T.P., Carvalho, G.R., Creer, S., Knapp, M., Yu, D.W., De Bruyn, M., 2014. Environmental DNA for wildlife biology and biodiversity monitoring. Trends Ecol. Evol. 29, 358–367.

Bont, R.D., 2015. Stations in the Field: A History of Place-Based Animal Research. University of Chicago Press, Chicago, pp. 1870–1930.

Borja, A., Elliott, M., Carstensen, J., Heiskanen, A.S., Van De Bund, W., 2010. Marine management—towards an integrated implementation of the European Marine Strategy Framework and the Water Framework Directives. Mar. Pollut. Bull. 60, 2175–2186.

Borja, A., Dauer, D.M., Gremare, A., 2012. The importance of setting targets and reference conditions in assessing marine ecosystem quality. Ecol. Indic. 12, 1–7.

Borja, A., Elliott, M., Andersen, J.H., Berg, T., Carstensen, J., Halpern, B.S., Heiskanen, A.-S., Korpinen, S., Lowndes, J.S.S., Martin, G., Rodriguez-Ezpeleta, N., 2016. Overview of integrative assessment of marine systems: the ecosystem approach in practice. Front. Mar. Sci. 3, 20.

Bourlat, S.J., Borja, A., Gilbert, J., Taylor, M.I., Davies, N., Weisberg, S.B., Griffith, J.F., Lettieri, T., Field, D., Benzie, J., Glockner, F.O., Rodriguez-Ezpeleta, N., Faith, D.P., Bean, T.P., Obst, M., 2013. Genomics in marine monitoring: new opportunities for assessing marine health status. Mar. Pollut. Bull. 74, 19–31.

Bush, A., Sollmann, R., Wilting, A., Bohmann, K., Cole, B., Balzter, H., Martius, C., Zlinszky, A., Calvignac-Spencer, S., Cobbold, C.A., Dawson, T.P., Emerson, B.C., Ferrier, S., Gilbert, M.T.P., Herold, M., Jones, L., Leendertz, F.H., Matthews, L., Millington, J.D.A., Olson, J.R., Ovaskainen, O., Raffaelli, D., Reeve, R., Rödel, M.-O., Rodgers, T.W., Snape, S., Visseren-Hamakers, I., Vogler, A.P., White, P.C.L., Wooster, M.J., Yu, D.W., 2017. Connecting Earth observation to high-throughput biodiversity data. Nat. Ecol. Evol. 1, 0176.

Butcher, R.W., 1946. The biological detection of pollution. J. Inst. Sew. Purif. 2, 92–97.

Callahan, B.J., Mcmurdie, P.J., Rosen, M.J., Han, A.W., Johnson, A.J.A., Holmes, S.P., 2016. DADA2: high-resolution sample inference from Illumina amplicon data. Nat. Methods 13, 581–583.

Callahan, B.J., Mcmurdie, P.J., Holmes, S.P., 2017. Exact sequence variants should replace operational taxonomic units in marker-gene data analysis. ISME J 11, 2639–2643.

CBD (Convention on Biological Diversity), 1992. 5 June 1992, 1760 UNTS 79; 31 ILM 818. Entered into force 29 Dec 1993. United Nations, New York, NY.

Choo, L.Q., Crampton-Platt, A., Vogler, A.P., 2017. Shotgun mitogenomics across body size classes in a local assemblage of tropical Diptera: phylogeny, species diversity and mitochondrial abundance spectrum. Mol. Ecol. 26, 5086–5098.

Cicconardi, F., Borges, P.A.V., Strasberg, D., Oromi, P., Lopez, H., Perez-Delgado, A.J., Casquet, J., Caujape-Castells, J., Fernandez-Palacios, J.M., Thebaud, C., Emerson, B.C., 2017. MtDNA metagenomics reveals large-scale invasion of belowground arthropod communities by introduced species. Mol. Ecol. 26, 3104–3115.

Cohn, F., 1853. Über lebende Organismen im Trinkwasser. Günsberg's Z. Klin. Med. 4, 229–237.

Cordier, T., Esling, P., Lejzerowicz, F., Visco, J., Ouadahi, A., Martins, C., Cedhagen, T., Pawlowski, J., 2017. Predicting the ecological quality status of marine environments from eDNA metabarcoding data using supervised machine learning. Environ. Sci. Technol. 51, 9118–9126.

Couto, C.R.D., Jurelevicius, D.D., Alvarez, V.M., Van Elsas, J.D., Seldin, L., 2016. Response of the bacterial community in oil-contaminated marine water to the addition of chemical and biological dispersants. J. Environ. Manage. 184, 473–479.

Crampton-Platt, A., Yu, D.W., Zhou, X., Vogler, A.P., 2016. Mitochondrial metagenomics: letting the genes out of the bottle. Gigascience 5, 15.

Creer, S., Deiner, K., Frey, S., Porazinska, D., Taberlet, P., Thomas, W.K., Potter, C., Bik, H.M., 2016. The ecologist's field guide to sequence-based identification of biodiversity. Methods Ecol. Evol. 7, 1008–1018.

de Vargas, C., Audic, S., Henry, N., Decelle, J., Mahe, F., Logares, R., Lara, E., Berney, C., Le Bescot, N., Probert, I., Carmichael, M., Poulain, J., Romac, S., Colin, S., Aury, J.M., Bittner, L., Chaffron, S., Dunthorn, M., Engelen, S., Flegontova, O., Guidi, L., Horak, A., Jaillon, O., Lima-Mendez, G., Lukes, J., Malviya, S., Morard, R., Mulot, M., Scalco, E., Siano, R., Vincent, F., Zingone, A., Dimier, C., Picheral, M., Searson, S., Kandels-Lewis, S., Acinas, S.G., Bork, P., Bowler, C., Gorsky, G., Grimsley, N., Hingamp, P., Iudicone, D., Not, F., Ogata, H., Pesant, S., Raes, J., Sieracki, M.E., Speich, S., Stemmann, L., Sunagawa, S., Weissenbach, J., Wincker, P., Karsenti, E., Coordinators, T.O., 2015. Eukaryotic plankton diversity in the sunlit ocean. Science 348, 1261605–1/11.

Deiner, K., Fronhofer, E.A., Machler, E., Walser, J.C., Altermatt, F., 2016. Environmental DNA reveals that rivers are conveyer belts of biodiversity information. Nat. Commun. 7, 12544.

Diaz, R.J., Solan, M., Valente, R.M., 2004. A review of approaches for classifying benthic habitats and evaluating habitat quality. J. Environ. Manage. 73, 165–181.

Dowle, E.J., Pochon, X., Banks, J.C., Shearer, K., Wood, S.A., 2016. Targeted gene enrichment and high-throughput sequencing for environmental biomonitoring: a case study using freshwater macroinvertebrates. Mol. Ecol. Resour. 16, 1240–1254.

Dudgeon, D., Arthington, A.H., Gessner, M.O., Kawabata, Z.I., Knowler, D.J., Leveque, C., Naiman, R.J., Prieur-Richard, A.H., Soto, D., Stiassny, M.L.J., Sullivan, C.A., 2006. Freshwater biodiversity: importance, threats, status and conservation challenges. Biol. Rev. 81, 163–182.

Egerton, F.N., 2014. History of ecological sciences, part 51: formalizing marine ecology, 1870s to 1920s. Bull. Ecol. Soc. Am. 95, 347–430.

Ekrem, T., Willassen, E., Stur, E., 2007. A comprehensive DNA sequence library is essential for identification with DNA barcodes. Mol. Phylogenet. Evol. 43, 530–542.

Elbrecht, V., Leese, F., 2015. Can DNA-based ecosystem assessments quantify species abundance? Testing primer bias and biomass—sequence relationships with an innovative metabarcoding protocol. PLoS One 10, e0130324.

Elbrecht, V., Leese, F., 2017. Validation and development of COI metabarcoding primers for freshwater macroinvertebrate bioassessment. Front. Environ. Sci. 5, 11.

Elbrecht, V., Peinert, B., Leese, F., 2017a. Sorting things out: assessing effects of unequal specimen biomass on DNA metabarcoding. Ecol. Evol. 7, 6918–6926.

Elbrecht, V., Vamos, E.E., Meissner, K., Aroviita, J., Leese, F., Yu, D., 2017b. Assessing strengths and weaknesses of DNA metabarcoding-based macroinvertebrate identification for routine stream monitoring. Methods Ecol. Evol. 8, 1265–1275.

Elliott, M., Snoeijs-Leijonmalm, P., Barnard, S., 2017. 'The dissemination diamond' and paradoxes of science-to-science and science-to-policy communication: lessons from large marine research programmes. Mar. Pollut. Bull. 125, 1–3.

Eren, A.M., Vineis, J.H., Morrison, H.G., Sogin, M.L., 2013. A filtering method to generate hgh quality short reads using Illumina paired-end technology. PLoS One 8, e66643.

Friberg, N., Bonada, N., Bradley, D.C., Dunbar, M.J., Edwards, F.K., Grey, J., Hayes, R.B., Hildrew, A.G., Lamouroux, N., Trimmer, M., Woodward, G., 2011. Biomonitoring of human impacts in freshwater ecosystems: the good, the bad and the ugly. Adv. Ecol. Res. 44 (44), 1–68.

Gibson, J.F., Shokralla, S., Curry, C., Baird, D.J., Monk, W.A., King, I., Hajibabaei, M., 2015. Large-scale biomonitoring of remote and threatened ecosystems via high-throughput sequencing. PLoS One 10, e0138432.

Gray, J.S., Elliott, M., 2009. Ecology of Marine Sediments. From Science to Management. Oxford University Press, New York.

Haase, P., Pauls, S.U., Schindehutte, K., Sundermann, A., 2010. First audit of macroinvertebrate samples from an EU Water Framework Directive monitoring program: human error greatly lowers precision of assessment results. J. N. Am. Benthol. Soc. 29, 1279–1291.

Hajibabaei, M., Shokralla, S., Zhou, X., Singer, G.A., Baird, D.J., 2011. Environmental barcoding: a next-generation sequencing approach for biomonitoring applications using river benthos. PLoS One 6, e17497.

Hajibabaei, M., Baird, D.J., Fahner, N.A., Beiko, R., Golding, G.B., 2016. A new way to contemplate Darwin's tangled bank: how DNA barcodes are reconnecting biodiversity science and biomonitoring. Philos. Trans. R. Soc. Lond. B Biol. Sci. 371, 20150330.

Hawkins, C.P., Norris, R.H., Hogue, J.N., Feminella, J.W., 2000. Development and evaluation of predictive models for measuring the biological integrity of streams. Ecol. Appl. 10, 1456–1477.

Hebert, P.D., Cywinska, A., Ball, S.L., 2003. Biological identifications through DNA barcodes. Proc. Biol. Sci. 270, 313–321.

Helmy, M., Awad, M., Mosa, K.A., 2016. Limited resources of genome sequencing in developing countries: challenges and solutions. Appl. Transl. Genom. 9, 15–19.

Hering, D., Johnson, R.K., Kramm, S., Schmutz, S., Szoszkiewicz, K., Verdonschot, P.F.M., 2006. Assessment of European rivers with diatoms, macrophytes, invertebrates and fish: a comparative metric-based analysis of organism response to stress. Freshw. Biol. 51, 1757–1785.

Hering, D., Carvalho, L., Argillier, C., Beklioglu, M., Borja, A., Cardoso, A.C., Duel, H., Ferreira, T., Globevnik, L., Hanganu, J., Hellsten, S., Jeppesen, E., Kodes, V., Solheim, A.L., Noges, T., Ormerod, S., Panagopoulos, Y., Schmutz, S., Venohr, M., Birk, S., 2015. Managing aquatic ecosystems and water resources under multiple stress—an introduction to the MARS project. Sci. Total Environ. 503, 10–21.

Jackson, M.C., Loewen, C.J.G., Vinebrooke, R.D., Chimimba, C.T., 2016a. Net effects of multiple stressors in freshwater ecosystems: a meta-analysis. Glob. Chang. Biol. 22, 180–189.

Jackson, M.C., Weyl, O.L.F., Altermatt, F., Durance, I., Friberg, N., Dumbrell, A.J., Piggott, J.J., Tiegs, S.D., Tockner, K., Krug, C.B., Leadley, P.W., Woodward, G., 2016b. Recommendations for the next generation of global freshwater biological monitoring tools. Adv. Ecol. Res. 55, 615–636.

Jerde, C.L., Chadderton, W.L., Mahon, A.R., Renshaw, M.A., Corush, J., Budny, M.L., Mysorekar, S., Lodge, D.M., 2013. Detection of Asian carp DNA as part of a Great Lakes basin-wide surveillance program. Can. J. Fish. Aquat. Sci. 70, 522–526.

Jörger, K.M., Schrödl, M., 2013. How to describe a cryptic species? Practical challenges of molecular taxonomy. Front. Zool. 10, 59.

Karr, J.R., 1991. Biological integrity—a long-neglected aspect of water-resource management. Ecol. Appl. 1, 66–84.

Keck, F., Rimet, F., Franc, A., Bouchez, A., 2016. Phylogenetic signal in diatom ecology: perspectives for aquatic ecosystems biomonitoring. Ecol. Appl. 26, 861–872.

Keck, F., Vasselon, V., Tapolczai, K., Rimet, F., Bouchez, A., 2017. Freshwater biomonitoring in the Information age. Front. Ecol. Environ. 15, 266–274.

Kelly, M.G., Schneider, S.C., King, L., 2015. Customs, habits, and traditions: the role of nonscientific factors in the development of ecological assessment methods. Wiley Inter. Rev. Water 2, 159–165.

Kermarrec, L., Franc, A., Rimet, F., Chaumeil, P., Humbert, J.F., Bouchez, A., 2013. Next-generation sequencing to inventory taxonomic diversity in eukaryotic communities: a test for freshwater diatoms. Mol. Ecol. Resour. 13, 607–619.

Kolkwitz, R., Marsson, M., 1902. Grundsätze für die biologische Beurteilung des Wassers nach seiner Flora und Fauna. Mitt. Prüfungsanst. Wasserversorg. Abwasserbeseit. 1, 33–72.

Kolkwitz, R., Marsson, M., 1908. Ökologie der pflanzlichen Saprobien. Ber. Deutsch. Bot. Ges. 26a, 505–519.

Kolkwitz, R., Marsson, M., 1909. Ökologie der tierischen Saprobien. Int. Rev. Gesamten Hydrobiol. 2, 126–152.

Krehenwinkel, H., Wolf, M., Lim, J.Y., Rominger, A.J., Simison, W.B., Gillespie, R.G., 2017. Estimating and mitigating amplification bias in qualitative and quantitative arthropod metabarcoding. Sci. Rep. 7, 17668.

Lange-Bertalot, H., Steindorf, A., 1996. Rote Liste der limnischen Kieselalgen (Bacillariophyceae) Deutschlands. Schrift. Vegetationsk 28, 633–677.

Leese, F., Altermatt, F., Bouchez, A., Ekrem, T., Hering, D., Meissner, K., Mergen, P., Pawlowski, J., Piggott, J., Rimet, F., Steinke, D., Taberlet, P., Weigand, A., Abarenkov, K., Beja, P., Bervoets, L., Björnsdóttir, S., Boets, P., Boggero, A., Bones, A., Borja, Á., Bruce, K., Bursić, V., Carlsson, J., Čiampor, F., Čiamporová-Zatovičová, Z., Coissac, E., Costa, F., Costache, M., Creer, S., Csabai, Z., Deiner, K., DelValls, Á., Drakare, S., Duarte, S., Eleršek, T., Fazi, S., Fišer, C., Flot, J., Fonseca, V., Fontaneto, D., Grabowski, M., Graf, W., Guðbrandsson, J., Hellström, M., Hershkovitz, Y., Hollingsworth, P., Japoshvili, B., Jones, J., Kahlert, M., Kalamujic Stroil, B., Kasapidis, P., Kelly, M., Kelly-Quinn, M., Keskin, E., Kõljalg, U., Ljubešić, Z., Maček, I., Mächler, E., Mahon, A., Marečková, M., Mejdandzic, M., Mircheva, G., Montagna, M., Moritz, C., Mulk, V., Naumoski, A., Navodaru, I., Padisák, J., Pálsson, S., Panksep, K., Penev, L., Petrusek, A., Pfannkuchen, M., Primmer, C., Rinkevich, B., Rotter, A., Schmidt-Kloiber, A., Segurado, P., Speksnijder, A., Stoev, P., Strand, M., Šulčius, S., Sundberg, P., Traugott, M., Tsigenopoulos, C., Turon, X., Valentini, A., van der Hoorn, B., Várbíró, G., Vasquez Hadjilyra, M., Viguri, J., Vitonytė, I., Vogler, A., Vrålstad, T., Wägele, W., Wenne, R., Winding, A., Woodward, G., Zegura, B., Zimmermann, J., 2016. DNAqua-Net: developing new genetic tools for bioassessment and monitoring of aquatic ecosystems in Europe. Res. Ideas Outcomes 2, e11321.

Leese, F., Hering, D., Wägele, J.-W., 2017. Potenzial genetischer Methoden für das Biomonitoring der Wasserrahmenrichtlinie. Wasserwirtschaft (7–8), 49–53.

Lejzerowicz, F., Esling, P., Pillet, L., Wilding, T.A., Black, K.D., Pawlowski, J., 2015. High-throughput sequencing and morphology perform equally well for benthic monitoring of marine ecosystems. Sci. Rep. 5, 13932.

Lin, X.-L., Stur, E., Ekrem, T., 2017. DNA barcodes and morphology reveal unrecognized species in Chironomidae (Diptera). Insect Syst. Evol. https://doi.org/10.1163/1876312X-00002172.

Louca, S., Jacques, S.M.S., Pires, A.P.F., Leal, J.S., Srivastava, D.S., Parfrey, L.W., Farjalla, V.F., Doebeli, M., 2016. High taxonomic variability despite stable functional structure across microbial communities. Nat. Ecol. Evol. 1, 15.

Lowndes, J.S.S., Best, B.D., Scarborough, C., Afflerbach, J.C., Frazier, M.R., O'hara, C.C., Jiang, N., Halpern, B.S., 2017. Our path to better science in less time using open data science tools. Nat. Ecol. Evol. 1, 0160.

Macher, J.N., Salis, R.K., Blakemore, K.S., Tollrian, R., Matthaei, C.D., Leese, F., 2016. Multiple-stressor effects on stream invertebrates: DNA barcoding reveals contrasting responses of cryptic mayfly species. Ecol. Indic. 61, 159–169.

Macher, J.N., Zizka, V., Weigand, A.M., Leese, F., 2017. A simple centrifugation protocol for metagenomic studies increases mitochondrial DNA yield by two orders of magnitude. Methods Ecol. Evol. 1–5.

Mächler, E., Deiner, K., Steinmann, P., Altermatt, F., 2014. Utility of environmental DNA for monitoring rare and indicator macroinvertebrate species. Freshw. Sci. 33, 1174–1183.

Mea, M., Newton, A., Uyarra, M.C., Alonso, C., Borja, A., 2016. From science to policy and society: enhancing the effectiveness of communication. Front. Mar. Sci. 3, 168.

Millennium Ecosystem Assessment, 2005. Ecosystems and Human Well-Being: Synthesis. Island Press, Washington, DC.

Morueta-Holme, N., Blonder, B., Sandel, B., Mcgill, B.J., Peet, R.K., Ott, J.E., Violle, C., Enquist, B.J., Jorgensen, P.M., Svenning, J.C., 2016. A network approach for inferring species associations from co-occurrence data. Ecography 39, 1139–1150.

Moss, D., Furse, M.T., Wright, J.F., Armitage, P.D., 1987. The prediction of the macroinvertebrate fauna of unpolluted running-water sites in great-Britain using environmental data. Freshw. Biol. 17, 41–52.

Mulder, C., Bennett, E.M., Bohan, D.A., Bonkowski, M., Carpenter, S.R., Chalmers, R., Cramer, W., Durance, I., Eisenhauer, N., Fontaine, C., Haughton, A.J., Hettelingh, J.P., Hines, J., Ibanez, S., Jeppesen, E., Krumins, J.A., Ma, A., Mancinelli, G., Massol, F., Mclaughlin, O., Naeem, S., Pascual, U., Penuelas, J., Pettorelli, N., Pocock, M.J.O., Raffaell, D., Rasmussen, J.J., Rusch, G.M., Scherber, C., Setala, H., Sutherland, W.J., Vacher, C., Voigt, W., Vonk, J.A., Wood, S.A., Woodward, G., 2015. 10 years later: revisiting priorities for science and society a decade after the millennium ecosystem assessment. Adv. Ecol. Res. 53, 1–53.

Oliveira, L.M., Knebelsberger, T., Landi, M., Soares, P., Raupach, M.J., Costa, F.O., 2016. Assembling and auditing a comprehensive DNA barcode reference library for European marine fishes. J. Fish Biol. 89, 2741–2754.

Pawlowski, J., Audic, S., Adl, S., Bass, D., Belbahri, L., Berney, C., Bowser, S.S., Cepicka, I., Decelle, J., Dunthorn, M., Fiore-Donno, A.M., Gile, G.H., Holzmann, M., Jahn, R., Jirku, M., Keeling, P.J., Kostka, M., Kudryavtsev, A., Lara, E., Lukes, J., Mann, D.G., Mitchell, E.a.D., Nitsche, F., Romeralo, M., Saunders, G.W., Simpson, A.G.B., Smirnov, A.V., Spouge, J.L., Stern, R.F., Stoeck, T., Zimmermann, J., Schindel, D., De Vargas, C., 2012. CBOL protist working group: barcoding eukaryotic richness beyond the animal, plant, and fungal kingdoms. PLoS Biol. 10, e1001419.

Pearson, T., Rosenberg, R., 1978. Macrobenthic succession in relation to organic enrichment and pollution of the marine environment. Oceanogr. Mar. Biol. Annu. Rev. 16, 229–311.

Pinol, J., Mir, G., Gomez-Polo, P., Agusti, N., 2015. Universal and blocking primer mismatches limit the use of high-throughput DNA sequencing for the quantitative metabarcoding of arthropods. Mol. Ecol. Resour. 15, 819–830.

Poikane, S., Zampoukas, N., Borja, A., Davies, S.P., Van De Bund, W., Birk, S., 2014. Intercalibration of aquatic ecological assessment methods in the European Union: lessons learned and way forward. Environ. Sci. Policy 44, 237–246.

Radom, M., Rybarczyk, A., Kottmann, R., Formanowicz, P., Szachniuk, M., Glockner, F.O., Rebholz-Schuhmann, D., Blazewicz, J., 2012. Poseidon: an information retrieval and extraction system for metagenomic marine science. Eco. Inform. 12, 10–15.

Rimet, F., Chaumeil, P., Keck, F., Kermarrec, L., Vasselon, V., Kahlert, M., Franc, A., Bouchez, A., 2016. R-Syst::diatom: an open-access and curated barcode database for diatoms and freshwater monitoring. Database 2016, baw016.

Rimet, F., Abarca, N., Bouchez, A., Kusber, W.-H., Jahn, R., Kahlert, M., Keck, F., Kelly, M.G., Mann, D.G., Piuz, A., Trobajo, R., Tapolczai, K., Vasselon, V., Zimmermann, J., 2018. The potential of high throughput sequencing (HTS) of natural samples as a source of primary taxonomic information for reference libraries of diatom barcodes. Fottea18/1.

Roslin, T., Majaneva, S., 2016. The use of DNA barcodes in food web construction-terrestrial and aquatic ecologists unite!. Genome 59, 603–628.

Rosser, N., 2017. Shortcuts in biodiversity research: what determines the performance of higher taxa as surrogates for species? Ecol. Evol. 7, 2595–2603.

Schmidt-Kloiber, A., Hering, D., 2015. www.freshwaterecology.info—an online tool that unifies, standardises and codifies more than 20,000 European freshwater organisms and their ecological preferences. Ecol. Indic. 53, 271–282.

Schmidt-Kloiber, A., Nijboer, R., 2004. The effect of taxonomic resolution on the assessment of ecological water quality classes. Hydrobiologia 516, 269–283.

Schuldt, A., Wubet, T., Buscot, F., Staab, M., Assmann, T., Bohnke-Kammerlander, M., Both, S., Erfmeier, A., Klein, A.M., Ma, K.P., Pietsch, K., Schultze, S., Wirth, C., Zhang, J.Y., Zumstein, P., Bruelheide, H., 2015. Multitrophic diversity in a biodiverse forest is highly nonlinear across spatial scales. Nat. Commun. 6, 10169.

Shokralla, S., Gibson, J., King, I., Baird, D., Janzen, D., Hallwachs, W., Hajibabaei, M., 2016. Environmental DNA barcode sequence capture: targeted, PCR-free sequence capture for biodiversity analysis from bulk environmental samples. bioRxiv 087437.

Sládeček, V., 1965. The future of the saprobity system. Hydrobiologia 25, 518–537.

Staats, M., Arulandhu, A.J., Gravendeel, B., Holst-Jensen, A., Scholtens, I., Peelen, T., Prins, T.W., Kok, E., 2016. Advances in DNA metabarcoding for food and wildlife forensic species identification. Anal. Bioanal. Chem. 408, 4615–4630.

Stein, E.D., White, B.P., Mazor, R.D., Jackson, J.K., Battle, J.M., Miller, P.E., Pilgrim, E.M., Sweeney, B.W., 2014. Does DNA barcoding improve performance of traditional stream bioassessment metrics? Freshw. Sci. 33, 302–311.

Stoeckle, M.Y., Soboleva, L., Charlop-Powers, Z., 2017. Aquatic environmental DNA detects seasonal fish abundance and habitat preference in an urban estuary. PLoS One 12, e0175186.

Taberlet, P., Coissac, E., Pompanon, F., Brochmann, C., Willerslev, E., 2012. Towards next-generation biodiversity assessment using DNA metabarcoding. Mol. Ecol. 21, 2045–2050.

Theissinger, K., Kästel, A., Elbrecht, V., Makkonen, J., Michiels, S., Schmidt, S., Allgeier, S., Leese, F., Brühl, C., 2018. Using DNA metabarcoding for assessing chironomid diversity and community change in mosquito controlled temporary wetlands. Metabarcoding Metagenomics 2, e21060.

Thomas, A.C., Deagle, B.E., Eveson, J.P., Harsch, C.H., Trites, A.W., 2016. Quantitative DNA metabarcoding: improved estimates of species proportional biomass using correction factors derived from control material. Mol. Ecol. Resour. 16, 714–726.

Thomsen, P.F., Kielgast, J., Iversen, L.L., Moller, P.R., Rasmussen, M., Willerslev, E., 2012. Detection of a diverse marine fish fauna using environmental DNA from seawater samples. PLoS One 7, e41732.

Townsend, C.R., Uhlmann, S.S., Matthaei, C.D., 2008. Individual and combined responses of stream ecosystems to multiple stressors. J. Appl. Ecol. 45, 1810–1819.

Tseng, C.H., Tang, S.L., 2014. Marine microbial metagenomics: from individual to the environment. Int. J. Mol. Sci. 15, 8878–8892.

United States, 1972. Federal Water Pollution Control Act Amendments of 1972. Pub.L. 92-500, October 18.

Valentini, A., Taberlet, P., Miaud, C., Civade, R., Herder, J., Thomsen, P.F., Bellemain, E., Besnard, A., Coissac, E., Boyer, F., Gaboriaud, C., Jean, P., Poulet, N., Roset, N., Copp, G.H., Geniez, P., Pont, D., Argillier, C., Baudoin, J.M., Peroux, T., Crivelli, A.J., Olivier, A., Acqueberge, M., Le Brun, M., Moller, P.R., Willerslev, E., Dejean, T., 2016. Next-generation monitoring of aquatic biodiversity using environmental DNA metabarcoding. Mol. Ecol. 25, 929–942.

Vamos, E., Elbrecht, V., Leese, F., 2017. Short COI markers for freshwater macroinvertebrate metabarcoding. Metabarcoding Metagenomics 1, e14625.

Vasselon, V., Domaizon, I., Rimet, F., Kahlert, M., Bouchez, A., 2017a. Application of high-throughput sequencing (HTS) metabarcoding to diatom biomonitoring: do DNA extraction methods matter? Freshwater Science 36, 162–177.

Vasselon, V., Rimet, F., Tapolczai, K., Bouchez, A., 2017b. Assessing ecological status with diatoms DNA metabarcoding: scaling-up on a WFD monitoring network (Mayotte island, France). Ecol. Indic. 82, 1–12.

Vivien, R., Wyler, S., Lafont, M., Pawlowski, J., 2015. Molecular barcoding of aquatic oligochaetes: implications for biomonitoring. PLoS One 10, e0125485.

Vörösmarty, C.J., Mcintyre, P.B., Gessner, M.O., Dudgeon, D., Prusevich, A., Green, P., Glidden, S., Bunn, S.E., Sullivan, C.A., Liermann, C.R., Davies, P.M., 2010. Global threats to human water security and river biodiversity. Nature 467, 555–561.

Woodward, G., Gray, C., Baird, D.J., 2013. Biomonitoring for the 21st century: new perspectives in an age of globalisation and emerging environmental threats. Limnetica 32, 159–173.

WWF, 2016. Living Planet Report 2016. Risk and Resilience in a New Era. WWW International, Gland, Switzerland.

Yu, D.W., Ji, Y.Q., Emerson, B.C., Wang, X.Y., Ye, C.X., Yang, C.Y., Ding, Z.L., 2012. Biodiversity soup: metabarcoding of arthropods for rapid biodiversity assessment and biomonitoring. Methods Ecol. Evol. 3, 613–623.

Zhou, X., Li, Y.Y., Liu, S.L., Yang, Q., Su, X., Zhou, L.L., Tang, M., Fu, R.B., Li, J.G., Huang, Q.F., 2013. Ultra-deep sequencing enables high-fidelity recovery of biodiversity for bulk arthropod samples without PCR amplification. Gigascience 2, 4.

Zimmermann, J., Glöckner, G., Jahn, R., Enke, N., Gemeinholzer, B., 2015. Metabarcoding vs. morphological identification to assess diatom diversity in environmental studies. Mol. Ecol. Resour. 15, 526–542.

CHAPTER THREE

Advances in Monitoring and Modelling Climate at Ecologically Relevant Scales

Isobel Bramer*, Barbara J. Anderson[†], Jonathan Bennie[‡],
Andrew J. Bladon[§], Pieter De Frenne[¶], Deborah Hemming[‖,#],
Ross A. Hill*, Michael R. Kearney**, Christian Körner[††],
Amanda H. Korstjens*, Jonathan Lenoir[‡‡], Ilya M.D. Maclean[‡],
Christopher D. Marsh*, Michael D. Morecroft[§§], Ralf Ohlemüller[¶¶],
Helen D. Slater*, Andrew J. Suggitt[‖‖], Florian Zellweger[##,***],
Phillipa K. Gillingham*,[1]

*Faculty of Science and Technology, Bournemouth University, Poole, Dorset, United Kingdom
[†]Manaaki Whenua Landcare Research, Biodiversity and Conservation Team, Dunedin, New Zealand
[‡]College of Life and Environmental Sciences, University of Exeter, Penryn, Cornwall, United Kingdom
[§]RSPB Centre for Conservation Science, The Lodge, Sandy, Bedfordshire, United Kingdom
[¶]Forest and Nature Lab, Ghent University, Ghent, Belgium
[‖]Met Office Hadley Centre, Exeter, Devon, United Kingdom
[#]Birmingham Institute of Forest Research, Birmingham University, Birmingham, United Kingdom
**School of BioSciences, The University of Melbourne, Melbourne, Australia
[††]Institute of Botany, University of Basel, Basel, Switzerland
[‡‡]UR "Ecologie et dynamique des systèmes anthropisés" (EDYSAN, UMR 7058 CNRS-UPJV), Université de Picardie Jules Verne, Amiens, France
[§§]Natural England c/o Mail Hub, County Hall, Worcester, Worcestershire, United Kingdom
[¶¶]University of Otago, Dunedin, New Zealand
[‖‖]University of York, York, Yorkshire, United Kingdom
[##]Forest Ecology and Conservation Group, University of Cambridge, Cambridge, Cambridgeshire, United Kingdom
***Swiss Federal Research Institute WSL, Birmensdorf, Switzerland
[1]Corresponding author: e-mail address: pgillingham@bournemouth.ac.uk

Contents

Advances in Ecological Research, Volume 58
ISSN 0065-2504
https://doi.org/10.1016/bs.aecr.2017.12.005

Abstract

Most ecological studies of the effects of climate on species are based on average conditions above ground level (measured by meteorological stations) averaged across 100 km^2
or larger areas. However, most terrestrial organisms experience conditions in a much
smaller area at the ground surface or within vegetation canopies, the climate of which
can be very different to large-scale averages. Therefore, to accurately characterise the
climatic conditions suitable for species, it is essential to include microclimate information.
Microclimates are affected by the shape of the landscape, including the steepness and
aspect of slopes, height above sea level, proximity to the sea or inland water, and whether
a site is in a valley or at the top of a hill. Plants also modify the conditions found within or
below their canopies, with the structure of vegetation playing an important role. The
recent increase in the availability of microsensors and remotely sensed data at appropriate
resolutions has led some ecologists to begin to include microclimate information within a
variety of contexts; however the field can be confusing and intimidating and mistakes are
often made along the way. In this chapter, we provide an overview of microclimatic processes and summarise the available methods of measuring and modelling microclimate
data for incorporation in ecological research. We highlight pitfalls to avoid emerging
novel methods and the limitations of some techniques. We also consider future research
directions and opportunities within this emerging field.

1. INTRODUCTION

Climate is key to the physiology and development of organisms, their
ecological interactions, and resulting geographical distributions. Focus on climate in ecological studies has increased in the context of anthropogenic climate change, the ecological impacts of which are becoming ever more
apparent (e.g. Lenoir and Svenning, 2015; Parmesan and Yohe, 2003;
Pauli et al., 2012; Pecl et al., 2017; Poloczanska et al., 2013; Settele et al.,
2014; Thackeray et al., 2016). Ecologists aiming to understand and predict
the influence of changing climate on species use distribution models based
on macroclimatic variables, which are generally measured by standard meteorological stations. However, organisms experience climate at a small scale,
where climatic conditions can significantly deviate from the macroclimate

(Potter et al., 2013). Ignoring microclimate effects within such studies could result in overprediction of the amount of climatically suitable area for species, under both current and future conditions (Gillingham et al., 2012a; Trivedi et al., 2008). It could also result in potential microrefugia being missed. If coarse resolution grid cells are misclassified as entirely unsuitable, this could result in overly pessimistic projections of the impacts of climate change (Gillingham et al., 2012b).

Microscale climates, or microclimates, have been defined in various ways depending on the discipline and context (e.g. Box 1). Broadly speaking, they are fine-scale climate variations which are, at least temporarily, decoupled from the background atmosphere (macroclimate, see Box 1). A wide range of variables, or combinations of variables, can be used to characterise microclimate, including temperature, precipitation, solar radiation, cloud cover, wind speed and direction, humidity, evaporation, and water availability. These are influenced by fine-resolution biotic and abiotic variations, including topography, soil type, land cover (especially vegetation), and proximity to the coast. The term *microclimate* is sometimes used interchangeably with topoclimate, although topoclimates are variations in climate solely as a function of topographical features such as elevation, slope, and aspect, and are generally considered to vary at a larger scale than microclimates, and occur higher off the ground (see Box 1; Barry and Blanken, 2016). Differences in the definition of microclimate may be a challenge in approaching microclimate research (see Box 2). Within the context of this chapter, we consider microclimates to typically have a spatial resolution of <100 m, and to be within a few metres of the vegetation canopy. The temporal resolution may vary depending on the process or application being studied, but generally timescales of hours (or higher frequency) are appropriate.

The ability to effectively and consistently measure and model microclimates is also important beyond ecology and climate science. Microclimates should be considered in civil applications such as architecture (Pérez Galaso et al., 2016; Terjung, 1974), and in urban design (Allegrini and Carmeliet, 2017; Yuan et al., 2017), forestry (Ma et al., 2010; Mason, 2015), agriculture (Lin, 2007; Waffle et al., 2017), and pest and disease epidemiology (Baker, 1980; Haider et al., 2017; Murdock et al., 2017). Although we focus here on microclimates within the context of conservation ecology, an understanding of microclimates could be vital to plant, animal and human health, food security, and sustainable development.

It has been established that the seasonal mean temperatures that species experience can deviate by as much as 5°C from the macroclimate (e.g. Scherrer and Körner, 2010; Suggitt et al., 2011), and deviations within

BOX 1 The definition of microclimate

Climatic observations and models are often grouped into spatial and/or temporal scales, typically based on the dimensions of the climate variations or processes they aim to represent. However, these groups are not precise and are much debated, resulting in widely varying definitions of what 'microclimate' is.

The World Meteorological Organisation (WMO) identifies the microscale as <100 m spatial resolution, with temporal resolution dependent on the application: 'minutes for aviation, hours for agriculture, and days for climate description' (WMO, 2014). Geiger et al. (2009) define microclimate as describing 'the climate of an individual site or station... characterised by rapid vertical and horizontal changes', while Barry and Blanken (2016) define microclimates as 'the layer of interface between the surface and atmosphere'. The tables below give the range of scales used to describe the different climatic groups by various sources.

Microclimate:

	Geiger et al. (2009)	Barry and Blanken (2016)	Littmann (2008)	Orlanski (1975); WMO (2010, 2014)
Horizontal scale	0.001–100 m	<~50 m (defined by vegetation canopy height)	10–100 m^2	<100 m
Vertical scale	−10 to 10 m	<A few 100 m		
Time scale	<10 s	<Minutes		

Topoclimate/local climate:

	Geiger et al. (2009)	Barry and Blanken (2016)	Littmann (2008)	Orlanski (1975); WMO (2010, 2014)
Horizontal scale	100 m to 10 km	100 m to ~10 km	100 m to ~2 km	100 m to 3 km
Vertical scale	5 m to 1 km	500 m to 1.5 km		
Time scale	10 s to h	Minutes to hours		

BOX 1 The definition of microclimate—cont'd
Mesoclimate:

	Geiger et al. (2009)	Barry and Blanken (2016)	Orlanski (1975); WMO (2010, 2014)
Horizontal scale	1 km to ~200 km	10 km to ~50 km	3–100 km
Vertical scale	500 m to 4 km		
Time scale	Hours to days	Hours to days	

Macroclimate:

	Geiger et al. (2009)	Barry and Blanken (2016)	Orlanski (1975); WMO (2010, 2014)
Horizontal scale	>200 km	>50 km	100–3000 km
Vertical scale	1–10 km		
Timescale	Days to weeks	>Hours	

landscapes can be as much as 20°C at any one time (Fig. 2; Scherrer and Körner, 2010). Even within a single inflorescence the temperature between the ovaries and the petals may differ by 10°C (Dietrich and Körner, 2014). However, the climate envelope models used by ecologists often do not account for such small-scale climate variations (Slavich et al., 2014). They generally use coarse-scale averages from meteorological stations with instruments sited a few metres above the earth's surface (WMO, 2014). These spread across grid scales which can be as much as 10,000-fold the body size of the species being studied (Potter et al., 2013), and thus they are not able to capture important climate-forcing factors and fine-scale climate variations, or weather extremes, which have strong limiting forces on species (Easterling et al., 2000; Parmesan et al., 2000). What is more, classic meteorological stations have guidance criteria which are designed to limit local climate influences (WMO, 2014). Utilising such observations to assess ecological processes that have a strong microclimate influence will decrease the accuracy of predictions of species' responses to climate change (Slavich et al., 2014) and hinder our ability to conserve biodiversity. For example,

BOX 2 Monitoring, modelling, and managing microclimates workshop

An open workshop supported by the British Ecological Society's Climate Change Ecology Special Interest Group was held in September 2017, attended by 25 microclimate scientists from 8 countries. Academics and conservation practitioners discussed the key challenges to microclimate research, and potential future directions.

Delegates were asked to first identify challenges for including microclimates in ecological research, and then from a master list choose the three that they considered to be the most challenging. These are listed below in order of identified importance:

- Investment of time and money, and challenges gaining funding.
- Lack of common data collection protocol and the resultant lack of comparability.
- Lacking knowledge of measurement methods.
- Lack of impact due to results not being generalisable to other systems.
- Lack of understanding of feedbacks, e.g., vegetation-climate, snow-climate.
- Difficulty of finding biologically relevant climate data in freely available datasets (no database dedicated to microclimate data).
- Lack of collaboration between meteorologists and ecologists.
- Availability/reliability/appropriateness of equipment.
- Insufficient computer power.
- Defining microclimate, both in general and specifically for the target system/species.
- Researchers are not often confronted with microclimate, so it is not considered important.
- What spatial resolution to measure microclimates at.
- By measuring the microclimate you may change it, e.g., flattening vegetation.
- Knowing which climatic variables to measure.
- It can be difficult and intimidating for people unfamiliar with the field.
- Knowing how often to measure the climatic variables.
- Balancing collecting enough data with having a manageable dataset.

Trivedi et al. (2008) showed that taking local and microclimate variations into account gave different projections of the future survival of plant species at a site in Scotland compared to using large-scale macroclimate data alone.

One of the reasons microclimates are ecologically important is that they can potentially buffer taxa against climate variability and longer term changes, thus providing microrefugia which allow species and populations to survive in locations which may be deemed unsuitable using low-resolution observations and models (De Frenne et al., 2013; Lenoir et al., 2017; Maclean et al., 2015; Slavich et al., 2014; Suggitt et al., 2015). Identifying and protecting refugia can be difficult (Ashcroft et al., 2012; Morelli et al., 2016), but it is becoming a more important aim for conservation science, and understanding the buffering role of microclimates may play a part in that.

Despite the long history of microclimatology (Geiger, 1927; Kraus, 1911), it is only more recently that consideration and understanding of microclimates have become widespread within the field of ecology. Developments in technology and advances in computing power have made taking simultaneous measurements over large areas much easier in the last two decades, and the mechanistic understanding of microclimates established in other scientific fields has begun to be considered by ecologists, allowing for a greater understanding of fine-scale climate–species interactions, which is vital for effective biomonitoring (Jones, 2013; Wang et al., 2013). The recent increase in expertise and subsequent research interest has resulted in the establishment of a strong knowledge base of microclimate ecology.

Topics that have been studied include the influence of terrain (Bennie et al., 2008; Finkel et al., 2001; Scherrer and Körner, 2010; Suggitt et al., 2011), forest structure (Chen et al., 1993; Pohlman et al., 2007; Pringle et al., 2003), and other vegetation types (Bauer and Kenyeres, 2007; Cavieres et al., 2007; D'Odorico et al., 2013; Suggitt et al., 2011) on microclimate. In addition, the effects of microclimate on species abundance (Checa et al., 2014; Curtis and Isaac, 2015; Gillingham et al., 2012a), diversity (Gómez-Cifuentes et al., 2017; Raabe et al., 2010), phenology (Weiss et al., 1993), distribution (Kelly et al., 2004; Martin, 2001), invasion success (Lembrechts et al., 2017), and behaviour (Cunningham et al., 2015; Hutchinson and Lacki, 2001; Kelly et al., 2004; Kleckova and Klecka, 2016; Willis and Brigham, 2005) have been quantified. Finally, the effects of habitat management on microclimates (Meyer et al., 2001; Ripley and Archibold, 1999), including water temperatures (Imholt et al., 2010), have also been investigated, as well as the potential for microclimate manipulation to be used as a means to offset the adverse impacts on biodiversity of climate change (Greenwood et al., 2016).

As interest in dedicated microclimate research has increased within ecology, so have attempts at microclimate modelling, with several different types of models in use (e.g. Gunton et al., 2015; Kearney et al., 2014; Maclean et al., 2017; Shi et al., 2016; see Section 5). Meanwhile, new technologies are becoming available, such as the use of unmanned aerial vehicles (UAVs) for remote sensing and monitoring of temperature and other environmental variables, presenting novel opportunities for future research (see Section 6).

Perhaps because the surge in interest is relatively recent, the methods used to measure, monitor, and model microclimates are varied, especially with regard to in situ measurement (see Section 4.2). This variability in data collection makes comparisons between datasets unreliable, the understanding of patterns in microclimate beyond individual field sites and case studies difficult, and the reliable identification of microrefugia challenging. For example, there are many types of relatively cheap miniature dataloggers and sensors available (see Table 1), but they vary widely in what they measure, their precision, reliability, and price. The WMO provides some practical guidance for observing microclimate, focussing on the siting of instruments for differing scales of representativeness (WMO, 2014, Annex 1.B). However, more broadly there is little to no guidance available as to which sensors are best for particular situations, how they should be placed and shielded, or even which factors of microclimate are important to measure.

We aim to tackle these challenges by summarising current methods in measuring and modelling microclimates, providing some guidance on the questions that should be asked when planning microclimate research, and how to go about deciding which approach to take. This should assist researchers and help to establish some consistency in the field, as well as presenting ideas for how the field may develop in coming years, laying the ground work for the next generation of climate biomonitoring.

2. FACTORS LEADING TO VARIABLE MICROCLIMATES

Microclimatic variation is driven primarily by the four components to the earth's heat budget, each representing the ways in which heat can be exchanged (Bennie et al., 2008; Geiger et al., 2009; Jones, 2013). In approximate order of importance these are (1) solar and thermal radiation, (2) latent-heat exchange, (3) sensible heat flux (heat convection), and (4) heat conduction (e.g. in soil).

2.1 Microclimatic Processes

2.1.1 Solar Radiation

Energy is received from the sun in the form of shortwave radiation, some of which reaches the surface directly, while some is reflected, scattered, and absorbed by particles in the atmosphere, to reach the ground as diffuse radiation. Small amounts of energy are also received at the ground as radiation is reflected from surrounding surfaces. The local intensity of radiation received at the surface, which influences local temperatures, is influenced largely by three factors (Hay and McKay, 1985). The first of these is the angle between the direction of the sun's rays and the earth's surface, and

Table 1 A Summary of Equipment That Has Been Used to Measure Microclimates in the Field, Quantified During a Systematic Review of the Literature (See Supplementary Methods for Details: Available on https://doi.org/10.1016/bs.aecr.2017.12.005)

Type/Brand of Datalogger Used	Details	Number of Studies	Price	Website
Onset's HOBO	Stand-alone dataloggers with internal sensors for a variety of environmental measurements. Most commonly used for temperature and humidity. Come in USB, WIFI, and Bluetooth compatible models	40	££	http://www.onsetcomp.com/
Maxim's iButton	Stand-alone miniature dataloggers with integral monitoring of temperature and humidity. USB compatible	37	£	https://www.maximintegrated.com/en.html
Campbell Scientific	Dataloggers compatible with a variety of external environmental sensors which can be bought separately	19	£££	https://www.campbellsci.co.uk/
Gemini's Tinytag	Both stand-alone and external sensor dataloggers for a variety of environmental measurements	7	£££	http://www.geminidataloggers.com/
Lascar	Stand-alone dataloggers for a variety of environmental measurements. Come in USB-, WIFI-, and Bluetooth-compatible models	2	££	https://www.lascarelectronics.com/data-loggers/
Logtag	Stand-alone and external sensor dataloggers for measuring temperature	2	££	http://logtagrecorders.com/
LI-COR	Dataloggers compatible with a variety of external environmental sensors which can be bought separately	2	£££	https://www.licor.com/env/

Continued

Table 1 A Summary of Equipment That Has Been Used to Measure Microclimates in the Field, Quantified During a Systematic Review of the Literature (See Supplementary Methods for Details: Available on https://doi.org/10.1016/bs.aecr.2017.12.005)—cont'd

Type/Brand of Datalogger Used	Details	Number of Studies	Price	Website
Testo	Both stand-alone and external sensor dataloggers for measuring temperature and humidity	1	££	http://www.testolimited.com/data-loggers-2
Volcraft	Stand-alone dataloggers for a variety of environmental measurements	1	££	http://www.voltcraft.com/environmental-measurement/
Decagon	Dataloggers compatible with a variety of external environmental sensors which can be bought separately	1	£££	https://www.decagon.com/en/
Kestrel handheld weather meters	Handheld instruments for a variety of environmental measurements, some with datalogger capabilities	2	£££	https://kestrelmeters.com/

Please note that this is a report of what was used in the literature and not a recommendation of what should be used; also most of the brands produce a variety of models with varying capabilities and prices and the model used has not always been reported. Price categories in £GBP: £ = ~£5 to ~£100; ££ = ~£50 to ~£200; £££ = > ~£200.

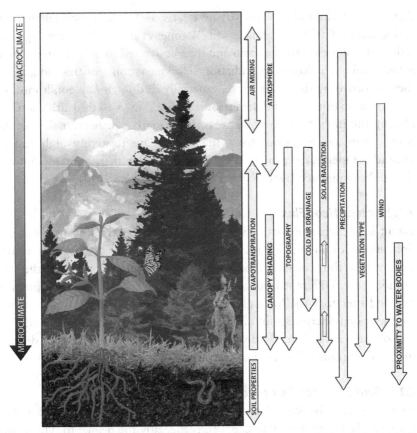

Fig. 1 An illustration of climate, microclimate, and the processes driving it. *Arrows* indicate the general extent, and in some cases direction, of the climate-forcing processes. See Section 2 for explanation of terms.

hence by latitudinal, seasonal, and diurnal changes in the position of the sun, as well as the slope and aspect of the ground surface (see Fig. 1). Second, the local intensity of radiation is also affected by various atmospheric constituents, namely gases, aerosols, and particularly clouds. Under turbid atmospheric conditions, direct radiation decreases, but diffuse radiation initially increases, and thus the effects of local topography on temperature are most pronounced under cloud-free conditions. Third, under vegetated canopies, canopy transmission not only decreases with canopy cover but is also affected by leaf structure; at low solar angles, radiation is lower when leaves are vertically oriented.

A considerable amount of solar radiation reaching the earth's surface is reflected. The fraction reflected is highly variable for different surfaces depending on their albedo and may be as large as 0.95 for freshly fallen snow and as small as 0.05 for a wet bare soil (Wanner et al., 1997). The energy not reflected is absorbed by the earth and then converted to heat energy, which is later emitted as longwave radiation. However, some of this radiation is then absorbed by the atmosphere, especially during cloudy conditions, or by vegetation canopies, and then reradiated back down to the earth's surface. During the night time in open areas, particularly under clear sky conditions, longwave emissions from the surface are relatively high and so ground temperatures can be considerably cooler than air temperatures above the boundary layer (Körner and Hiltbrunner, 2017).

2.1.2 Latent-Heat Flux

Latent-heat exchange, particularly due to surface evaporation and transpiration, affects local temperatures because heat is required to evaporate water. This process thus consumes some of the energy budget and cools the surrounding air. Different land cover types influence the evapotranspiration rate and therefore the local climate (Zeng et al., 2017). Similarly, when ice melts or undergoes sublimation, energy is required, and the converse is also true: energy is released when water condenses or undergoes deposition as frost (Lunardini, 1981).

2.1.3 Sensible Heat Flux (Heat Convection)

Heat convection between the earth's surface and the atmosphere is affected by boundary layer processes (Stull, 2012), particularly the degree to which heat and moisture are transported by eddy diffusion, which itself has two causes: frictional and convective exchange (Oke, 2002). The former is determined by the roughness of the ground surface, whereas the latter is a function of wind speed (see Fig. 1). By day, particularly during warm weather, convective mixing also contributes to air heat transport (Geiger et al., 2009).

2.1.4 Heat Conduction

Heat conduction, in this case the transfer of heat from the air to the soil and vice versa, is dependent on air–soil temperature gradients. Temperatures typically increase with soil depth at night and in winter, but on warm, sunny days in summer, the converse is true. Conduction is also dependent on the thermal properties of soil. These in turn depend on soil moisture content as well as on substrate type: rocks typically have a much higher thermal conductivity than clay and sand, thus animals basking on warm rocks will warm up more quickly than if basking on sand (Johansen, 1977; Fig. 2).

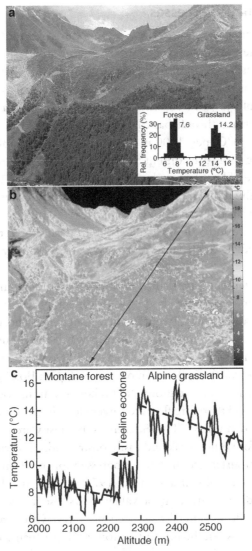

Fig. 2 Temperatures experienced across a heterogeneous landscape. (A) Image of an Alpine slope in Switzerland, (B) thermal infrared image of the same slope, and (C) temperature of pixels down the transect shown in (B), showing the effect of vegetation on temperature. Although there is a general decrease in temperature with elevation within habitat types, the short alpine meadow at higher elevations shows much higher temperatures than the wooded area at lower elevations. The relative frequency of pixels at given temperatures within the two habitats is summarised in (A) (Körner, 2007).

2.2 Mesoclimatic Processes

Over mesoclimatic extents, vertical temperature gradients and horizontal air movement become increasingly important. Arguably the most important effect is that of altitude. At higher altitudes, the expansion of air caused by lower atmospheric pressure consumes energy, thus reducing the heat available per unit volume. By this process, dry air cools at a predictable rate, termed the adiabatic lapse rate (see Box 3 for the explanation of terms), of 9.8°C per 1000 m altitude (Barry and Blanken, 2016). With increasing moisture content the latent heat released by condensation reduces this rate of heat loss in the air, to about 5.0°C per 1000 m altitude depending on the temperature. However, the additional effects of mixing of air masses mean that environmental lapse rates are often variable, and, while typically around 5.0–6.0°C per 1000 m (Blandford et al., 2008), can be hard to predict. Such lapse rates also vary seasonally (Kollas et al., 2014).

Airmass movement is particularly important in mountainous and coastal areas, or close to large water bodies. On hot sunny days in mountainous areas, greater heating over adjacent lowlands can create a horizontal pressure gradient that produces a movement of air towards the mountain. A return flow often occurs at night with the reversal of gradient. Under certain climatic conditions, katabatic flow can occur (see Fig. 1; Box 3). This happens when air in contact with the ground at higher elevations cools, increases in specific gravity, and therefore flows to lower elevations and typically escapes by funnelling down valleys. The differences in density are marginal and the airflow is therefore quite weak and easily obstructed, occurring only during still nights when surface cooling has progressed for some time (Haiden and Whiteman, 2005; Manins and Sawford, 1979). During the day, when slope flow is reversed, the direction of valley flows also reverses (Whiteman, 2000). These air flows generally reduce lapse rates, with smaller differences in temperature between the top and bottom of slopes than would otherwise be expected. During the day, the transport of air to higher elevations results in warming, reducing the temperature gradient. At night, the sinking of cold air can result in the usual decrease in temperature with height reversing, causing a temperature inversion. In extreme cases, frost hollows form in which the vegetation cover can differ substantially from surrounding areas (Davidson and Reid, 1985).

Another important landscape-scale determinant of surface temperature results from the presence of large waterbodies, particularly the sea. Water has a particularly high specific heat capacity relative to many substances,

BOX 3 Glossary of terms

Adiabatic processes	Process that do not involve heat or matter exchange
Free atmosphere	The portion of the earth's atmosphere, above the planetary boundary layer, in which the effect of surface friction on air motion is negligible, and in which the air behaves like an ideal fluid
Heat budget	The balance between incoming and outgoing heat
Hyperspectral data	Used in the context of remote sensing to describe imagery in which reflectance is measured across a range of the electromagnetic spectra
Irradiance	The flux of radiant energy per unit area
Katabatic wind/flow	Airflow that carries high-density air from a higher elevation down a slope under the force of gravity
Lapse rate	The rate at which air temperature falls with increasing altitude
Latent heat	Energy absorbed by or released from a substance during a phase change from a gas to a liquid or a solid or vice versa
Microrefugia	Small areas outside the core distribution area where species persist despite the surroundings being inhospitable
Photogrammetry	Gaining accurate measurements from photography
Physiographic	The physical factors of the earth
Refugia	An area in which a population can survive during a period of unfavourable conditions
Remote sensing	Obtaining information about objects or areas from a distance, typically from aircraft or satellites
Sensible heat	The energy required to change the temperature of a substance with no phase change
Shortwave radiation	Energy radiating from the sun and received by the earth in the visible, near-ultraviolet, and near-infrared spectral region ($\sim$0.1–5 µm)
Stevenson screen	A standard shelter to shield meteorological instruments against radiation and precipitation, while still allowing air to circulate freely
Thermodynamic	Of or related to the actions of heat and other types of energy

including the various compounds that comprise terrestrial landmasses, and therefore tends to heat up and cool down more slowly. During the afternoon, when land is typically warmer than water, the resulting vertical expansion of the air above land causes compensating onshore winds, while at night the air over water is warmer and the situation is reversed (Wexler, 1946). For this reason, both seasonal and daily temperature variations are often much lower near the sea, an effect noticeable even over relatively short distances of a few kilometres (Maclean et al., 2017).

2.3 Fine-Scale Variation in Water Availability

Hydrological conditions vary steeply in space and over time (Maclean et al., 2012), and to understand these processes, it is helpful to consider a single hydrological basin. Water reaches a basin as precipitation, and then either infiltrates into the ground or runs off and collects as surface water in low-lying areas. Infiltration rates are dependent on the physical properties of soil (Pepper and Morrissey, 1985; see Fig. 1). Infiltration occurs until the soil is saturated, but thereafter any remaining water entering the basin runs off and accumulates as surface water. Water can leave a basin at its lowest boundary and similarly can enter a basin from adjacent basins. Variations within a basin are driven predominantly by topography, with the lowest areas generally accumulating the most water, and flat areas being typically wetter than steep slopes due to lower rates of surface run-off.

Water also leaves a basin as a result of evapotranspiration. Rates of evapotranspiration are strongly weather dependent and are caused by the motion of water molecules at any temperature above absolute zero (Burman et al., 1987). Direct solar radiation and, to a lesser extent, the ambient temperature of the air provide the energy required to change the state of the molecules of water from liquid to vapour. The rate of removal of water vapour from the evaporating surface is driven by the difference between the water vapour pressure at the evaporating surface and that of the surrounding atmosphere. As evaporation proceeds, the surrounding air becomes gradually saturated and the process will slow down and stop if wet air is not transferred to the atmosphere. Consequently, evapotranspiration is higher from surfaces receiving more solar energy and from more exposed locations, where wind speeds are higher. The transpiration component of evapotranspiration is influenced by the nature of vegetation in addition to weather. Plants predominately lose their water through stomata: small openings in the leaves through which water vapour can pass. Different vegetation types have different transpiration rates, but in general, transpiration increases with the leaf area, the roughness of vegetated surfaces, and the size of stomatal apertures (Jarvis and McNaughton, 1986; see Fig. 1). This process, coupled with shading and microscale leaf surface topography, is responsible for driving very fine-scale variation in temperature across a leaf's surface, which in turn can influence the distribution of microorganisms on the plant (Vacher et al., 2016). Detection of these differences using thermal imaging can be used to monitor and assess plant health (Chaerle and Van Der Straeten, 2001).

As theory suggests, dry conditions typically occur on steep slopes, particularly those that face the sun, and, rainfall aside, valley bottoms are typically wetter than mountain tops. The effects of vegetation on water availability are, however, more complex. On the one hand, greater evapotranspiration is associated with higher vegetation cover, due to the effects of leaf area. On the other hand, canopy shading can reduce evapotranspiration, leading to damper conditions in the understory. The processes are also dynamic: drought conditions can cause vegetation moisture stress and possibly die back, which influences evapotranspiration rates and hence water availability (da Costa et al., 2010).

3. ORGANISMS AND THEIR ENVIRONMENT
3.1 Individuals and Microclimate

The fundamental reason for the importance of microclimates to organisms is that they drive energy, mass, and momentum exchange which, in turn, set the thermodynamic bounds on life (Bird et al., 2002; Porter and Gates, 1969). For example, the temperature of a terrestrial organism at a given moment is the result of the balance of energy fluxes, from incoming and reflected solar radiation, and heat from its metabolism, as well as the heat lost or gained by the processes of convection, conduction, and evaporation, and any storage of heat in the body (Porter and Gates, 1969). All of these processes except for metabolism and storage explicitly involve one or more aspects of microclimate: radiation, air temperature, precipitation, wind speed, humidity, and substrate temperature. Similarly, the mass budget of an organism in terms of water is the outcome of water gained from the environment, from its metabolism, as well as that lost by evaporation and potentially excretion/faeces. The evaporation process is microclimate sensitive, depending on the vapour pressure difference between the relevant surfaces of the organism and the air immediately surrounding the organism, as well as the air temperature and wind speed (Gates, 1980; Tracy, 1976). Finally, movement costs for mobile organisms and mechanical stress on sessile organisms depend on the force exerted by wind or water, which itself depends on fluid temperature and speed (e.g. Denny, 2014; Telewski, 1995).

The actual microclimate experienced by an organism in a particular habitat is the outcome of the interaction between the organism and its environment (Lewontin, 2000). The microclimate of all organisms is in part a function of their morphology and physiology. For example, the simple change in height above the ground with growth and development can have a strong effect on microclimate exposure (e.g. Körner and Hiltbrunner, 2017). Even different parts of the same organism may be at different temperatures: tree leaf temperature is different to that of the trunk and different again to root temperature (Kollas et al., 2014). Behaviour is an important factor for animals and, to a lesser but not unimportant degree, for plants (Huey et al., 2002). Ectotherms have long been noted for their ability to regulate body temperature precisely by behavioural means (Cowles and Bogert, 1944), but plants can also alter radiative microclimates via leaf angle (Darwin and Darwin, 1880) and their internal environment by opening and closing their stomata (Michaletz et al., 2015), while endotherms have sophisticated behavioural and postural means for regulating the microclimates they experience (Bennett et al., 1984; Briscoe et al., 2014a,b; Cunningham et al., 2015; Kelly et al., 2004).

The interaction between an organism and the environment can generate microclimatic effects that structure entire landscapes (D'Odorico et al., 2013; Rietkerk et al., 2002). Variations in climate at different scales can interact to give an overall impact of climate on species (Frey et al., 2016). Organisms are always subject to the influence of microclimates, but these effects can be mediated by biotic interactions that either change the microclimate where an organism occurs (e.g. plant competition for light, see Fig. 1) or induce an organism to move from one microclimate to another (e.g. due to competitive or predatory interactions). Organisms can also change the microclimate of the location they occupy; for example, the metabolic heat production of bats can measurably influence cave microclimates (Baudunette et al., 1994). Microclimates may also be important indirectly through their impact on resources such as food, micronutrients, water, and nesting materials.

For these reasons, the way in which microclimates are studied, measured, and simulated is much more organism specific than in biological considerations of macroclimate. In different cases, different variables may assume varying degrees of importance. For example, humidity may be of far greater importance to the body temperature of a frog than to a lizard, and soil moisture may be more directly influential to the survival of a germinating seed than it is to a hatchling bird. The relevant spatial and temporal scale of microclimatic observations is also intimately connected to the biological entity or phenomenon in question. For example, fine temporal resolution may be important for assessing exposure to stress but not for assessing climatic influences on life history (Helmuth et al., 2010; Kearney et al., 2012).

Correlative Species Distribution Models (SDMs; see Box 4 for all acronyms) can benefit from the inclusion of microclimatic variables, although they are not a prerequisite for the generation of high-quality models (Bennie et al., 2013). For example, Slavich et al. (2014) found that models run with 25 m topographic variables projected fewer species becoming critically endangered as a result of climate change than models run with 50 m resolution macroclimatic variables, with a difference of as much as 18%. Topoclimate also accounted for 6.1% more variance in the distribution of fern species than macroclimate, and 3.6% more variance in grasses. In addition, Flint and Flint (2012) found that 270 m resolution models highlighted the importance of topography in predicted redwood distribution as a result of water availability, while 4 km resolution models did not show any influence of topography at all.

For truly mechanistic SDMs that are based on measured physiological responses as a function of, e.g., body temperature, organism-scale microclimatic forcing is a prerequisite. For example, the CLIMEX modelling system is ideally based on measured physiological responses of individual organisms, but is often applied with macroclimatic data and 'tuned' to the distribution data, as in the case of the invasive cane toad in Australia (Sutherst et al., 1996). Such inverse fitting of mechanistic models converts the results into a correlative prediction, albeit mechanistically inspired. The explicit inclusion of microclimates can allow a stronger connection between individual physiology and environmental conditions (e.g. for the cane toad, see Kearney et al., 2008).

4. MEASURING MICROCLIMATES

Ultimately, the question facing any ecologist interested in studying microclimates is: 'What microclimate do I want to measure or estimate?' This may be the actual body or organ temperature of an organism, or the microclimate of the air or soil surrounding an organism. The answer will be unique to the experiment being planned and will lie somewhere on the spectrum between a full international standard meteorological setup (ISO, 2015) and the true set of conditions that the target organism experiences in real time. Some consideration should also be given to if/how the

BOX 4 Acronyms explained

Acronym	Full form	Description
ALS	Airborne Laser Scanning	An active remote sensing method that acquires 3D point clouds of objects using LiDAR instrumentation mounted on an aircraft
DEM	Digital Elevation Model	Encompasses both DTMs and DSMs
DGVM	Dynamic Global Vegetation Model	Shifts in global vegetation and associated processes
DSM	Digital Surface Model	The elevation of the land surface plus surface features such as vegetation and buildings
DTM	Digital Terrain Model	The elevation of a land surface, typically above sea level
CHM	Canopy Height Model	Canopy height calculated from the difference between a DSM and a DTM
GDD5	Growing Degree Days above 5 degrees	Essentially the amount of time available for most plants to grow
GIS	Geographic Information System	A system for managing and present spatial or geographic data
GNSS	Global Navigation Satellite System	Satellite navigation systems that provide autonomous geospatial positioning with global coverage
GPS	Global Positioning System	A type of GNSS using the US constellation of satellites
IBM	Individual Based Model	Simulation models that describe autonomous individual organisms
LAI	Leaf Area Index	The one-sided green leaf area per unit ground surface area
LiDAR	Light Detection and Ranging	A remote sensing method utilising a pulsed laser to measure variable distances to backscattering objects, such as vegetation elements and the bare ground
PAR	Photosynthetically Active Radiation	The light available for photosynthesis, in the 400–700 nm wavelength range
SDM	Species Distribution Model	Species occurrence or abundance in response to environmental factors
TDR	Time Domain Reflectometry	A measurement technique used to determine the characteristics of electrical lines by observing reflected waveforms
TLS	Terrestrial Laser Scanning	An active remote sensing method that acquires 3D point clouds of objects using LiDAR instrumentation on the ground
UAV	Unmanned Arial Vehicles	Aircraft without a pilot aboard, aka drones

sensor might affect this target organism, i.e., by changing its behaviour (Willis et al., 2009) or its body temperature. Thorough background research on a putative sensor type should reveal such issues where they occur.

As the field of microclimate science matures, it is also useful to consider whether measurements can be taken in such a way as to be more broadly useful to others, to allow sharing of data between researchers and provide ground-truthing data for modellers. In order to achieve this, it is important to thoroughly record every aspect of the method used. For example, many papers do not report the model of climate sensor used, but it is vital to do so if data are to be comparable. A list of sensors which have been used in microclimate research can be seen in Table 1, while Table 2 presents some issues that may be encountered when utilising different types of sensors.

The scope of microclimate monitoring can range from the placement of portable meteorological stations in locations uncaptured by standard networks, to bespoke microsensors placed on (or even in) an individual organism of interest, to the simultaneous capture of temperature information across a landscape using remote sensing. Any measure of an atmospheric condition (temperature, moisture, or others) is an approximation of reality, including those data collected by meteorological station networks. The data that are derived are therefore a function of the physical properties of the monitoring device, its emissivity, its albedo, its mass, its susceptibility to radiation and wind, and, most critically, where it is placed: the level of shade and orientation to the sun, and thus exposure to direct solar radiation. Because microclimate by definition varies at small spatial scales, it is a challenge for ecologists to develop a consistent, standard methodology for measuring microclimatic variables. Slight variation in the position of sensors, in their exposure to sunlight and wind, in their height above the ground, or in the physical characteristics of the sensors themselves can lead to large discrepancies in measurements, but attempting to reduce these discrepancies (for example by mounting sensors in a radiation screen at a standard height and at a standard aspect), will inevitably lead to a reduction in the variability observed, including those aspects which may be of most interest.

4.1 Measuring Microclimate In Situ

The in situ study of microclimates has a long history that can broadly be divided into two eras: before and after automation. Prior to automation, most instruments would take instantaneous measurements of the microclimate (see Monteith, 1972; Unwin, 1980 for useful guides). Although the

Table 2 A Summary of the Variables Measured in Ecological Studies of Microclimates in the Field, Quantified During a Systematic Review of the Literature (See Supplementary Methods for Details: Available on https://doi.org/10.1016/bs.aecr.2017.12.005)

Variables Measured	Number of Studies	Potential Problems
Temperature	59	If left unshielded, temperature sensors absorb radiation: either direct shortwave radiation from the sun or indirect short- and longwave radiation via other components of the local environment (e.g. trees, ground surface). This can result in recorded temperatures being higher than the ('true') air temperature, particularly in clear-sky conditions in unshaded locations. Sensors also emit longwave radiation, so under conditions where the net radiation balance of the device is negative—such as at night, or in shady habitats in daytime—they can also be warmer than the environment (Unwin, 1980). The size of any radiation effect varies based on the sensor design (see Section 4.2.1)
Combined temperature and humidity	79	Humidity sensors are vulnerable to getting wet. If water gets on the sensor, it results in inaccurate humidity readings (see Section 4.2.2)
PAR/LUX	31	There are dataloggers available which measure both light and temperature; however, this presents the problem of radiative warming of the sensor discussed above. In these cases, one logger should be used to record light, and a shielded logger should be used to record temperature
Wind speed/ direction	11	There have been a lack of small self-contained anemometers which can be *left* in situ. Some models are now available which can be left logging for 3 days
Soil moisture	28	Soil moisture sensors need to be calibrated for the soil texture and organic matter concentration that determines the electric conductivity; otherwise the results are uncertain

Please note that there are a variety of types and brands available for each sensor group.

level of replication in these experiments was limited by obvious practical considerations, such as the numbers of instruments to be employed, and the number of personnel available to attend them, the preautomation era led to much of the fundamental physics of boundary layer climatology being resolved (e.g. Geiger, 1927; Homén, 1897; Kraus, 1911). Because the sensor technology behind early, analogue instruments and their modern, digital equivalents has changed little, this early microclimatological work has continued relevance today.

The falling price of the microchip during the 1980s and 1990s heralded a new wave of instruments that could autonomously collect and store microclimate measurements in the field (see Table 1). Large levels of replication became available to scientists interested in how microclimates vary across space and through time. This technology has improved our understanding of how and when the boundary layer microclimate differs from measurements derived from meteorological station networks, such as the difference between recording at screen or other sensor heights (1.5–2.0 m) and close to the ground (e.g. Scherrer et al., 2011; Suggitt et al., 2011), how vegetation types differ in their microclimate (e.g. Morecroft et al., 1998), or even the contexts in which simplifying assumptions, such as lapse rate adjustments, break down (e.g. Bennie et al., 2010; Pepin et al., 1999).

The WMO provides guidance on the desirable characteristics, standards, and uncertainties of standard meteorological instruments (WMO, 2014), some of which are applicable for micrometeorological observations. However, some of this guidance will only apply to regional and synoptic processes, so a thorough consideration of their suitability for micrometeorological studies should be made before they are applied.

Sections 4.1.1–4.1.8 briefly describe what we consider to be the more important considerations when choosing and using each type of in situ sensor.

4.1.1 Temperature

Numerous sensor types have been used to measure temperature (Table 1), and the merits and drawbacks of each are typically associated with cost and accuracy: more expensive devices record temperatures more accurately. However, by far the greatest consideration for accurate temperature recording is what temperature is actually of interest. This will dictate where sensors should be placed and how measurements should be taken (see Fig. 3 for illustration). Aerodynamically well-coupled organisms such as adult trees and bushes are likely to experience temperatures close to those represented by standard meteorological station data obtained 1.5–2 m above the ground. Where finer spatial-scale data are needed (e.g. at the seedling life stage for trees), sensors appropriately shaded from solar radiation will likely provide a reasonable proxy. In other circumstances obtaining measurements 1–2 cm below the ground, where many smaller plants have

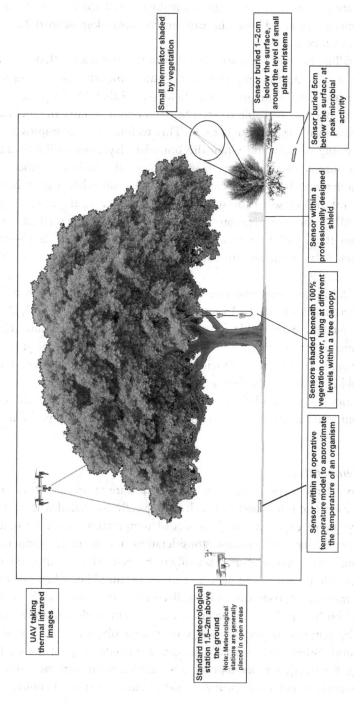

Small thermistor shaded by vegetation

Sensor buried 1–2cm below the surface, around the level of small plant meristems

Sensor buried 5cm below the surface, at peak microbial activity

Sensor within a professionally designed shield

Sensors shaded beneath 100% vegetation cover, hung at different levels within a tree canopy

UAV taking thermal infrared images

Standard meteorological station 1.5–2m above the ground
Note: Meteorological stations are generally placed in open areas

Sensor within an operative temperature model to approximate the temperature of an organism

Fig. 3 An illustration of the potential methods of temperature measurement.

their meristems, or 5 cm below the ground, where microbial activity peaks, may be desirable. To obtain such measurements burying sensors with appropriate damp proofing is likely to be the most sensible option. It is when trying to measure the temperature experienced by small organisms close to the ground that the temperature measurement becomes most problematic. Here, the temperature of an organism is influenced not only by solar radiation but by the temperature of the ambient air with which it exchanges heat. Air temperatures close to the earth's surface, even when shaded from direct sunlight, can often differ substantially from that within the free atmosphere due to convective heating or cooling from the ground. It is thus necessary to approximate the effects of both higher ambient air temperature and radiative heating.

In an attempt to approximate near-ground air temperatures, many ecologists have opted to shield their sensors with a radiation shield (or even two), electing for either a 'homemade' or 'off-the-shelf' version, which benefit from being low-cost (e.g. De Frenne et al., 2011; Holden et al., 2013; Lundquist and Huggett, 2008). However, we caution against doing so without rigorous testing. Different types of shield result in substantially different measurements of temperature (see Fig. 4), and low-cost nonstandardised screens tend to reduce ventilation and result in overestimation of the temperatures in comparison to expensive purpose-built screens (Hubbart et al., 2005; Fig. 4). Differences in temperature recorded under homemade screens are more extreme during the daytime, especially under clear sky conditions. In Fig. 4 the data are averaged across the testing period and so the differences shown are conservative compared to what would be seen if only clear days were included. Mechanical ventilation ('aspiration'), ensuring a unit flow of air passes the sensor per unit time, can minimise this effect, but the power required to implement it limits its use in most field ecology contexts. If the aim is to obtain a representative local air temperature that permits comparisons with weather service data, measuring air temperatures either under 100% tree shade or in soils in deep shade may provide reliable results (Körner and Paulsen, 2004; Lundquist and Huggett, 2008). In contrast, if fully sunlit life conditions are to be understood, shielding the device may actually mask the very component of microclimate that is of interest. Because the thermophysics of the device will almost always differ from that of the organism of interest, the best choice for accurate air temperature close to the ground in open terrain would be very thin (<0.1 mm), unshielded, freely exposed thermocouple junctions. Many temperature sensors are often housed in weather-proof

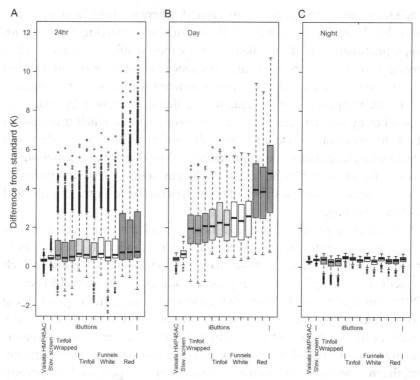

Fig. 4 The effect of different shielding methods (inverted funnels covered in tinfoil, painted white or red) on in situ temperature measurements (see Supplementary Methods available on https://doi.org/10.1016/bs.aecr.2017.12.005)—iButtons in custom-made shields were on average warmer and experienced more extreme temperatures, particularly higher temperatures, than standard sensors and compared to the iButton in the Stevenson Screen (A). However, this was very dependent on radiative warming. Measurements taken during the two hottest hours of the day (14:00–16:00) tended to be more extreme (B) with the biggest difference being driven by colour (*red funnels* vs everything else). In comparison measurements taken between midnight and 02:00 showed little discernible difference between treatments or from the standard sensors (C). Tinfoil-wrapped iButtons slightly underestimate temperature at night possibly due to cooling from wetting or condensation (C).

casing and are thus essentially shielded to some degree. Therefore the surface albedo of the device and degree of ventilation the sensors experience may be quite different to those experienced by an organism in its environment. There are several possible ways forward for measuring air temperature close to the ground:

1. Use dataloggers installed close to the ground within a professionally designed radiation shield. While a full Stevenson screen (see Box 3) is impractical for microclimate work, many operating meteorological stations (e.g. Campbell Scientific) use radiation shelters that are 15–20 cm in height/diameter, and these can simply be installed close to the ground.

2. Use very small external sensors, which due to their low surface area/volume ratio are closely coupled to air temperature. Simple thermocouples or small thermistors are usually too small to heat up above the air temperature. Such probes should not be shaded because reradiation from the screen removes the benefit of convective coupling of such thin sensors.

3. Use improvised shading methods with extreme caution, and only if the design is fully tested against more preferred methods and the errors present in strong radiation have been quantified.

4. Use sensor/shield combinations that have similar physical properties (size, thermal capacity, albedo) to the organism of interest, and measure a standardised effective organism temperature (operative temperature, see below) rather than air temperature.

5. Consider the use of thermal imagery obtained under varying radiation heat flux conditions (see Section 4.1), coupled with mechanistic models that permit temperature to be predicted from radiation (see Section 5.2).

Above all, it is important to remember that the temperature of the sensor is reported, not that of the air or the organism. Any assumptions about the strength of coupling to air temperature/organism temperature must be carefully considered and ideally tested within a formal experiment (e.g. Fig. 4; see Supplementary Methods available on https://doi.org/10.1016/bs.aecr.2017.12.005), as the effect of irradiance on reported temperatures is often nonlinear and difficult to predict a priori (Barozzi et al., 2016).

4.1.1.1 Operative Temperature Models

The most organism-specific empirical microclimatic measurements are made using 'operative environmental temperature' thermometers, which approximate the steady-state temperature an organism would come to give its size, shape, and radiative thermal properties if it had zero thermal mass (Bakken, 1981, 1992; Bakken and Angilletta, 2014). For ectotherms, these are typically made out of hollow copper objects (for rapid heat transfer), often by making a mould of the specimen (Hertz, 1992; Porter et al., 1973) but also by simply using hollow copper tubes (Shine and Kearney, 2001). For endotherms, heating elements inside taxidermic mounts can be used to estimate metabolic heating requirements to maintain a constant body temperature, or to include basal metabolic heat generation for estimating heat loads (Bakken, 1981; Bakken et al., 1983; for a critical review, see Walsberg and Wolf, 1996). Of course, such models cannot replicate the reality of organism's thermoregulatory behaviour, where even subtle behaviours such as changes in the angle of an individuals' fur can greatly affect the degree of insulation (Walsberg and Wolf, 1996). Depending on the question being asked, attention may need to be paid to fine details of posture, shape, and reflectivity of the models; even minor differences in size and colour of models can result in measurements differing by 2–4°C (Bakken and Angilletta, 2014). The approach has typically been used for small- to medium-sized animals (1–1000 g), but the lower size of application is limited by sensor dimensions (e.g. 40 gauge thermocouple wire is ∼0.2 mm diameter). Bakken and Angilletta (2014) provide a useful overview of the use of operative temperatures in thermal ecological studies.

4.1.2 Humidity

Atmospheric humidity is critical to organisms as, along with temperature, wind speed, and the radiation balance, it determines the potential rate of water loss from a surface. For ecological studies, in the vast majority of cases the most appropriate measure of humidity is vapour pressure deficit (VPD), which scales in an almost linear manner with the rate of evaporation from a surface. VPD is the difference between the saturation vapour pressure within free air and the actual vapour pressure at a particular temperature; when the air is saturated and dew forms on surfaces, the VPD is zero. Therefore, in order to accurately record VPD, a measure of temperature is also needed, and the same considerations concerning the effects of radiation and shelter housing that apply to temperature measurement will also apply to measurement of humidity. Many forms of hygrometer will record relative humidity (RH), rather than VPD—in this case the temperature of the air must be known in order to convert to VPD. Because the meaning of RH with respect to evaporative forcing differs greatly for different air temperatures (50% RH at around 0°C and 50% RH at around 20°C correspond to an approximately 10-fold difference in VPD), it is best to use VPD in ecological studies rather than RH, which can be converted to VPD if the air temperature is known. Electronic humidity sensors typically record both RH and temperature by default.

Several different types of hygrometer, employing different physical principles, are used to measure humidity. The most widespread in meteorological stations is the psychometric method. A psychrometer consists of two thermometers exposed side by side, with the sensing surface of one covered by a film of water (the wet bulb), and the other dry (the dry bulb). The cooling effect of evaporation from the wet surface is used to estimate humidity. Although accurate, this method has several drawbacks for measuring fine-scale humidity, particularly for automated measurements in remote areas. The size of standard psychrometer units is often fairly large, making it difficult to make measurements in confined spaces or close to the ground, near to an organism or microenvironment of interest. Also, regular maintenance is needed to ensure that water reservoirs are kept topped up.

Electrical (capacitive or resistive) humidity sensors are increasingly used to provide a cheap, lightweight, and flexible alternative to traditional methods, particularly for remote applications (Table 2). Electrical sensors exploit the properties of hygroscopic materials that change their electrical properties (resistance or capacitance) with a change in the ambient relative humidity, with a small temperature dependence. Several manufacturers produce electrical relative humidity sensors with miniature dataloggers, designed to be deployed outdoors and subsequently downloaded to a PC or a laptop.

Another frequently encountered problem with electronic humidity data sensors is the trade-off between exposing the sensor to freely circulating air and reducing the probability that moisture condenses on the sensor itself (or gets wet from rainfall or spray in damp environments), typically leading to false measurements of zero VPD (100% RH) or even damage to their circuitry. While zero VPD is frequently observed in many environments, particularly at night when the atmosphere cools below the dew point and the air is saturated, if liquid water remains on the sensor, then erroneous measurements are likely to persist after the atmospheric humidity is reduced. Radiation shielding, good ventilation, and protection from rain are imperative in any attempt at arriving at reliable air humidity data. Where the practicalities of housing a humidity sensor for measuring VPD prevent accurate temperature measurement (for example, where sensing housing heats the sensor), the vapour pressure can be calculated and the VPD at a given temperature (and therefore saturation vapour pressure) can be recalculated. An alternative or complementary approach to humidity sensors in sporadically damp environments is leaf wetness sensor, which measure the dielectric constant of the surface of the sensor itself (usually shaped like a leaf), and can detect small amounts of water or ice on the 'leaf' surface.

4.1.3 Radiation

This section refers to those parts of the electromagnetic spectrum which are biologically relevant and measureable, essentially the thermal infrared and visible parts of the spectrum. Note that visible, photosynthetically active part (wavelength 400–700 nm) will be dealt with separately below.

The total solar irradiance per unit horizontal area is called global radiation. In terms of its energy ($W\ m^{-2}$) roughly half falls in a long wavelength part, often referred to as thermal radiation, while the other half is largely visible and termed 'photosynthetically active radiation', PAR (or PhAR). Radiation sensors are calibrated in such a way that the data reflect total incoming radiation. Global radiation is typically measured by a solarimeter, and some devices also separate the direct and diffuse components (by shading the sensor in some way). As its name suggests, a 'net' radiometer will additionally capture outgoing longwave radiation, and thus the overall radiation budget can be discerned.

Although in theory quite possible to design, a battery-integrated radiation sensor with full spectral response would in practice suffer from a short battery life due to the power required to operate the sensor(s). The power requirement is increased for deployments where the weather conditions (snow, ice, dew) might obscure the path of radiation to the sensor, necessitating the use of a preventative heater.

Separate monitoring of radiation (in addition to temperature) may be required to understand the thermal opportunities available to organisms in situations where the timing of a change in the radiation budget may not be apparent in recorded temperatures, but could easily be apparent in ecological findings (such as activity levels or metabolic processes).

4.1.4 Visible Light and PAR

Light may be considered either as a flux of energy in the form of waves or as a flux of photons, with the amount of energy a property of the photon. Since the energy carried by a photon is higher in shortwave (blue) compared to longwave (red) light, physical (energy-) sensors will always give more weight to blue light. It is possible, by using appropriate filters, to account for this, so that all photons will have roughly the same weight. Such sensors measure the photon flux density PFD (or PPFD the photosynthetically active photon flux density), and the units applied are μmol photons $m^{-2}\ s^{-1}$. PAR, the energy-based parameter, and PFD, the photon flux-based parameter, should not be confused. Since the only sensors on the market that are tailored to plant's photon demand measure PFD, the only appropriate unit is μmol m^{-2} s^{-1}, and thus the term PAR is best avoided.

Although the wavelength range of 'visible' light and that of PFD or PAR are roughly the same, the typical response to light within these bounds (McCree, 1971) is quite different to that of the human eye. PAR sensors are therefore calibrated to a different power distribution, although in practice integrated sensing units are available that will report PAR, visible light, and total radiation (including direct and diffuse) simultaneously. Because photon flux readings can be highly variable under a vegetation canopy (as mentioned in discussion of solarimeters, Section 4.1.3), a number of sensors deployed over a long 'bar' can capture an integrated PFD far better than a point sensor. Sensors designed to capture visible light can also be calibrated in lux. All these units can be converted into each other, as long as the spectrum does not change (as it does, if an electric lamp and the sun is compared). Field of view or cosine correction is important—many PFD or lux meters measure on a plane (and thus require cosine correction) rather than from all directions, as an organism would experience.

When the aim is to obtain subcanopy PFD conditions from freely exposed PFD signals in open terrain, one can either apply the Beer–Lambert law of light extinction as described by Monsi and Saeki (Hirose, 2004) or apply hemispherical photographs. Both methods permit calculating the Leaf Area Index (LAI), the ratio of leaf area to ground area, and in the case of hemispheric photography, surprisingly accurate estimates of LAI ($R^2 \sim 0.95$) are possible if exposure levels are correctly set to maximise the contrast between the vegetation and the sky (Zhang et al., 2005). Red/far-red sensors measure the fraction of red light that is absorbed by green leaves vs the fraction of red that is not absorbed by chlorophyll (730 nm), thereby quantifying the density of the green cover above any point on the ground or in a plant canopy in relative terms (and so quite independent of incoming radiation).

4.1.5 Wind

Anemometers, devices used to measure wind speed, generally fall into three categories. The cheapest and most widely used in ecological studies are cup or propeller anemometers, in which rotating cups or a propeller is driven by the wind. In the latter case, the device must either be held perpendicular to the direction of the wind or be mounted on a vane; for automated measurements such devices typically measure direction as well as speed. While mechanical anemometers are still in use in meteorological stations, ultrasonic anemometers are becoming more popular. Handheld anemometers are frequently used by ecologists for short-term measurements, and they are less often installed with dataloggers to measure microclimate (see Table 2). In part, this is due to the relatively large size of many units, making it difficult to measure within plant canopies or close to the ground, where the wind is decoupled from the background atmosphere. However, miniature propeller anemometers with integral dataloggers and the ability to measure temperature and humidity as well are now available at a relatively low cost (e.g. Samson and Hunt, 2012) and may be appropriate for some microclimate applications.

The second form of anemometers used in ecological studies are hot-wire anemometers, in which an electrically heated wire element is cooled by the wind, and the wind speed calculated by the rate of heat loss. Unlike mechanical anemometers, these devices have a rapid response time and the lack of moving parts allows them to be installed close to the ground or within vegetation, so they have the potential for measuring small-scale eddies and microclimatic effects. However, the relative expense of the units is prohibitive for many ecological applications.

More complex ultrasonic anemometers measure wind speed in three directions based on the time of flight of sonic pulses between pairs of transducers. While expensive, such sonic anemometers are suitable for measuring turbulent air flow with a very high temporal resolution and are typically used in conjunction with infrared or laser-based gas analysers to measure ecosystem fluxes using the eddy covariance method (Burba and Anderson, 2007). The eddy covariance micrometeorological technique, which involves high-speed measurements of fluxes of water, gas, heat, and momentum within the atmospheric boundary layer, is widely used by micrometeorologists across the globe.

4.1.6 Soil Moisture

Soil moisture is a key component of microclimate that both directly affects water availability to plants and indirectly influences the temperature and humidity close to the ground, via its effect on the surface water and energy balance. While measurements can be made by extracting soil cores, drying, and weighing them, this form of measurement is destructive and not suitable for long-term monitoring (but is essential to validate instrumental readings). Automated in situ measurements can be made by tensiometers, by various types of electrical resistance sensors, or by Time-Domain Reflectometry (TDR). TDR sensors send an electrical signal via metal rods into the soil and measure the signal return. This has the potential for spatial surveys of soil moisture as the probe does not need to equilibrate with the soil, readings are fast and accurate, and many readings can be taken using a single probe within a short space of time. However, the calibration of TDR probes is sensitive to soil characteristics, and so may need separate calibrations for different soil types (see Table 2). All three probe types can be connected to loggers where a network of probes collecting time-series data is required.

As below-ground microclimates (such as soil moisture or temperature) typically change on a slower timescale than those above ground (which are more subject to rapid changes in the radiation balance and air movement) for many applications, it may be possible to sample them less frequently, or to take measurements across a spatial domain using a single sensor within a specified time period. However, soil moisture can change rapidly during and following precipitation events, particularly near the soil surface, so care must be taken when planning spatial sampling strategies.

The above sensors typically measure soil moisture in the close vicinity of the probe itself (within centimetres), making them suitable for high spatial resolution measurement or investigating fine-scale variation. However, as soil moisture varies over fine scales, many point-based measurements may be needed to characterise a domain if measurements representative of a wider area are needed. In this case, another technique, cosmic ray soil neutron sensing, can be used to measure integrated soil moisture over a footprint with a diameter of up to 600 m (IAEA, 2017). However, for most studies with a focus on microclimate, this technique is likely to integrate over too large an area to be useful.

4.1.7 Ground Truthing and Sensor Calibration

An important consideration when planning a project is the need to ensure that the data collected by each of the numerous sensors and/or dataloggers are truly comparable. Most units come with a guaranteed degree of accuracy, but this will vary between designs, and there can be differences between units. For less accurate units, one solution is to deploy multiple sensors together and calculate the average for a more accurate reading. A simple solution is to calibrate the units, by deploying them together in the same environment for a few days prior to use in the field. The recorded values can be modelled as a function of unit and time, and the deviation of each unit from the mean calculated. This offers a simple but effective 'correction value' which can be applied to the field data collected on each unit. Another method is to deploy devices overnight in an ice–water mixture (more ice volume than water) in a thermos flask, which also obtains information on the absolute accuracy (0.0°C). The process can be repeated following field-work in order to test for changes in relative measurement accuracy over time. For long-term studies ongoing cross-checks to a reference weather station that is subject to regular calibration checks is valuable.

4.1.8 Sampling Design

Whether placing sensors to allow statistical interpolation of microclimates or to provide ground-truthing data for models or remotely sensed data, it is nec-essary to carefully consider the sampling design. The level of replication will depend to some extent on the number of loggers available and the heteroge-neity and extent of the area to be sampled. Care should be taken to sample entire gradients of microclimatic conditions prevailing in a study system. For example, Suggitt et al. (2011) used a Digital Elevation Model (DEM) to categorise a topographically heterogeneous landscape into categories of slope steepness, aspect, and elevation, and then within a Geographic Informa-tion Systems (GIS) generated three random sample locations within each cat-egory to ensure that the full range of these drivers was sampled at the earth's surface, with sensors placed >100 m apart to reduce the impact of spatial auto-correlation in these variables. In addition, measurements should be taken for a sufficient amount of time to capture the full range of weather conditions at a site, including unusual situations such as those that drive cold-air drainage (see Section 3). In practice the number of available sensors may limit the number of processes that can be investigated, and the placement of sensors themselves can result in large differences in measured variables (especially with temperature sensors) and this should also be quantified where possible. In some cases, experimental manipulations of the factors driving microclimates have been used, such as removal of vegetation to examine effects of the interactions between slope aspect and vegetation cover (Lembrechts et al., 2017).

Other applications may require sampling below the soil surface (e.g. Edwards et al., 2017) or within vegetation canopies (e.g. Graae et al., 2012). In these cases, it is necessary to consider whether absolute height or depth is important, or distance from a feature of interest (e.g. distance from the top of the canopy). This is likely to depend on the target organism and should approximate the location of their behaviour or life stage of interest. For example, to measure the climatic experiences of tree dwellers (like primates or tree frogs, e.g., Scheffers et al., 2013), it is important to capture variation of climatic conditions within the canopy, close to the ground, and above the canopy where direct sunlight greatly increases experienced temperatures but where the animals often have to go for food.

4.2 Ex Situ Sensing of Microclimate

Satellite remote sensing (RS), although not traditionally available at resolutions suitable for monitoring microclimate, has advanced rapidly and is now providing products at spatial and temporal resolutions that are useful for some microclimate-related applications. Over the last 50 years, satellite RS has played a major role in the monitoring and understanding of large-scale (global and regional) environmental changes (Pettorelli et al., 2014), including land cover (Elias et al., 2015), crop condition (Atzberger, 2013), land and marine productivity (Brewington et al., 2014; Guay et al., 2014; Myneni et al., 1997), phenology (Parmesan and Yohe, 2003), desertification (Ibrahim et al., 2015), and forest fires (Cuomo et al., 2001). During this time, advances in satellite and sensor technologies have enabled considerable improvements in the spatial, temporal, spectral, and radiometric resolution of the products available. Moderate to very high resolution (VHR) satellite imagery is now routinely used to monitor and assess relatively small-scale (<100 m) or rapid (<1 day) processes relevant to, e.g., operational meteorology (Søraas et al., 2017), soil moisture/drought (Mishra et al., 2017), forest fires (Wooster et al., 2013), pest outbreaks (Hicke and Logan, 2009), permafrost layer dynamics (National Research Council, 2014), and species/wildlife tracking (Yang et al., 2014).

The spatial and temporal coverage of satellite sensors and their resultant datasets varies considerably, mainly due to the application for which the data are to be used and technical limitations of the satellites and sensors. Geostationary satellites have the advantage of high temporal resolution coverage over a large area of the earth's surface. However, polar-orbiting (sun synchronous) satellites, being at lower altitudes, are generally able to sense higher spatial resolutions. Fig. 5 highlights some of the applications and sensors available for satellite RS across a range of spatial and temporal scales.

The term 'micro' suggests that the magnitude of the synergies between RS and microclimate research is constrained by the level of detail, i.e., the spatial resolution or grain size, of the RS data. The greatest potential lies in very high resolution (VHR) RS instrumentation providing environmental data at a level of detail that matches the scale at which the focal species experiences its environment or habitat.

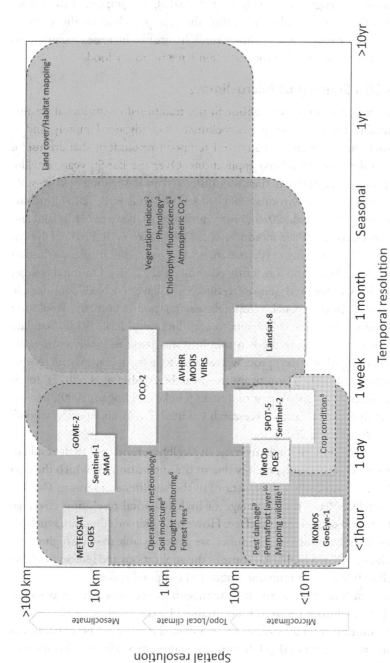

Fig. 5 Temporal and spatial resolutions of primary data available from selected satellites used for environmental monitoring. Polar orbiting (hashed background), geostationary (grey background). *Coloured boxes* highlight various environmental applications utilising satellite RS data, and *superscript numbers* refer to publications as follows: 1, Tsendbazar et al. (2017) and Russell et al. (2014); 2, Myneni et al. (1997); 3, Sun et al. (2017); 4, Eldering et al. (2017); 5, Søraas et al. (2017); 6, Mishra et al. (2017); 7, Wooster et al. (2013); 8, Atzberger (2013); 9, Hicke and Logan (2009); 10, National Research Council (2014); 11, Yang et al. (2014). See Table 3 for satellite details.

Table 3 Satellites From Which Data for Environmental Monitoring Is Available, Corresponding to Fig. 5

Acronym	Name	Organisation	Website
OCO-2	Orbiting Carbon Observatory	NASA/JPL, USA	https://oco.jpl.nasa.gov/
AVHRR	Advanced Very High Resolution Radiometer	NOAA, USA	http://noaasis.noaa.gov/NOAASIS/ml/avhrr.html
MODIS	Moderate-Resolution Imaging Spectroradiometer	NASA, USA	https://modis.gsfc.nasa.gov/
VIIRS	Visible Infrared Imaging Radiometer Suite	NASA, USA	https://ncc.nesdis.noaa.gov/VIIRS/
	Landsat-8	NASA/USGS, USA	https://landsat.usgs.gov/landsat-8
GOME-2	Global Ozone Monitoring Experiment-2	ESA, Europe	http://gome.aeronomie.be/
	Sentinel 1 and 2	ESA, Europe	https://sentinel.esa.int/web/sentinel/missions/sentinel-1 https://sentinel.esa.int/web/sentinel/missions/sentinel-2
SMAP	Soil Moisture Active/Passive	NASA, USA	https://smap.jpl.nasa.gov/
	METEOSAT	EUMETSAT, Europe	https://www.eumetsat.int/website/home/Satellites/CurrentSatellites/Meteosat/index.html
GOES	Geostationary Operational Environmental Satellites	NOAA, USA	https://www.nasa.gov/content/goes
	IKONOS	DigitalGlobe, USA	https://www.satimagingcorp.com/satellite-sensors/ikonos/
SPOT-5	Satellite Pour l'Observation de la Terre-5	CNES, France	https://spot.cnes.fr/en/SPOT/index.htm
	MetOp	EUMETSAT/ESA, Europe	https://www.eumetsat.int/website/home/Satellites/CurrentSatellites/Metop/index.html
POES	Polar Operational Satellites	NOAA, USA	https://poes.gsfc.nasa.gov/index.html
	GeoEye-1	DigitalGlobe, USA	https://www.satimagingcorp.com/satellite-sensors/geoeye-1/

4.2.1 Remote Sensing of Microclimate Variables

One option for taking temperature measurements simultaneously across a landscape is to take thermal infrared images using a specialised camera (e.g. Scherrer and Körner, 2010, who took images every minute for 24 h of an opposing hillside). This is becoming more affordable with the advent of devices that can be attached to mobile phones, although care must be taken to correctly calibrate and ground truth the data. Many of these devices are point and click and appear easy to use for the novice, but parameters like emissivity and distance to target can affect the readings obtained (Vollmer and Möllmann, 2017). Consideration also must be given to the wavelengths measured: for different applications, different parts of the infrared spectrum might be appropriate.

Thermal imagers can also be mounted to UAVs to give a 'bird's-eye view' data layer for the top of a vegetation canopy which can be imported for use in GIS for subsequent analysis (e.g. Zarco-Tajeda et al., 2012). To use these data with distribution records or field observations, it is necessary to measure the location of ground control points with a Global Navigational Satellite System (GNSS) to 'tie' the image to the correct part of the earth's surface (Greenwood, 2015). The type of GNSS should be selected based on the spatial accuracy required for the application. Handheld devices have a spatial accuracy of 3–5 m at best, so for fine spatial resolution data layers (e.g. <5 m) a differential GNSS should be used (spatial accuracy of 1–3 cm). However, some UAVs have inbuilt navigational GNSS devices that outperform handheld GNSS (Greenwood, 2015; Turner et al., 2014).

Thermal images have been available from satellites at a moderate resolution of 60 m × 60 m since 1999 (ETM+ on LANDSAT 7). For finer resolution images, aircraft with thermal imagers attached can capture thermal images at resolutions <5 m × 5 m, depending on the height of the flight above the earth's surface. These have already been used in studies of urban heat islands (Zhao and Wentz, 2016) and their increased availability could represent an opportunity for microclimate research in nonurban areas in future.

There are caveats when using thermal images, in that what is measured is the temperature of the surface itself, not the atmospheric conditions that many would consider to be the microclimate. However, thermal imagers can directly measure organism temperatures, which could be useful for some applications (e.g. Dietrich and Körner, 2014; Töpfer and Gedeon, 2014). In addition, thermal images provide a snapshot of temperature at the time recorded and do not necessarily represent microclimatic conditions as a whole. Indeed, mean, minimum, and maximum temperatures within landscapes do not necessarily correlate, the warmest places on average may not experience the hottest maximum temperatures, and the coolest places on average may not experience the coldest minimum temperatures (Suggitt et al., 2011; see Section 2). In order to fully represent microclimates, images should be recorded at different times of day and night, across seasons, and the full range of mesoclimatic conditions experienced at a site.

4.2.2 Remote Sensing of Proxy Variables

Microclimatic gradients and the associated availability of ecological microniches are to a large extent determined by three-dimensional (3D) structure, e.g., microtopography and/or horizontal and vertical vegetation structure (Section 2; Bazzaz and Wayne, 1994; Scherrer and Körner, 2010). Thus, RS data that accurately assess the 3D structure of natural or human made systems are expected to benefit microclimate research the most. In this respect, the recent surge of airborne RS generates many useful datasets and tools. Full-waveform Airborne Laser Scanning (ALS or airborne Light Detection and Ranging LiDAR), for example, provides contiguous and highly detailed information about the 3D structure of the land surface environment, which allows the derivation of high-end Digital Terrain Models (DTMs) as well as detailed data on horizontal and vertical vegetation structure, including Canopy Height Models (CHMs; Lefsky et al., 2002; Morsdorf et al., 2006). ALS data are thus suited to parameterise and map principle ecological phenomena, such as small-scale variations of temperature due to microtopographic variability, as well as temperature buffering mediated by canopy structure, providing crucial input data to microclimate modelling (see Section 5.2; Frey et al., 2016; Leempoel et al., 2015; Lenoir et al., 2017).

As an alternative to ALS, detailed landscape and vegetation structure data can also be derived from airborne radar or photogrammetry and image matching (i.e. structure from motion; Ginzler and Hobi, 2015; Gruen, 2012). However, photogrammetry suffers from a number of limitations such as difficulties with deriving accurate terrain and detailed vertical vegetation structure information beneath tree canopies due to occlusion effects, as well as insufficient handling of shadow areas and cloud effects (Gruen, 2012). Detailed data across large areas are normally recorded from airplanes or helicopters, whereas UAVs are increasingly used to map smaller areas, with the advantage of increased spatial and temporal resolutions (Anderson and Gaston, 2013). Terrestrial Laser Scanning (TLS) constitutes a further promising data source, as it provides 3D environmental data at an even higher level of detail than UAVs and ALS, but not contiguously across landscapes, being spatially restricted to sample plots (Liang et al., 2016; Telling et al., 2017). However, an advantage of TLS is that it reliably depicts understorey vegetation and within-canopy structure, which can complement ALS structure data in landscape-scale analysis (Hancock et al., 2017). A promising ecological application of detailed 3D vegetation structure data is to approximate below-canopy light regimes and associated microclimatic conditions prevailing across horizontal and vertical forest structure gradients (e.g. Moeser et al., 2014).

Air- and space-borne systems recording hyperspectral data provide information that is complementary to vegetation structure measurements because such data reveal insights into vegetation functional trait composition that may affect microclimatic gradients, e.g., via effects of leaf traits on the light regime beneath forest canopies (Asner et al., 2015; Kimmins, 2004). In addition, measuring surface albedo can be important for microclimate research as it affects the radiation balance (see Section 2) and using remotely sensed data to assess cloud cover could greatly help in downscaling radiation. As with directly measured microclimate variables, there must be accurate spatial coregistration of in situ measurements and the RS-derived environmental data.

5. MODELLING MICROCLIMATES
5.1 Why Model Microclimates?

Modelling allows the different factors which affect microclimate to be explored and the relative sensitivities within the study system to factors such as aspect, elevation, and canopy cover (see Section 3), to be tested. It also allows a scaling up from measurements at a network of locations to larger scales, for example, to predict variations in temperature over a heterogeneous landscape. In turn it allows relationships with organisms' distribution (Gillingham et al., 2012a), characteristics, and behaviour to be identified. Modelling also allows predictions and projections of ecological impacts, including projection of the impacts of climate change on species distributions (Gillingham et al., 2012b). There are a wide range of approaches, which incorporate a variety of the factors and processes that influence microclimates and result in a variety of predicted variables (Lenoir et al., 2017; see Table 4 for an overview). Model spatial resolution is driven by a trade-off between the availability of fine-resolution DEMs and other input data, and available computing power to run the model for the extent of interest. Fig. 6A shows the spatial resolution of a variety of models examining microclimate ecology, while Fig. 6B shows the temporal scale of models and

Table 4 A Summary of Models Utilised in Microclimate Research

Model Type	Climate Data Input	Other Predictors	Processes Modelled	Response Variables	References
Interpolated	Microsensors	Topographic from DEM, Canopy cover	Radiation, cold-air drainage	Temperature	Ashcroft et al. (2012)
Interpolated	Microsensors	Topographic from DEM, Canopy cover (remotely sensed), distance from coast	Radiation, cold-air drainage, distance to coast, latent heat	Temperature, humidity (if recorded by sensors)	Ashcroft and Gollan (2012, 2013), Gollan et al. (2014), Slavich et al. (2014)
Interpolated	Microsensors	Topographic and vegetation structure both from LiDAR	Radiation	Temperature	Frey et al. (2016)
Interpolated	Weather stations and microsensors	Topographic from DEM	Radiation, cold-air drainage, soil moisture	Temperature	Fridley (2009)
Interpolated	Microsensors	Topographic, distance from coast	Radiation, topographic wetness index	Temperature	Vanwalleghem and Meentemeyer (2009)
Interpolated	Weather stations with satellite data to ground-truth	Topographic	Radiation, cold-air drainage	Temperature (note this is 'free air' at 2 m temperature, not near the ground)	Dobrowski et al. (2009)
Downscaled	Macroclimatic grid with microsensors to ground-truth	Topographic from DEM	Radiation, cold-air drainage, PET (driven by radiation)	Temperature, rainfall, climate water deficit	Flint and Flint (2012), Dingman et al. (2013), McCullough et al. (2016)

Continued

Table 4 A Summary of Models Utilised in Microclimate Research—cont'd

Model Type	Climate Data Input	Other Predictors	Processes Modelled	Response Variables	References
Downscaled	Weather stations	Topographic	Radiation, cool air drainage, topographic wetness index, distance to water bodies	Temperature	Meineri and Hylander (2017)
Mechanistic	Weather station or macroclimatic grids with microsensors to ground-truth	Topographic from DEM	Radiation, local wind speed	Temperature	Bennie et al. (2008, 2013), Gillingham et al. (2012a,b)
Mechanistic	Weather station or macroclimatic grid with microsensors to ground-truth	Topographic, soil properties, vegetation shading	Radiation, convection, conduction, evaporation, latent heat, water infiltration, and evapotranspiration (of soil)	Air and soil temperature, air humidity, wind velocity, soil moisture, snow	Kearney et al. (2014), Kearney and Porter (2017)
Mechanistic	Weather station or macroclimatic grids with microsensors to ground-truth	Topographic from DEM, sea-surface temperatures, albedo from aerial imagery	Radiation, local wind speed, coastal effects, elevation, cold-air drainage, latent-heat exchange	Temperature	Maclean et al. (2017)
Mechanistic	Weather station or macroclimatic grids with field measurements to ground-truth	Topographic from DEM	Soil and surface water conditions	Surface water depth, soil moisture fraction	Maclean et al. (2012)

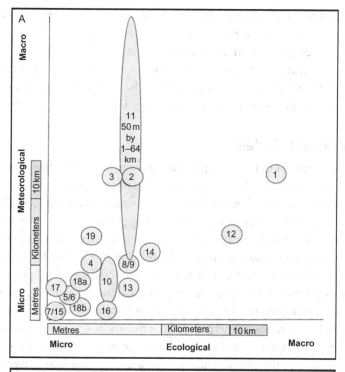

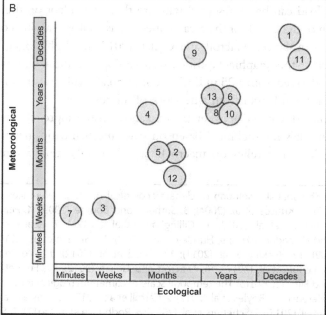

Fig. 6 See legend on next page.

studies on the same subject, which may be useful as a reference to assist in deciding the appropriate resolution and scales to model at. In some cases, such as the direct computation of heat, mass, and momentum budgets, one needs estimates of actual microclimatic conditions (e.g. Porter and Gates, 1969), whereas in correlative models, proxies such as topographic indices or radiation metrics may suffice (e.g. Gillingham et al., 2012a,b).

5.2 Statistical Models

Two main types of statistical approaches have been used in the scientific literature to model microclimate in a spatially explicit manner: directly through spatial interpolations of microclimatic measurements from georeferenced microsensors (e.g. Ashcroft et al., 2012; Suggitt et al., 2011) or indirectly by downscaling macroclimate from georeferenced synoptic weather stations or coarse-grained climatic grids (e.g. WorldClim or CHELSA data: Hijmans et al., 2005; Karger et al., 2017). Whatever the overall statistical approach used (interpolating microclimate or downscaling macroclimate), high-resolution DEMs are needed to generate meaningful predictor variables that are subsequently linked to micro- (interpolation) or macroclimate (downscaling) measurements for mapping climatic conditions at very fine spatial resolutions. The resolution of the output and thus the accuracy that factors such as solar radiation load can be represented with are thus dependent on the available resolution of the DEM for the area in question as well as available computing power. For instance, Ashcroft and Gollan (2013) used physiographic variables, such as topographic exposure, relative slope position, and distance to the coast, derived from a 25 m DEM, to interpolate daily maximum air temperature at 5 cm height from a network of microsensors.

The only difference between the two above-listed approaches is that one directly models microclimate based on true microclimatic measurements, whereas the other relies on macroclimate data only, thus assuming that

---

Fig. 6 (A) The spatial resolution of climate/ecological models. References: 1, Kriticos et al. (2012); 2, Kriticos et al. (2004); 3, Sutherst and Bourne (2009); 4, Pearson et al. (2014); 5, Bennie et al. (2013); 6, Gillingham et al. (2012a); 7, Kearney (2013); 8, Meineri and Hylander (2017); 9, Huntley et al. (2017); 10, Slavich et al. (2014); 11, Seo et al. (2009); 12, Buckley et al. (2011); 13, Graae et al. (2012); 14, De Frenne et al. (2013); 15, Lenoir et al. (2017); 16, Ashcroft and Gollan (2012); 17, Frey et al. (2016); 18, Flint and Flint (2012). (B) The temporal scale of climate/ecological models and field studies. References: 1, Boyles et al. (2017); 2, Carroll et al. (2017); 3, Agosta et al. (2017); 4, Tampucci et al. (2017); 5, Scheffers et al. (2013); 6, Rodhouse et al. (2017); 7, Sporn et al. (2009); 8, Graae et al. (2012); 9, De Frenne et al. (2013); 10, Lenoir et al. (2017); 11, Ashcroft and Gollan (2012); 12, Frey et al. (2016).

the set of predictor variables used to model climatic conditions at finer spatial resolutions will capture microclimatic processes. For example, Dobrowski et al. (2009) used a 30 m DEM to generate physiographically informed variables accounting for processes such as elevation-based lapse rates, solar insolation, and cold-air drainage effects. This set of topographic variables was then used as predictors and regressed against the residuals of a model relating free-air temperature measurements from synoptic weather stations to remotely sensed free-air temperature estimates from radiosondes and satellites that provide high temporal (3 h) and coarse spatial (32 km) resolution data on macroclimate (cf. regional free-air temperature conditions). This two-step statistical approach based on the general model of Lundquist et al. (2008) does not require any microclimatic measurements to model temperature conditions at reasonably fine (30 m) spatial resolution. Although very appealing, this indirect approach has strong limitations since it still represents free-air temperature conditions, and cannot be extrapolated to habitat types other than the ones equipped with weather stations. Unfortunately, most weather stations to date are installed in open habitats: standard meteorological recording protocols standardise measures above closely cut grass and thus (deliberately) fail to capture the impact of vegetation on the microclimate near the ground. Others are installed for very specific purposes, such as on mountain tops in ski resorts, which are also unlikely to capture the impacts of vegetation. Very few long-term weather stations have been installed within forest habitats (De Frenne and Verheyen, 2016) although forest microclimate has often been monitored as part of studies of forest ecophysiology, carbon, and water fluxes.

Statistical approaches have a drawback in that relationships are established over a relatively short period. It is well appreciated that models based on statistical methods can be unreliable when used to predict beyond the realm of existing data (e.g. Rice, 2004).

5.3 Mechanistic Models

Models that are based on the physical processes underpinning local climatic variation are more likely to provide reliable predictions under novel conditions (Evans and Westra, 2012). Mechanistic models seek to model microclimate using a mathematical representation of the processes involved in determining it. For example, the spatially explicit grid-based model of Bennie et al. (2008) calculates direct and indirect solar radiation from the slope and aspect of a location (themselves calculated from a DEM), given the average conditions experienced at nearby weather stations. Porter et al. (1973) and Kearney and Porter (2017) have developed a point-based model that makes hourly calculations of solar and infrared radiation intensities, above-ground profiles of air temperature, wind velocity and relative humidity at user-defined heights, and soil temperature and moisture profiles at 10 depths from the surface down to a maximum depth specified by the user. The fractional shade, slope and aspect, and horizon angles can be specified to capture topographic effects, and depth-specific soil thermal and hydraulic properties can also be user-defined. Mechanistic models still require inputs from weather stations or climate models (such as total radiation, wind speed, and temperature), but crucially the downscaling process is based on known mechanisms rather than being statistical or using interpolation algorithms.

6. LOOKING TO THE FUTURE OF MICROCLIMATE ECOLOGY

There are currently some clear gaps in microclimate research within ecology, due to trends in research focus or lack of technological or computer capabilities. For example, there is a lack of small-scale, self-contained wind speed sensors which can be left in situ alongside temperature and moisture dataloggers, resulting in this important factor being understudied. Also, more data on the vertical temperature profile near the earth's surface would be beneficial (see Lenoir et al., 2017), and there are currently large geographical gaps, particularly in Africa, Asia, and the Polar regions.

It is understood that microclimates are important to species distributions (Briscoe et al., 2014a,b; Kelly et al., 2004); however, there has been little research into the effects of microclimate on phenology, and how important it is in influencing phenological responses to climate change. It is also vital to understand how species use microclimates, and what factors influence their ability to fully utilise potential microclimatic niches. There has been a focus on thermal microclimates, driven by the fact that these appear easier to measure. However, for many species and in many ecosystems, hydro- and hygromicroclimates might be more relevant. Increasing our efforts in measuring and modelling water- and humidity-driven variation in microclimates will offer new opportunities in terrestrial ecological research.

It would be useful to understand to what extent habitats and species are buffered by and reliant upon microclimates (Wakelin et al., 2013). For this it is important to have long-term datasets. At present microclimate research is generally done on the short term, with measurements being taken for a few months to a year at a time. In order to understand the decoupling of microclimates from macroclimate, we need multiple year studies (Lenoir et al., 2017). This will allow us to better understand and model how that decoupling is affected by climate change (for example, does the difference in temperature increase or decrease with climate change?). There is also the potential for changes in vegetation structure and species composition caused by climate change to modify microclimate. The most obvious effect would be at treelines, where a taller canopy has major impacts on surface and soil temperatures; however the extent to which canopy height is limited by temperature or wind speeds will modify this.

As yet there has been little consideration of climatic extremes in microclimate research. As it has been established that extremes are important limiting factors to organisms (Cunningham et al., 2013), this is a topic that cannot be overlooked if microclimate research is to be useful in protecting species. Relevant to this is the understanding of how microclimates buffer not just against general climate change but against extreme events such as heat waves and droughts. Morecroft et al. (1998) found that in one winter, the presence of a tree canopy at one site in Britain prevented the incidence of ground frost which was frequently experienced at a grassland site nearby. It would be useful to gain a good understanding of how habitat management practises influence microclimates, in order to establish the best ways to manage habitats for the protection and creation of microclimatic refuges. There will hopefully be opportunities to work with conservation practitioners in order to increase awareness of the importance of microclimates to species and allow them to integrate awareness of microclimates into their management practises.

In order to achieve these goals and develop an understanding of microclimates which is widely applicable and useful to conservationists, it is necessary to have data widely available to researchers and the resulting information available to practitioners. An online global dataset of freely available microclimate data would make the establishment of patterns in microclimates around the world and comparisons between studies far easier. In order for such data to be comparable, it is necessary to have some level of standardisation in data collection methods, which would be facilitated by standardisation of sensor and datalogger technology (although the continuing proliferation of sensor designs available suggests that such standardisation makes this less and less likely), and a high level of metadata describing the methods of measurement and study sites. Freely available data would also assist if a Microclimate Model Intercomparison Project (MMIP) was established. Such a project would facilitate the development of ever more effective microclimate models, as well as potentially establishing the importance of microclimate in the wider climate modelling community.

Recent advances in technology and data availability present interesting opportunities for future microclimate research. For example, the recent development of dataloggers from which data can be downloaded in the field via Bluetooth presents a potential area for further advancement. Sensors from which data are downloaded automatically in real time would allow for more long-term studies, with reduced need for large memory storage, and could potentially allow for microclimate models which are continually evaluated and refined via data automatically fed into them from an array of sensors in the field (e.g. as proposed for DNA sampling by Bohan et al., 2017). Another sector that is developing rapidly is remotely sensed data, which is becoming more widely available online. Ongoing plans for satellite RS, particularly through the recent EU/ESA Copernicus programme (Sentinel satellites), as well as continued high-resolution monitoring programmes, e.g., the NASA/USGS Landsat programme, will ensure that the quality and availability of high-resolution satellite RS data continue to improve, providing new and improved opportunities for studying microclimate from space. Meanwhile, the development and increasing use of UAVs for agriculture may result in this technology becoming more accessible to ecologists in the near future. An increasing number of apps which may be of use in microclimate research are also being developed, such as apps to allow thermal cameras to be attached to a smartphone, and apps which capture canopy cover measurements using a cheap fisheye lens attachment.

Considering these developments, strengthening the cooperation between RS experts and ecologists is likely to reveal mutual benefits (Turner et al., 2003) and constitutes an important way forward to advance in microclimate research. This is a subject where new researchers could greatly benefit from supervisors and mentors in a variety of fields, so that they may have access to the necessary knowledge to work with varied methods early in their career. This would also reduce the intimidation factor of these methods which can seem overwhelming.

In short, the future of microclimate ecology may lie in the use of cutting edge technology in order to obtain standardised and freely accessible data for use in large-scale, collaborative studies, and the development of further microclimate models facilitated by the establishment of an MMIP.

7. CONCLUSIONS

It is clear that microclimates are closely tied to habitats and are important to organisms at a fine scale. As such microclimatic monitoring

BOX 5 Summary of important points

When planning microclimate research, there are a variety of questions that should be asked in order to develop an appropriate methodology.

Before designing methods, consider:

- Does the coarse-scale climate account for the most important variation?
- How might microclimate influence your species/community? (See Section 3.)
- Which microclimatic factors are the most biologically relevant to your species/community, and which are the most important for you to measure/model? E.g. air temperature or operative temperature, extremes, or averages.
- Which processes are most important in influencing the microclimate your study species/community experiences? (See Section 2.)
- At what scales does your study species/community experience microclimate?
- Does in situ measurement or modelling make more sense for your research question? (See Sections 4 and 5.)
- How much data do you need vs how much you can manage?

When designing methods, consider:

In situ measurement

- Do you want to measure the environment or what the individual's experience?
- What equipment should you use for that purpose?
- What could go wrong with that equipment? Do you need to consider shielding?
- What calibrations do you need to do?

Modelling

- Do you want to be spatially implicit or explicit?
- Do you want to be temporally implicit or explicit?
- What input data is best for your research question?
- Do you need ground-truth data?

at taxon-specific resolutions is vital to the routine monitoring of the environment, ecosystems, and species, and how these will be affected in a changing world. While it is important to consider the potential influence of microclimates when considering climatic interactions with ecology, and conservation efforts such as the protection of microrefugia, it is a complex topic with many different potential methods available. There are a variety of questions that need to be asked at the very beginning of research planning, and which need to continue to be checked through the design process. Box 5 provides a quick reference of some of the most important questions in order to design an effective, appropriate, replicable study of microclimate ecology.

REFERENCES

Agosta, S.J., Hulshof, C.M., Staats, E.G., 2017. Organismal responses to habitat change: herbivore performance, climate and leaf traits in regenerating tropical dry forests. J. Anim. Ecol. 86 (3), 590–604.

Allegrini, J., Carmeliet, J., 2017. Coupled CFD and building energy simulations for studying the impacts of building height topology and buoyancy on local urban microclimates. Urban Clim. 21, 278–305.

Anderson, K., Gaston, K.J., 2013. Lightweight unmanned aerial vehicles will revolutionize spatial ecology. Front. Ecol. Environ. 11 (3), 138–146.

Ashcroft, M.B., Gollan, J.R., 2012. Fine-resolution (25 m) topoclimatic grids of near-surface (5 cm) extreme temperatures and humidities across various habitats in a large (200 × 300 km) and diverse region. Int. J. Climatol. 32 (14), 2134–2148. John Wiley & Sons, Ltd.

Ashcroft, M., Gollan, J., 2013. Moisture, thermal inertia, and the spatial distributions of near-surface soil and air temperatures: understanding factors that promote microrefugia. Agric. For. Meteorol. 176, 77–89.

Ashcroft, M.B., Gollan, J.R., Warton, D.I., Ramp, D., 2012. A novel approach to quantify and locate potential microrefugia using topoclimate, climate stability, and isolation from the matrix. Glob. Chang. Biol. 18 (6), 1866–1879.

Asner, G.P., Martin, R.E., Anderson, C.B., Knapp, D.E., 2015. Quantifying forest canopy traits: imaging spectroscopy versus field survey. Remote Sens. Environ. 158, 15–27.

Atzberger, C., 2013. Advances in remote sensing of agriculture: context description, existing operational monitoring systems and major information needs. Remote Sens. (Basel) 5 (2), 949–981.

Baker, C.R.B., 1980. Some problems in using meteorological data to forecast the timing of insect life cycles. OEPP/EPPO Bull. 10 (2), 83–91. Blackwell Publishing Ltd.

Bakken, G.S., 1981. How many equivalent black-body temperatures are there? J. Therm. Biol. 6 (1), 59–60.

Bakken, G.S., 1992. Measurement and application of operative and standard operative temperatures in ecology. Am. Zool. 32 (2), 194–216.

Bakken, G.S., Angilletta, M.J., 2014. How to avoid errors when quantifying thermal environments. Funct. Ecol. 28 (1), 96–107. Edited by M. Konarzewski.

Bakken, G.S., Erskine, D.J., Santee, W.R., 1983. Construction and operation of heated taxidermic mounts used to measure standard operative temperature. Ecology 64 (6), 1658–1662. Ecological Society of America.

Barozzi, B., Bellazzi, A., Pollastro, M.C., 2016. The energy impact in buildings of vegetative solutions for extensive green roofs in temperate climates. Buildings 6 (3), 33.

Barry, R.G., Blanken, P.D., 2016. Microclimate and Local Climate. Cambridge University Press.

Baudunette, R.V., Wells, R.T., Sanderson, K.J., Clark, B., 1994. Microclimatic conditions in maternity caves of the bent-wing bat, Miniopterus schreibersii: an attempted restoration of a former maternity site. Wildl. Res. 21, 607–619.

Bauer, N., Kenyeres, Z., 2007. Seasonal changes of microclimatic conditions in grasslands and its influence on orthopteran assemblages. Biologia 62 (6), 742–748.

Bazzaz, F.A., Wayne, P.M., 1994. Coping with environmental heterogeneity: the physiological ecology of tree seedling regeneration across the gap–understory continuum. In: Caldwell, M.M., Pearcy, R.W. (Eds.), Exploitation of Environmental Heterogeneity by Plants; Ecophysiological Processes Above and Below Ground. Academic Press, New York, pp. 349–390.

Bennett, A.F., Huey, R.B., John-Alder, H., Nagy, K.A., 1984. The parasol tail and thermoregulatory behavior of the Cape ground squirrel Xerus inauris. Physiol. Zool. 57 (1), 57–62.

Bennie, J., Huntley, B., Wiltshire, A., Hill, M.O., Baxter, R., 2008. Slope, aspect and climate: spatially explicit and implicit models of topographic microclimate in chalk grassland. Ecol. Model. 216 (1), 47–59.

Bennie, J.J., Wiltshire, A.J., Joyce, A.N., Clark, D., Lloyd, A.R., Adamson, J., Parr, T., Baxter, R., Huntley, B., 2010. Characterising inter-annual variation in the spatial pattern of thermal microclimate in a UK upland using a combined empirical–physical model. Agric. For. Meteorol. 150 (1), 12–19.

Bennie, J., Hodgson, J.A., Lawson, C.R., Holloway, C.T., Roy, D.B., Brereton, T., Thomas, C.D., Wilson, R.J., 2013. Range expansion through fragmented landscapes under a variable climate. Ecol. Lett. 16 (7), 921–929. Edited by N. Haddad.

Bird, R.B., Stewart, W.E., Lightfoot, E.N., 2002. Transport Phenomena, second ed. Wiley and Sons, New York.

Blandford, T.R., Humes, K.S., Harshburger, B.J., Moore, B.C., Walden, V.P., Ye, H., 2008. Seasonal and synoptic variations in near-surface air temperature lapse rates in a mountainous basin. J. Appl. Meteorol. Climatol. 47 (1), 249–261.

Bohan, D.A., Vacher, C., Tamaddoni-Nezhad, A., Raybould, A., Dumbrell, A.J., Woodward, G., 2017. Next-generation global biomonitoring: large-scale, automated reconstruction of ecological networks. Trends Ecol. Evol. 32 (7), 477–487.

Boyles, J.G., Boyles, E., Dunlap, R.K., Johnson, S.A., Brack, V., 2017. Long-term microclimate measurements add further evidence that there is no "optimal" temperature for bat hibernation. Mamm. Biol. 86, 9–16.

Brewington, L., Frizzelle, B.G., Walsh, S.J., Mena, C.F., Sampedro, C., 2014. Remote sensing of the marine environment: challenges and opportunities in the Galapagos Islands of Ecuador. In: Denkinger, J., Vinueza, L. (Eds.), The Galapagos Marine Reserve. Springer, pp. 109–136.

Briscoe, N.J., Handasyde, K.A., Griffiths, S.R., Porter, W.P., Krockenberger, A., Kearney, M.R., 2014a. Tree-hugging koalas demonstrate a novel thermoregulatory mechanism for arboreal mammals. Biol. Lett. 10 (6), 20140235.

Briscoe, N.J., Handasyde, K.A., Griffiths, S.R., Porter, W.P., Krockenberger, A., Kearney, M.R., 2014b. Tree-hugging koalas demonstrate a novel thermoregulatory mechanism for arboreal mammals Ecology. Ecol. Soc. Am. 92 (12), 2214–2221.

Buckley, L.B., Waaser, S.A., MacLean, H.J., Fox, R., 2011. Does including physiology improve species distribution model predictions of responses to recent climate change? Ecology 92 (12), 2214–2221.

Burba, G., Anderson, D., 2007. Introduction to the Eddy Covariance Method: General Guidelines and Conventional Workflow. Li-Cor Biosciences, p. 141.

Burman, R.D., Jensen, M., Allen, R.G., 1987. Thermodynamic factors in evapotranspiration. In: James, L.G., English, M.J. (Eds.), Irrigation Systems for the 21st Century. ASCE, pp. 140–148.

Carroll, J.M., Davis, C.A., Elmore, R.D., Fuhlendorf, S.D., 2017. Using a historic drought and high-heat event to validate thermal exposure predictions for ground-dwelling birds. Ecol. Evol. 7 (16), 6413–6422.

Cavieres, L.A., Badano, E.I., Sierra-Almeida, A., Molina-Montenegro, M.A., 2007. Microclimatic modifications of cushion plants and their consequences for seedling survival of native and non-native herbaceous species in the high Andes of central Chile. Arct. Antarct. Alp. Res. 39 (2), 229–236. Allen Press Publishing Services Inc.

Chaerle, L., Van Der Straeten, D., 2001. Seeing is believing: imaging techniques to monitor plant health. Biochim. Biophys. Acta 1519 (3), 153–166.

Checa, M.F., Rodriguez, J., Willmott, K.R., Liger, B., 2014. Microclimate variability significantly affects the composition, abundance and phenology of butterfly communities in a highly threatened Neotropical dry forest. Fla. Entomol. 97 (1), 1–13.

Chen, J., Franklin, J.F., Spies, T.A., 1993. Contrasting microclimates among clearcut, edge, and interior of old-growth Douglas-fir forest. Agric. For. Meteorol. 63 (3–4), 219–237.

Cowles, R.B., Bogert, C.M., 1944. A preliminary study of the thermal requirements of desert reptiles. Bull. AMNH 83, 261–296.

Cunningham, S.J., Kruger, A.C., Nxumalo, M.P., Hockey, P.A., 2013. Identifying biologically meaningful hot-weather events using threshold temperatures that affect life-history. PLoS One 8 (12), 1. Public Library of Science.

Cunningham, S.J., Martin, R.O., Hockey, P.A.R., 2015. Can behaviour buffer the impacts of climate change on an arid-zone bird? Ostrich 86 (1–2), 119–126. Taylor & Francis.

Cuomo, V., Lasaponara, R., Tramutoli, V., 2001. Evaluation of a new satellite-based method for forest fire detection. Int. J. Remote Sens. 22 (9), 1799–1826.

Curtis, R.J., Isaac, N.J.B., 2015. The effect of temperature and habitat quality on abundance of the Glanville fritillary on the Isle of Wight: implications for conservation management in a warming climate. J. Insect Conserv. 19 (2), 217–225.

da Costa, A.C.L., Galbraith, D., Almeida, S., Portela, B.T.T., da Costa, M., de Athaydes Silva Junior, J., Braga, A.P., de Gonçalves, P.H., de Oliveira, A.A., Fisher, R., Phillips, O.L., 2010. Effect of 7 yr of experimental drought on vegetation dynamics and biomass storage of an eastern Amazonian rainforest. New Phytol. 187 (3), 579–591.

Darwin, C., Darwin, F., 1880. The Power of Movement in Plants. John Murray, London.

Davidson, N.J., Reid, J.B., 1985. Frost as a factor influencing the growth and distribution of subalpine eucalypts. Aust. J. Bot. 33 (6), 657–667.

De Frenne, P., Verheyen, K., 2016. Weather stations lack forest data. Science 351 (6270), 234.

De Frenne, P., Brunet, J., Shevtsova, A., Kolb, A., Graae, B.J., Chabrerie, O., Cousins, S.A., Decocq, G., De Schrijver, A.N., Diekmann, M., Gruwez, R., 2011. Temperature effects on forest herbs assessed by warming and transplant experiments along a latitudinal gradient. Glob. Chang. Biol. 17 (10), 3240–3253.

De Frenne, P., Rodríguez-Sánchez, F., Coomes, D.A., Baeten, L., Verstraeten, G., Vellend, M., Bernhardt-Römermann, M., Brown, C.D., Brunet, J., Cornelis, J., Decocq, G.M., 2013. Microclimate moderates plant responses to macroclimate warming. Proc. Natl. Acad. Sci. U.S.A. 110 (46), 18561–18565.

Denny, M., 2014. Biology and the Mechanics of the Wave-Swept Environment. Princeton University Press.

Dietrich, L., Körner, C., 2014. Thermal imaging reveals massive heat accumulation in flowers across a broad spectrum of alpine taxa. Alp. Bot. 124 (1), 27–35.

Dingman, J.R., Sweet, L.C., McCullough, I., Davis, F.W., Flint, A., Franklin, J., Flint, L.E., 2013. Cross-scale modeling of surface temperature and tree seedling establishment in mountain landscapes. Ecol. Process. 2 (1), 30.

Dobrowski, S.Z., Abatzoglou, J.T., Greenberg, J.A., Schladow, S.G., 2009. How much influence does landscape-scale physiography have on air temperature in a mountain environment? Agric. For. Meteorol. 149 (10), 1751–1758.

D'Odorico, P., He, Y., Collins, S., De Wekker, S.F., Engel, V., Fuentes, J.D., 2013. Vegetation–microclimate feedbacks in woodland–grassland ecotones. Glob. Ecol. Biogeogr. 22 (4), 364–379.

Easterling, D.R., Meehl, G.A., Parmesan, C., Changnon, S.A., Karl, T.R., Mearns, L.O., 2000. Climate extremes: observations, modeling, and impacts. Science 289 (5487), 2068–2074.

Edwards, F.A., Finan, J., Graham, L.K., Larsen, T.H., Wilcove, D.S., Hsu, W.W., Chey, V.K., Hamer, K.C., 2017. The impact of logging roads on dung beetle assemblages in a tropical rainforest reserve. Biol. Conserv. 205, 85–92.

Eldering, A., Wennberg, P.O., Crisp, D., Schimel, D.S., Gunson, M.R., Chatterjee, A., Liu, J., Schwandner, F.M., Sun, Y., O'dell, C.W., Frankenberg, C., 2017. The Orbiting Carbon Observatory-2 early science investigations of regional carbon dioxide fluxes. Science 358 (6360), eaam5745.

Elias, M., Hensel, O., Richter, U., Hülsebusch, C., Kaufmann, B., Wasonga, O., 2015. Land conversion dynamics in the Borana rangelands of Southern Ethiopia: an integrated assessment using remote sensing techniques and field survey data. Environments 2 (1), 1–31.

Evans, J.P., Westra, S., 2012. Investigating the mechanisms of diurnal rainfall variability using a regional climate model. J. Climate 25 (20), 7232–7247.

Finkel, M., Fragman, O., Nevo, E., 2001. Biodiversity and interslope divergence of vascular plants caused by sharp microclimatic differences at "Evolution Canyon II", Lower Nahal Keziv, Upper Galilee, Israel. Isr. J. Plant Sci. 49 (4), 285–295.

Flint, L.E., Flint, A.L., 2012. Downscaling future climate scenarios to fine scales for hydrologic and ecological modeling and analysis. Ecol. Process. 1 (1), 2.

Frey, S.J., Hadley, A.S., Johnson, S.L., Schulze, M., Jones, J.A., Betts, M.G., 2016. Spatial models reveal the microclimatic buffering capacity of old-growth forests. Sci. Adv. 2 (4), e1501392.

Fridley, J.D., 2009. Downscaling climate over complex terrain: high finescale (<1000 m) spatial variation of near-ground temperatures in a montane forested landscape (Great Smoky Mountains). J. Appl. Meteorol. Climatol. 48 (5), 1033–1049.

Gates, D.M., 1980. Biophysical Ecology. Springer Verlag, New York.

Geiger, R., 1927. Das Klima der bodennahen Luftschicht. Vieweg & Sohn, Brunswick.

Geiger, R., Aron, R.H., Todhunter, P., 2009. The Climate Near the Ground. Rowman & Littlefield.

Gillingham, P.K., Huntley, B., Kunin, W.E., Thomas, C.D., 2012a. The effect of spatial resolution on projected responses to climate warming. Divers. Distrib. 18 (10), 990–1000.

Gillingham, P.K., Palmer, S.C., Huntley, B., Kunin, W.E., Chipperfield, J.D., Thomas, C.D., 2012b. The relative importance of climate and habitat in determining the distributions of species at different spatial scales: a case study with ground beetles in Great Britain. Ecography 35 (9), 831–838.

Ginzler, C., Hobi, M., 2015. Countrywide stereo-image matching for updating digital surface models in the framework of the Swiss National Forest Inventory. Remote Sens. (Basel) 7 (4), 4343–4370.

Gollan, J.R., Ramp, D., Ashcroft, M.B., 2014. Assessing the distribution and protection status of two types of cool environment to facilitate their conservation under climate change. Conserv. Biol. 28 (2), 456–466.

Gómez-Cifuentes, A., Munevar, A., Gimenez, V.C., Gatti, M.G., Zurita, G.A., 2017. Influence of land use on the taxonomic and functional diversity of dung beetles (Coleoptera: Scarabaeinae) in the southern Atlantic forest of Argentina. J. Insect Conserv. 21 (1), 147–156.

Graae, B.J., De Frenne, P., Kolb, A., Brunet, J., Chabrerie, O., Verheyen, K., Pepin, N., Heinken, T., Zobel, M., Shevtsova, A., Nijs, I., 2012. On the use of weather data in ecological studies along altitudinal and latitudinal gradients. Oikos 121 (1), 3–19. Blackwell Publishing Ltd.

Greenwood, F., 2015. How to make maps with drones. In: Greenwood, F., Kakaes, K. (Eds.), Drones and Aerial Observation: New Technologies for Property Rights, Human Rights, and Global Development. New America, pp. 35–47.

Greenwood, O., Mossman, H.L., Suggitt, A.J., Curtis, R.J., Maclean, I.M.D., 2016. Using in situ management to conserve biodiversity under climate change. J. Appl. Ecol. 53 (3), 885–894.

Gruen, A., 2012. Development and status of image matching in photogrammetry. Photogramm. Rec. 27 (137), 36–57.

Guay, K.C., Beck, P.S., Berner, L.T., Goetz, S.J., Baccini, A., Buermann, W., 2014. Vegetation productivity patterns at high northern latitudes: a multi-sensor satellite data assessment. Glob. Chang. Biol. 20 (10), 3147–3158.

Gunton, R.M., Polce, C., Kunin, W.E., 2015. Predicting ground temperatures across European landscapes. Methods Ecol. Evol. 6 (5), 532–542.

Haiden, T., Whiteman, C.D., 2005. Katabatic flow mechanisms on a low-angle slope. J. Appl. Meteorol. 44 (1), 113–126.

Haider, N., Kirkeby, C., Kristensen, B., Kjær, L.J., Sørensen, J.H., Bødker, R., 2017. Microclimatic temperatures increase the potential for vector-borne disease transmission in the Scandinavian climate. Sci. Rep. 7 (1), 1–12.

Hancock, S., Anderson, K., Disney, M., Gaston, K.J., 2017. Measurement of fine-spatial-resolution 3D vegetation structure with airborne waveform lidar: calibration and validation with voxelised terrestrial lidar. Remote Sens. Environ. 188, 37–50.

Hay, J.E., McKay, D.C., 1985. Estimating solar irradiance on inclined surfaces: a review and assessment of methodologies. Int. J. Sol. Energy 3 (4–5), 203–240.

Helmuth, B., Broitman, B.R., Yamane, L., Gilman, S.E., Mach, K., Mislan, K.A.S., Denny, M.W., 2010. Organismal climatology: analyzing environmental variability at scales relevant to physiological stress. J. Exp. Biol. 213 (6), 995.

Hertz, P.E., 1992. Temperature regulation in Puerto Rican Anolis lizards: a field test using null hypotheses. Ecology 73 (4), 1405–1417.

Hicke, J.A., Logan, J., 2009. Mapping whitebark pine mortality caused by a mountain pine beetle outbreak with high spatial resolution satellite imagery. Int. J. Remote Sens. 30 (17), 4427–4441.

Hijmans, R.J., Cameron, S.E., Parra, J.L., Jones, P.G., Jarvis, A., 2005. Very high resolution interpolated climate surfaces for global land areas. Int. J. Climatol. 25 (15), 1965–1978. John Wiley & Sons, Ltd.

Hirose, T., 2004. Development of the Monsi–Saeki theory on canopy structure and function. Ann. Bot. 95 (3), 483–494.

Holden, Z.A., Klene, A.E., Keefe, R.F., Moisen, G.G., 2013. Design and evaluation of an inexpensive radiation shield for monitoring surface air temperatures. Agric. For. Meteorol. 180, 281–286.

Homén, T., 1897. Der tägliche Wärmeumsatz im Boden und die Wärmestrahlung zwischen Himmel und Erde. Engelmann, Leipzig.

Hubbart, J., Link, T., Campbell, C., Cobos, D., 2005. Evaluation of a low-cost temperature measurement system for environmental applications. Hydrol. Process. 19 (7), 1517–1523. John Wiley & Sons, Ltd.

Huey, R.B., Carlson, M., Crozier, L., Frazier, M., Hamilton, H., Harley, C., Hoang, A., Kingsolver, J.G., 2002. Plants versus animals: do they deal with stress in different ways? Integr. Comp. Biol. 42 (3), 415–423.

Huntley, B., Allen, J.R., Bennie, J., Collingham, Y.C., Miller, P.A., Suggitt, A.J., 2017. Climatic disequilibrium threatens conservation priority forests. Conserv. Lett. 1–9.

Hutchinson, J.T., Lacki, M.J., 2001. Possible microclimate benefits of roost site selection in the Red Bat, Lasiurus borealis, in mixed mesophytic forests of Kentucky. Can. Field Nat. 115 (2), 205–209.

IAEA, 2017. Cosmic Ray Neutron Sensing: Use, Calibration and Validation for Soil Moisture Estimation. IAEA-TECDOC-1809. IAEA, Vienna.

Ibrahim, Y.Z., Balzter, H., Kaduk, J., Tucker, C.J., 2015. Land degradation assessment using residual trend analysis of GIMMS NDVI3g, soil moisture and rainfall in Sub-Saharan West Africa from 1982 to 2012. Remote Sens. (Basel) 7 (5), 5471–5494.

Imholt, C., Gibbins, C.N., Malcolm, I.A., Langan, S., Soulsby, C., 2010. Influence of riparian cover on stream temperatures and the growth of the mayfly Baetis rhodani in an upland stream. Aquat. Ecol. 44 (4), 669–678.

ISO, 2015. Siting Classifications for Surface Observing Stations on Land (ISO 19289:2015). ISO, Geneva.

Jarvis, P.G., McNaughton, K.G., 1986. Stomatal control of transpiration: scaling up from leaf to region. Adv. Ecol. Res. 15, 1–49.

Johansen, O., 1977. Thermal Conductivity of Soils. Cold Regions Research and Engineering Laboratory, Hanover, NH.

Jones, H.G., 2013. Plants and Microclimate: A Quantitative Approach to Environmental Plant Physiology. Cambridge University Press.

Karger, D.N., Conrad, O., Böhner, J., Kawohl, T., Kreft, H., Soria-Auza, R.W., Zimmermann, N.E., Linder, H.P., Kessler, M., 2017. Climatologies at high resolution for the earth's land surface areas. Sci. Data 4.

Kearney, M., 2013. Activity restriction and the mechanistic basis for extinctions under climate warming. Ecol. Lett. 16 (12), 1470–1479. Edited by L. Buckley.

Kearney, M.R., Porter, W.P., 2017. NicheMapR—an R package for biophysical modelling: the microclimate model. Ecography 40 (5), 664–674.

Kearney, M.R., Matzelle, A., Helmuth, B., 2012. Biomechanics meets the ecological niche: the importance of temporal data resolution. J. Exp. Biol. 215 (6), 922.

Kearney, M., Phillips, B.L., Tracy, C.R., Christian, K.A., Betts, G., Porter, W.P., 2008. Modelling species distributions without using species distributions: the cane toad in Australia under current and future climates. Ecography 31 (4), 423–434.

Kearney, M.R., Shamakhy, A., Tingley, R., Karoly, D.J., Hoffmann, A.A., Briggs, P.R., Porter, W.P., 2014. Microclimate modelling at macro scales: a test of a general microclimate model integrated with gridded continental-scale soil and weather data. Methods Ecol. Evol. 5 (3), 273–286.

Kelly, A., Godley, B.J., Furness, R.W., 2004. Magpies, Pica pica, at the southern limit of their range actively select their thermal environment at high ambient temperatures. Zool. Middle East 32 (1), 13–26. Taylor & Francis.

Kimmins, J.P., 2004. Forest Ecology—A Foundation for Sustainable Forest Management and Environmental Ethics in Forestry. Prentice Hall, Upper Saddle River, New Jersey.

Kleckova, I., Klecka, J., 2016. Facing the heat: thermoregulation and behaviour of lowland species of a cold-dwelling butterfly genus, Erebia. PLoS One 11 (3), e0150393.

Kollas, C., Randin, C.F., Vitasse, Y., Körner, C., 2014. How accurately can minimum temperatures at the cold limits of tree species be extrapolated from weather station data? Agric. For. Meteorol. 184, 257–266.

Körner, C., 2007. Climatic treelines: conventions, global patterns, causes. Erdkunde 61 (4), 316–324.

Körner, C., Hiltbrunner, E., 2017. The 90 ways to describe plant temperature. Perspect. Plant Ecol. Evol. Syst. https://doi.org/10.1016/j.ppees.2017.04.004.

Körner, C., Paulsen, J., 2004. A world-wide study of high altitude treeline temperatures. J. Biogeogr. 31 (5), 713–732.

Kraus, G.C.M., 1911. Boden und Klima auf kleinstem raum. Gustav Fischer, Jena.

Kriticos, D.J., Lamoureaux, S., Bourdôt, G.W., Pettit, W., 2004. Nassella tussock: current and potential distributions in New Zealand. N. Z. Plant Prot. 57, 81.

Kriticos, D.J., Webber, B.L., Leriche, A., Ota, N., Macadam, I., Bathols, J., Scott, J.K., 2012. CliMond: global high-resolution historical and future scenario climate surfaces for bioclimatic modelling. Methods Ecol. Evol. 3 (1), 53–64. Blackwell Publishing Ltd.

Leempoel, K., Parisod, C., Geiser, C., Daprà, L., Vittoz, P., Joost, S., 2015. Very high-resolution digital elevation models: are multi-scale derived variables ecologically relevant? Methods Ecol. Evol. 6 (12), 1373–1383.

Lefsky, M.A., Cohen, W.B., Parker, G.B., Harding, D.J., 2002. Lidar remote sensing for ecosystem studies. Bioscience 52 (1), 19–30.

Lembrechts, J.J., Lenoir, J., Nuñez, M.A., Pauchard, A., Geron, C., Bussé, G., Milbau, A., Nijs, I., 2017. Microclimate variability in alpine ecosystems as stepping stones for non-native plant establishment above their current elevational limit. Ecography https://doi.org/10.1111/ecog.03263. in press.

Lenoir, J., Svenning, J.C., 2015. Climate-related range shifts—a global multidimensional synthesis and new research directions. Ecography 38 (1), 15–28.

Lenoir, J., Hattab, T., Pierre, G., 2017. Climatic microrefugia under anthropogenic climate change: implications for species redistribution. Ecography 40 (2), 253–266. Blackwell Publishing Ltd.

Lewontin, R.C., 2000. The Triple Helix: Genes, Organism and Environment. Harvard University Press, Cambridge, MA.

Liang, X., Kankare, V., Hyyppä, J., Wang, Y., Kukko, A., Haggrén, H., Yu, X., Kaartinen, H., Jaakkola, A., Guan, F., Holopainen, M., 2016. Terrestrial laser scanning in forest inventories. ISPRS J. Photogramm. Remote Sens. 115, 63–77.

Lin, B.B., 2007. Agroforestry management as an adaptive strategy against potential microclimate extremes in coffee agriculture. Agric. For. Meteorol. 144 (1–2), 85–94.

Littmann, T., 2008. Topoclimate and microclimate. In: Breckle, S.W., Yair, A., Veste, M. (Eds.), Arid Dune Ecosystems. Springer, Berlin, Heidelberg, pp. 175–182.

Lunardini, V.J., 1981. Heat Transfer in Cold Climates. Van Nostrand Reinhold Company.

Lundquist, J.D., Huggett, B., 2008. Evergreen trees as inexpensive radiation shields for temperature sensors. Water Resour. Res. 44 (4), W00D04.

Lundquist, J.D., Pepin, N., Rochford, C., 2008. Automated algorithm for mapping regions of cold-air pooling in complex terrain. J. Geophys. Res. 113 (D22), D22107.

Ma, S., Concilio, A., Oakley, B., North, M., Chen, J., 2010. Spatial variability in microclimate in a mixed-conifer forest before and after thinning and burning treatments. For. Ecol. Manage. 259 (5), 904–915.

Maclean, I.M.D., Bennie, J.J., Scott, A.J., Wilson, R.J., 2012. A high-resolution model of soil and surface water conditions. Ecol. Model. 237, 109–119.

Maclean, I.M.D., Hopkins, J.J., Bennie, J., Lawson, C.R., Wilson, R.J., 2015. Microclimates buffer the responses of plant communities to climate change. Glob. Ecol. Biogeogr. 24 (11), 1340–1350.

Maclean, I.M.D., Suggitt, A.J., Wilson, R.J., Duffy, J.P., Bennie, J.J., 2017. Fine-scale climate change: modelling spatial variation in biologically meaningful rates of warming. Glob. Chang. Biol. 23 (1), 256–268.

Manins, P.C., Sawford, B.L., 1979. A model of katabatic winds. J. Atmos. Sci. 36 (4), 619–630.

Martin, T.E., 2001. Abiotic vs. biotic influences on habitat selection of coexisting species: climate change impacts? Ecology 82 (1), 175–188.

Mason, W.L., 2015. Implementing continuous cover forestry in planted forests: experience with sitka spruce (Picea sitchensis) in the British Isles. Forests 6 (4), 879–902.

McCree, K.J., 1971. The action spectrum, absorptance and quantum yield of photosynthesis in crop plants. Agric. Meteorol. 9 (Suppl. C), 191–216.

McCullough, I.M., Davis, F.W., Dingman, J.R., Flint, L.E., Flint, A.L., Serra-Diaz, J.M., Syphard, A.D., Moritz, M.A., Hannah, L., Franklin, J., 2016. High and dry: high elevations disproportionately exposed to regional climate change in Mediterranean-climate landscapes. Landsc. Ecol. 31 (5), 1063–1075.

Meineri, E., Hylander, K., 2017. Fine-grain, large-domain climate models based on climate station and comprehensive topographic information improve microrefugia detection. Ecography 40 (8), 1003–1013.

Meyer, C.L., Sisk, T.D., Wallace Covington, W., 2001. Microclimatic changes induced by ecological restoration of ponderosa pine forests in Northern Arizona. Restor. Ecol. 9 (4), 443–452.

Michaletz, S.T., Weiser, M.D., Zhou, J., Kaspari, M., Helliker, B.R., Enquist, B.J., 2015. Plant thermoregulation: energetics, trait–environment interactions, and carbon economics. Trends Ecol. Evol. 30 (12), 714–724.

Mishra, A., Vu, T., Veettil, A.V., Entekhabi, D., 2017. Drought monitoring with soil moisture active passive (SMAP) measurements. J. Hydrol. 552, 620–632.

Moeser, D., Roubinek, J., Schleppi, P., Morsdorf, F., Jonas, T., 2014. Canopy closure, LAI and radiation transfer from airborne LiDAR synthetic images. Agric. For. Meteorol. 197, 158–168.

Monteith, J.L., 1972. Survey of Instruments for Micrometeorology. Blackwell Scientific, London.

Morecroft, M.D., Taylor, M.E., Oliver, H.R., 1998. Air and soil microclimates of deciduous woodland compared to an open site. Agric. For. Meteorol. 90 (1), 141–156.

Morelli, T.L., Daly, C., Dobrowski, S.Z., Dulen, D.M., Ebersole, J.L., Jackson, S.T., Lundquist, J.D., Millar, C.I., Maher, S.P., Monahan, W.B., Nydick, K.R., 2016. Managing climate change refugia for climate adaptation. PLoS One 11 (8), e0159909.

Morsdorf, F., Kötz, B., Meier, E., Itten, K.I., Allgöwer, B., 2006. Estimation of LAI and fractional cover from small footprint airborne laser scanning data based on gap fraction. Remote Sens. Environ. 104 (1), 50–61.

Murdock, C.C., Evans, M.V., McClanahan, T.D., Miazgowicz, K.L., Tesla, B., 2017. Fine-scale variation in microclimate across an urban landscape shapes variation in mosquito population dynamics and the potential of Aedes albopictus to transmit arboviral disease. PLoS Negl. Trop. Dis. 11 (5), e0005640.

Myneni, R.B., Keeling, C.D., Tucker, C.J., Asrar, G., Nemani, R.R., 1997. Increased plant growth in the northern high latitudes from 1981 to 1991. Nature 386 (6626), 698–702.

National Research Council, 2014. Opportunities to Use Remote Sensing in Understanding Permafrost and Related Ecological Characteristics: Report of a Workshop. National Academies Press.

Oke, T.R., 2002. Boundary Layer Climates. Routledge.

Orlanski, I., 1975. A rational subdivision of scales for atmospheric processes. Bull. Am. Meteorol. Soc. 56, 527–530.

Parmesan, C., Yohe, G., 2003. A globally coherent fingerprint of climate change impacts across natural systems. Nature 421 (6918), 37–42.

Parmesan, C., Root, T.L., Willig, M.R., 2000. Impacts of extreme weather and climate on terrestrial biota. Bull. Am. Meteorol. Soc. 81 (3), 443–450.

Pauli, H., Gottfried, M., Dullinger, S., Abdaladze, O., Akhalkatsi, M., Alonso, J.L.B., Coldea, G., Dick, J., Erschbamer, B., Calzado, R.F., Ghosn, D., 2012. Recent plant diversity changes on Europe's mountain summits. Science 336 (6079), 353–355.

Pearson, R.G., Stanton, J.C., Shoemaker, K.T., Aiello-Lammens, M.E., Ersts, P.J., Horning, N., Fordham, D.A., Raxworthy, C.J., Ryu, H.Y., McNees, J., Akçakaya, H.R., 2014. Life history and spatial traits predict extinction risk due to climate change. Nat. Clim. Chang. 4 (3), 217–221. Nature Publishing Group.

Pecl, G.T., Araújo, M.B., Bell, J.D., Blanchard, J., Bonebrake, T.C., Chen, I.C., Clark, T.D., Colwell, R.K., Danielsen, F., Evengård, B., Falconi, L., 2017. Biodiversity redistribution under climate change: impacts on ecosystems and human well-being. Science 355 (6332), p. eaai9214.

Pepin, N., Benham, D., Taylor, K., 1999. Modeling lapse rates in the Maritime Uplands of Northern England: implications for climate change. Arct. Antarct. Alp. Res. 31, 151–164.

Pepper, R.G., Morrissey, J.G., 1985. Soil properties affecting runoff. J. Hydrol. 79 (3–4), 301–310.

Pérez Galaso, J.L., Ladrón de Guevara López, I., Boned Purkiss, J., 2016. The influence of microclimate on architectural projects: a bioclimatic analysis of the single-family detached house in Spain's Mediterranean climate. Energ. Effic. 9 (3), 621–645.

Pettorelli, N., Laurance, W.F., O'Brien, T.G., Wegmann, M., Nagendra, H., Turner, W., 2014. Satellite remote sensing for applied ecologists: opportunities and challenges. J. Appl. Ecol. 51 (4), 839–848.

Pohlman, C.L., Turton, S.M., Goosem, M., 2007. Edge effects of linear canopy openings on tropical rain forest understory microclimate. Biotropica 39 (1), 62–71.

Poloczanska, E.S., Brown, C.J., Sydeman, W.J., Kiessling, W., Schoeman, D.S., Moore, P.J., Brander, K., Bruno, J.F., Buckley, L.B., Burrows, M.T., Duarte, C.M., 2013. Global imprint of climate change on marine life. Nat. Clim. Chang. 3 (10), 919–925. https://doi.org/10.1038/nclimate1958.

Porter, W.P., Gates, D.M., 1969. Thermodynamic equilibria of animals with environment', ecological monographs. Ecol. Soc. Am. 39 (3), 227–244.

Porter, W.P., Mitchell, J.W., Beckman, W.A., DeWitt, C.B., 1973. Behavioural implications of mechanistic ecology. Oecologia 13 (1), 1–54.

Potter, K.A., Arthur Woods, H., Pincebourde, S., 2013. Microclimatic challenges in global change biology. Glob. Chang. Biol. 19 (10), 2932–2939.

Pringle, R.M., Webb, J.K., Shine, R., 2003. Canopy structure, microclimate, and habitat selection by a nocturnal snake, Hoplocephalus bungaroides. Ecology 84 (10), 2668–2679. Ecological Society of America.

Raabe, S., Müller, J., Manthey, M., Dürhammer, O., Teuber, U., Göttlein, A., Förster, B., Brandl, R., Bässler, C., 2010. Drivers of bryophyte diversity allow implications for forest management with a focus on climate change. For. Ecol. Manage. 260 (11), 1956–1964.

Rice, K., 2004. Sprint research runs into a credibility gap. Nature 432 (7014), 147.

Rietkerk, M., Boerlijst, M.C., van Langevelde, F., HilleRisLambers, R., de Koppel, J.V., Kumar, L., Prins, H.H., de Roos, A.M., 2002. Self-organization of vegetation in arid ecosystems. Am. Nat. 160 (4), 524–530. The University of Chicago Press.

Ripley, E.A., Archibold, O.W., 1999. Effects of burning on prairie aspen grove microclimate. Agric. Ecosyst. Environ. 72 (3), 227–237.

Rodhouse, T.J., Hovland, M., Jeffress, M.R., 2017. Variation in subsurface thermal characteristics of microrefuges used by range core and peripheral populations of the American pika (Ochotona princeps). Ecol. Evol. 7 (5), 1514–1526.

Russell, G., Congalton, R.G., Gu, J., Yadav, K., Thenkabail, P., Ozdogan, M., 2014. Global land cover mapping: a review and uncertainty analysis. Remote Sens. 6 (12), 12070–12093.

Samson, D.R., Hunt, K.D., 2012. A thermodynamic comparison of arboreal and terrestrial sleeping sites for dry-habitat chimpanzees (Pan troglodytes schweinfurthii) at the Toro-Semliki Wildlife Reserve, Uganda. Am. J. Primatol. 74 (9), 811–818.

Scheffers, B.R., Phillips, B.L., Laurance, W.F., Sodhi, N.S., Diesmos, A., Williams, S.E., 2013. Increasing arboreality with altitude: a novel biogeographic dimension. Proc. R. Soc. Lond. B Biol. Sci. 280 (1770), p. 20131581.

Scherrer, D., Körner, C., 2010. Infra-red thermometry of alpine landscapes challenges climatic warming projections. Glob. Chang. Biol. 16 (9), 2602–2613. Blackwell Publishing Ltd.

Scherrer, D., Schmid, S., Körner, C., 2011. Elevational species shifts in a warmer climate are overestimated when based on weather station data. Int. J. Biometeorol. 55 (4), 645–654.

Seo, C., Thorne, J.H., Hannah, L., Thuiller, W., 2009. Scale effects in species distribution models: implications for conservation planning under climate change. Biol. Lett. 5 (1), 39.

Settele, J., Scholes, R., Betts, R.A., Bunn, S., Leadley, P., Nepstad, D., Overpeck, J.T., Taboada, M.A., Fischlin, A., Moreno, J.M., Root, T., 2014. Terrestrial and inland water systems. In: Field, C.B., Barros, V.R. (Eds.), Climate Change 2014 Impacts, Adaptation and Vulnerability: Part A: Global and Sectoral Aspects. Cambridge University Press.

Shi, H.J., Paull, D., Rayburg, S., 2016. Spatial heterogeneity of temperature across alpine boulder fields in New South Wales, Australia: multilevel modelling of drivers of microhabitat climate. Int. J. Biometeorol. 60 (7), 965–976.

Shine, R., Kearney, M., 2001. Field studies of reptile thermoregulation: how well do physical models predict operative temperatures? Funct. Ecol. 15 (2), 282–288.

Slavich, E., Warton, D.I., Ashcroft, M.B., Gollan, J.R., Ramp, D., 2014. Topoclimate versus macroclimate: how does climate mapping methodology affect species distribution models and climate change projections? Divers. Distrib. 20 (8), 952–963.

Søraas, F., Sandanger, M.I., Smith-Johnsen, C., 2017. NOAA POES and MetOp particle observations during the 17 March 2013 storm. J. Atmos. Sol. Terr. Phys. https://doi.org/10.1016/j.jastp.2017.09.004.

Sporn, S.G., Bos, M.M., Hoffstätter-Müncheberg, M., Kessler, M., Gradstein, S.R., 2009. Microclimate determines community composition but not richness of epiphytic understory bryophytes of rainforest and cacao agroforests in Indonesia. Funct. Plant Biol. 36 (2), 171–179.

Stull, R.B., 2012. An Introduction to Boundary Layer Meteorology. Springer Science & Business Media.

Suggitt, A.J., Gillingham, P.K., Hill, J.K., Huntley, B., Kunin, W.E., Roy, D.B., Thomas, C.D., 2011. Habitat microclimates drive fine-scale variation in extreme temperatures. Oikos 120 (1), 1–8.

Suggitt, A.J., Wilson, R.J., August, T.A., Fox, R., Isaac, N.J., Macgregor, N.A., Morecroft, M.D., Maclean, I.M.D., 2015. Microclimate affects landscape level persistence in the British Lepidoptera. J. Insect Conserv. 19 (2), 237–253.

Sun, Y., Frankenberg, C., Wood, J.D., Schimel, D.S., Jung, M., Guanter, L., Drewry, D.T., Verma, M., Porcar-Castell, A., Griffis, T.J., Gu, L., 2017. OCO-2 advances photosynthesis observation from space via solar-induced chlorophyll fluorescence. Science 358 (6360), eaam5747.

Sutherst, R.W., Bourne, A.S., 2009. Modelling non-equilibrium distributions of invasive species: a tale of two modelling paradigms. Biol. Invasions 11 (6), 1231–1237.

Sutherst, R.W., Floyd, R.B., Maywald, G.F., 1996. The potential geographical distribution of the cane toad. Bufo marinus L. in Australia. Conserv. Biol. 10 (1), 294–299.

Tampucci, D., Gobbi, M., Marano, G., Boracchi, P., Boffa, G., Ballarin, F., Pantini, P., Seppi, R., Compostella, C., Caccianiga, M., 2017. Ecology of active rock glaciers and surrounding landforms: climate, soil, plants and arthropods. Boreas 46 (2).

Telewski, F.W., 1995. Wind-induced physiological and developmental responses in trees. In: Coutts, M.P., Grace, J.E. (Eds.), Wind and Trees. Cambridge University Press, pp. 237–263.

Telling, J., Lyda, A., Hartzell, P., Glennie, C., 2017. Review of earth science research using terrestrial laser scanning. Earth Sci. Rev. 169, 35–68.

Terjung, W.H., 1974. Urban climatology: with reference to the interrelationship between external weather and the microclimate in houses and buildings. In: Progress in Biometeorology. Tromp, S.W., Swets and Zeitlinger, Amsterdam, vol. 1(1 A), p. 168–180, Pt. 1, Chapter 4, Section 5.

Thackeray, S.J., Henrys, P.A., Hemming, D., Bell, J.R., Botham, M.S., Burthe, S., Helaouet, P., Johns, D.G., Jones, I.D., Leech, D.I., Mackay, E.B., 2016. Phenological sensitivity to climate across taxa and trophic levels. Nature 535 (7611), 241–245.

Töpfer, T., Gedeon, K., 2014. In: Facial skin provides thermoregulation in Stresemann's Bush-crow Zavattariornis stresemanni. Presented at the 26th International Ornithological Congress, p. 24.

Tracy, C.R., 1976. A model of the dynamic exchanges of water and energy between a terrestrial amphibian and its environment. Ecol. Monogr. 46 (3), 293–326. Ecological Society of America.

Trivedi, M.R., Berry, P.M., Morecroft, M.D., Dawson, T.P., 2008. Spatial scale affects bioclimate model projections of climate change impacts on mountain plants. Glob. Chang. Biol. 14 (5), 1089–1103.

Tsendbazar, N.-E., de Bruin, S., Herold, M., 2017. Integrating global land cover datasets for deriving user-specific maps. Int. J. Digital Earth 10 (3), 219–237.

Turner, W., Spector, S., Gardiner, N., Fladeland, M., Sterling, E., Steininger, M., 2003. Remote sensing for biodiversity science and conservation. Trends Ecol. Evol. 18 (6), 306–314.

Turner, D., Lucieer, A., Malenovský, Z., King, D.H., Robinson, S.A., 2014. Spatial co-registration of ultra-high resolution visible, multispectral and thermal images acquired with a micro-UAV over Antarctic moss beds. Remote Sens. (Basel) 6 (5), 4003.

Unwin, D.M., 1980. Microclimate Measurement for Ecologists. Cambridge University Press, Cambridge.

Vacher, C., Hampe, A., Porté, A.J., Sauer, U., Compant, S., Morris, C.E., 2016. The phyllosphere: microbial jungle at the plant–climate interface. Annu. Rev. Ecol. Evol. Syst. 47, 1–24.

Vanwalleghem, T., Meentemeyer, R.K., 2009. Predicting forest microclimate in heterogeneous landscapes. Ecosystems 12 (7), 1158–1172.

Vollmer, M., Möllmann, K.-P., 2017. Infrared Thermal Imaging: Fundamentals, Research and Applications. John Wiley & Sons.

Waffle, A.D., Corry, R.C., Gillespie, T.J., Brown, R.D., 2017. Urban heat islands as agricultural opportunities: an innovative approach. Landsc. Urban Plan. 161, 103–114.

Wakelin, J., Wilson, A.-L., Downs, C.T., 2013. Ground cavity nest temperatures and their relevance to Blue Swallow Hirundo atrocaerulea conservation. Ostrich 84 (3), 221–226.

Walsberg, G.E., Wolf, B.O., 1996. An appraisal of operative temperature mounts as tools for studies of ecological energetics. Physiol. Zool. 69 (3), 658–681.

Wang, C., Jones, R., Perry, M., Johnson, C., Clark, P., 2013. Using an ultrahigh-resolution regional climate model to predict local climatology. Q. J. R. Meteorol. Soc. 139 (677), 1964–1976.

Wanner, W., Strahler, A.H., Hu, B., Lewis, P., Muller, J.-P., Li, X., Schaaf, C.L.B. m Barnsley M. J., 1997. Global retrieval of bidirectional reflectance and albedo over land from EOS MODIS and MISR data: theory and algorithm. J. Geophys. Res. Atmos. 102 (D14), 17143–17161.

Weiss, S.B., Murphy, D.D., Ehrlich, P.R., Metzler, C.F., 1993. Adult emergence phenology in checkerspot butterflies: the effects of macroclimate, topoclimate, and population history. Oecologia 96 (2), 261–270.

Wexler, R., 1946. Theory and observations of land and sea breezes. Bull. Am. Meteor. Soc 27, 272–287.

Whiteman, C.D., 2000. Mountain Meteorology: Fundamentals and Applications. Pacific Northwest National Laboratory, Richland, WA (US).

Willis, C.K.R., Brigham, R.M., 2005. Physiological and ecological aspects of roost selection by reproductive female hoary bats (Lasiurus cinereus). J. Mammal. 86 (1), 85–94.

Willis, C.K.R., Jameson, J.W., Faure, P.A., Boyles, J.G., Brack Jr., V., Cervone, T.H., 2009. Thermocron iButton and iBBat temperature dataloggers emit ultrasound. J. Comp. Physiol. B 179 (7), 867–874.

Wooster, M.J., Roberts, G., Smith, A.M.S., Johnston, J., Freeborn, P., Amici, S., Hudak, A.T., 2013. Thermal remote sensing of active vegetation fires and biomass burning events. In: Kuenzer, C., Dech, S. (Eds.), Thermal Infrared Remote Sensing. In: Remote Sensing and Digital Image Processing, vol. 17. Springer, Dordrecht, pp. 347–390.

World Meteorological Organisation (WMO), 2010. Manual on the Global Observing System. WMO-No. 544. Geneva.

World Meteorological Organisation (WMO), 2014. Guide to Meteorological Instruments and Methods of Observation. WMO-No.8. Geneva.

Yang, Z., Wang, T., Skidmore, A.K., de Leeuw, J., Said, M.Y., Freer, J., 2014. Spotting East African mammals in open savannah from space. PLoS One 9 (12), e115989.

Yuan, J., Emura, K., Farnham, C., 2017. Is urban albedo or urban green covering more effective for urban microclimate improvement? A simulation for Osaka. Sustain. Cities Soc. 32, 78–86.

Zarco-Tajeda, P.J., González-Dugo, V., Berni, J.A.J., 2012. Fluorescence, temperature and narrow-band indices acquired from a UAV platform for water stress detection using a micro-hyperspectral imager and a thermal camera. Remote Sens. Environ. 117, 322–337.

Zeng, Z., Piao, S., Li, L.Z.X., Zhou, L., Ciais, P., Wang, T., Li, Y., Lian, X., Wood, E.F., Friedlingstein, P., Mao, J., Estes, L.D., Myneni, R.B., Peng, S., Shi, X., Seneviratne, S.I., Wang, Y., 2017. Climate mitigation from vegetation biophysical feedbacks during the past three decades. Nat. Clim. Chang. 7, 432–436.

Zhang, Y., Chen, J.M., Miller, J.R., 2005. Determining digital hemispherical photograph exposure for leaf area index estimation. Agric. For. Meteorol. 133 (1), 166–181.

Zhao, Q., Wentz, E., 2016. A MODIS/ASTER airborne simulator (MASTER) imagery for urban heat island research. Data 1 (1), 7.

FURTHER READING

Kearney, M., 2018. Microclimate ecology. In: Gibson, D. (Ed.), Oxford Bibliographies in Ecology. Oxford University Press, New York.

Nakamura, R., Mahrt, L., 2005. Air temperature measurement errors in naturally ventilated radiation shields. J. Atmos. Ocean. Technol. Am. Meteorol. Soc. 22 (7), 1046–1058.

Uvarov, B.P., 1931. Insects and climate. Trans. R. Entomol. Soc. Lond. 79 (1), 1–232.

CHAPTER FOUR

Challenges With Inferring How Land-Use Affects Terrestrial Biodiversity: Study Design, Time, Space and Synthesis

Adriana De Palma*[,1], Katia Sanchez-Ortiz*[,†], Philip A. Martin[‡], Amy Chadwick[§], Guillermo Gilbert*, Amanda E. Bates[¶], Luca Börger[∥], Sara Contu*, Samantha L.L. Hill*, Andy Purvis*[,†]

*Natural History Museum, London, United Kingdom
[†]Grand Challenges in Ecosystems and the Environment, Imperial College London, Ascot, United Kingdom
[‡]Conservation Science Group, University of Cambridge, Cambridge, United Kingdom
[§]University College London, London, United Kingdom
[¶]Ocean and Earth Science, National Oceanography Centre Southampton, University of Southampton, Southampton, United Kingdom
[∥]College of Science, Swansea University, Swansea, United Kingdom
[1]Corresponding author: e-mail address: a.de-palma@nhm.ac.uk

Contents

Advances in Ecological Research, Volume 58
ISSN 0065-2504
https://doi.org/10.1016/bs.aecr.2017.12.004

Abstract

Land use has already reshaped local biodiversity on Earth, with effects expected to increase as human populations continue to grow in both numbers and prosperity. An accurate depiction of the state of biodiversity on our planet, combined with identifying the mechanisms driving local biodiversity change, underpins our ability to predict how different societal priorities and actions will influence biodiversity trajectories. Quantitative syntheses provide a fundamental tool by taking information from multiple sources to identify generalisable patterns. However, syntheses, by definition, combine data sources that have fundamentally different purposes, contexts, designs and sources of error and bias; they may thus provide contradictory results, not because of the biological phenomena of interest, but due instead to combining diverse data. While much attention has been focussed on the use of space-for-time substitution methods to estimate the impact of land-use change on terrestrial biodiversity, we show that the most common study designs all face challenges—either conceptual or logistical—that may lead to faulty inferences and ultimately mislead quantitative syntheses. We outline these study designs along with their advantages and difficulties, and how quantitative syntheses can combine the strengths of each class of design.

1. INTRODUCTION

Land-use change and intensification are among the most widespread pressures facing terrestrial biodiversity worldwide. Only 39% of land has never been converted to human use; around $265,000 \, km^2$ of unaltered landscapes is lost each year, while around $290,000 \, km^2$ is abandoned and reverting to secondary vegetation (Hurtt et al., 2017). Loss of habitat to human use is the most commonly identified direct driver of global extinction risk to terrestrial species (e.g. Brummitt et al., 2015; Estrada et al., 2017). As well as global loss, local biodiversity loss is also of concern, as it can harm ecosystem function and jeopardise delivery of ecosystem services (Cardinale et al., 2012; Díaz et al., 2006; Hooper et al., 2012). Although there is consensus that global biodiversity is declining (McGill et al., 2015), there is disagreement about the average trend in local biodiversity (Dornelas et al., 2014; Gonzalez et al., 2016; McGill et al., 2015; Newbold et al., 2015; Vellend et al., 2013). Given that sustainable development depends on using natural resources sustainably (Sustainable

Development Goals; Griggs et al., 2013), there is a pressing need to synthesise available evidence on how local biodiversity is changing in the face of land-use change and intensification.

One complication facing such syntheses is that many ecological aspects of the study system are likely to influence how the intensity of a given change in anthropogenic pressure affects different measures of site-level biodiversity. Observed effects can depend on—among other things—the taxonomic or functional group studied (e.g. Gibson et al., 2011; Lawton et al., 1998; Murphy and Romanuk, 2014), how long the landscape has experienced strong human impacts (Balmford, 1996), the habitat stratum sampled (Dumbrell and Hill, 2005) and the location of the study (e.g. Gibson et al., 2011). Our main focus here, however, is on a further possible source of disagreement that has recently become prominent (França et al., 2016; Gonzalez et al., 2016; Vellend et al., 2017)—the spatiotemporal design of the study. Some studies directly observe biodiversity change over time (by repeated sampling), others make spatial comparisons to infer temporal changes (space-for-time substitution), while others combine spatial and temporal comparisons. Global syntheses of time series data of assemblage diversity show no overall temporal trend (Vellend et al., 2013; Dornelas et al., 2014; but see Gonzalez et al., 2016), whereas global meta-analyses of spatial comparisons between sites varying in land use suggest a net decline in diversity (Gibson et al., 2011; Murphy and Romanuk, 2014; Newbold et al., 2015), and sampling both before and after a disturbance at both control and impact (i.e. undisturbed vs disturbed) sites can yield even larger estimates of decline (França et al., 2016).

Another aspect of study design that can affect inferences about land-use impacts on biodiversity is the experimental approach used, from manipulative experiments to correlational 'quasi-' or mensurative experiments (Block et al., 2001; McGarigal and Cushman, 2002). In the former, land-use change treatments are assigned by the researchers in order to minimise effects of confounding variables (e.g. randomly), while the latter approach makes use of externally imposed variation in land use. Manipulative experiments are rare, especially at a large enough spatial and temporal scale to fully capture the real-world effects of land-use change (see Ewers et al., 2011 for a notable exception), because land-use change is seldom within a researcher's control. Though much more common, quasi-experiments face the risk that the land-use variation incorporated as a predictor can also reflect other factors that may confound inferences (McGarigal and Cushman, 2002) and can rarely disentangle effects of different land-use pressures that are inherently linked (e.g. habitat loss vs fragmentation effects; Ewers and Didham, 2006).

Synthetic analyses of trends in local terrestrial biodiversity may not only be confounded by ecological differences, but also the study designs can influence the trends implied by different data sets. Recent extensive, thorough and careful syntheses of temporal and spatial comparisons disagree whether average assemblage-level diversity is declining (Newbold et al., 2015; Vellend et al., 2017). This disagreement makes it harder for global and regional assessments, such as those of the Intergovernmental Science-Policy Platform on Biodiversity and Ecosystem Services (Díaz et al., 2015), to deliver clear messages about human impacts on biodiversity. To clarify how methodology can influence our capacity to detect responses of local terrestrial biodiversity to land-use change, we explicitly consider the advantages and disadvantages of the study designs most commonly used to achieve this objective. We highlight when each study design is particularly likely to mislead and suggest a new synthetic modelling framework in which all relevant designs can be incorporated. We also address the broader question of whether syntheses of published data can provide an accurate global picture of how terrestrial biodiversity is responding to human impacts.

2. DESIGNS OF STUDIES FOR ASSESSING BIOTIC IMPACTS OF LAND-USE CHANGE

Many different sampling designs have been used to estimate the effect of land-use change on local biodiversity. Because the costs of sampling increase with numbers of sites, distances between sites and numbers of individual sampling campaigns, a study design represents different compromises between temporal data (before and after a change in land-use pressure) and spatial data (in both control and impact sites). Here we report the most common study designs in Fig. 1, according to an ongoing survey of the literature we are conducting.

2.1 Control–Impact (Also Known as Space-for-Time Substitution)

The quickest, usually cheapest, and therefore the most frequently used design is Control–Impact (Fig. 1A), which assumes space-for-time substitution (Fukami and Wardle, 2005; Pickett, 1989); in lieu of repeated sampling over time, nearby sites that differ in land use but are hopefully otherwise

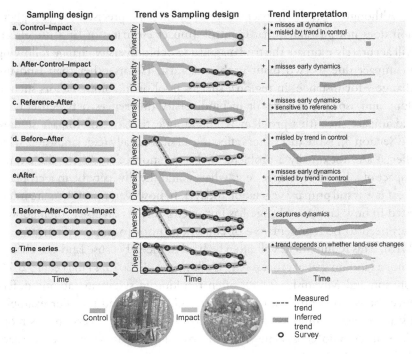

Fig. 1 Schematic showing the essence of different sampling designs and how they can affect the inferred change in diversity in response to a land-use change. Sampling designs (A)–(G) differ in their spatial and temporal patterns of survey events (*open black circles*). Not all designs include the predisturbance state or equal sampling effort in the impacted (i.e. treatment site: *orange*) and control (i.e. reference: *blue*) land uses. Thus, against a backdrop of ongoing diversity change, the measured (*black dotted lines*) and inferred response of diversity (*grey line*) to a disturbance can differ markedly from the counterfactual or baseline, which in turn varies according to study design. For example, in (E) and (G), the first time point sampled is used as the baseline; panel (E) therefore suggests an increase in diversity as the initial loss of diversity is missed, whereas the inference from (G) will depend on whether the site's land use changes (hence the three alternative inferences).

comparable are surveyed in a single field season using a consistent sampling approach. The difference in biodiversity between land-use classes is assumed to accurately capture the effect of land-use change from one class to another. Since resources are focused on one field season, Control–Impact studies may have higher replication (more sites in more land-use classes) than multiple-season designs; such studies can therefore provide quite precise estimates of land-use impacts on biodiversity.

Although Control–Impact studies may increase precision, high replication does not guarantee that Control–Impact estimates of land-use effects will accurately estimate the true magnitude of biodiversity change following an impact due to two main problems. First, factors other than land-use change—for instance, in regional species pool or in abiotic factors such as slope and aspect—may result in biodiversity differences between control and impact sites (this problem is not unique to Control–Impact studies: see Section 3). Note that factors affecting all sites equally (e.g. landscape-wide effects) are not usually a problem, if the question of interest is the overall impact of a land-use change (e.g. what would the diversity be in a cropland site if it was still primary vegetation today?). If, however, researchers are interested in how land-use change has impacted previous diversity, then control sites need to be representative of conditions prior to impact (França et al., 2016). This requirement will often be hard to meet, because land-use change is not a random spatial process (e.g. Dullinger et al., 2013; França et al., 2016; Johnson and Miyanishi, 2008), often resulting in systematic abiotic differences between control and impact sites other than land use. For example, a study of agricultural expansion in Brazil found that conversion was more likely in cerrado ecosystems with shallow slopes that were close to roads (Jasinski et al., 2005).

The second problem is that Control–Impact studies can provide only a snapshot of what is in reality a dynamic process of biodiversity change because sampling is undertaken within a single field season; land-use impacts may take many years—even many decades—to fully unfold (a phenomenon termed 'biotic lag', Box 1; Kuussaari et al., 2009; Wearn et al., 2012). Transient changes are likely to be missed by Control–Impact studies (Pickett, 1989), and the data will show only differences in population means, rather than variability (Blüthgen et al., 2016; Underwood, 1994). The effects seen in Control–Impact studies therefore depend on the length of time since the land-use change, a variable that is often not known and thus cannot be accounted for, which complicates effect-size interpretation. Faced with this problem, many analyses treat the observed biodiversity difference as being the equilibrium effect of land-use change (e.g. Gibson et al., 2011; Murphy and Romanuk, 2014; Newbold et al., 2015). However, Control–Impact studies may provide the only feasible way to assess historical land-use transitions; for example, the ongoing recovery of secondary forest over a century after it had been cleared (see Box 2 for more detail).

BOX 1 Biotic lag

Although land-use change, habitat loss and fragmentation often have some immediate effect on assemblage composition, the full biotic response may take many years to unfold, a phenomenon that can be termed *biotic lag*. Biotic lag has two components. The *extinction debt* (Tilman et al., 1994) is the number of species that will be lost from the assemblage but whose extinction is not immediate. In a study of forest fragments that had become isolated in Thailand, Gibson et al. (2013) estimated that on average it would take 13.9 years for 50% of resident small mammal species to disappear after fragmentation, while Vellend et al. (2006) found that forest plants in the United Kingdom and Belgium still showed evidence of an extinction debt more than a century after habitat fragmentation. *Immigration credit* (Jackson and Sax, 2010) arises because arrival of new species into the assemblage is likewise not immediate. For instance, dispersal limitation slows the arrival of plant species into patches of relatively young forest in introduction experiments (Flinn and Vellend, 2005). Mechanisms for biotic lag have been reviewed recently by Kuussaari et al. (2009) and Essl et al. (2015). The probability of a long lag time is lessened at the species level by shorter generation times and high dispersal ability (Essl et al., 2015; Hylander and Ehrlén, 2013) and at the landscape level through greater site connectivity and availability of large, stable habitat patches (Kuussaari et al., 2009).

Studies that assume space-for-time substitution cannot fully appreciate lagged responses, although some have tested for biotic lag by assessing whether the variation in biodiversity among sites is best explained by current or past pressures (e.g. Dullinger et al., 2013). Such efforts may be biased if the quality of pressure data has changed over time, and anyway provide limited insight into the dynamics of local biodiversity change over time.

BOX 2 Chronosequences

Chronosequences, by assuming space-for-time substitution, aim to infer temporal dynamics from measurements at sites of different ages but similar land-use histories. Chronosequences are particularly useful when investigating post-disturbance recovery of systems that take decades to centuries to recover (Walker et al., 2010). However, their value has been questioned, primarily because they rely on the poorly tested assumption that all sites share similar biotic and abiotic conditions, and (apart from time since change) disturbance histories (Capers et al., 2005; Chazdon et al., 2007; Johnson and Miyanishi, 2008).

If the assumptions of chronosequence methods are met, however, this approach can make useful predictions, depending on the characteristics of the ecosystem. Changes in communities are most easily predicted where regional species pools are relatively small (e.g. Lebrija-Trejos et al., 2010). With larger regional species pools, there is more opportunity for priority effects to cause trajectories to

Continued

BOX 2 Chronosequences—cont'd

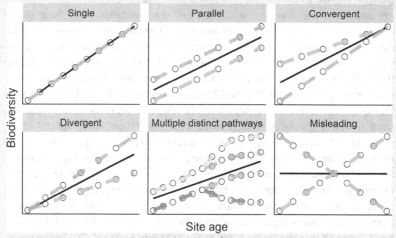

Fig. 2 True patterns of biodiversity dynamics following ecosystem recovery compared with inferences from chronosequences. *Dashed lines* represent the true dynamics of a system, with coloured lines representing different potential pathways. *Points* represent the sites sampled as part of a chronosequence; *solid lines* show the statistical relationship between biodiversity and site age, which is inferred to also be how site-level biodiversity changes over time. Single and parallel pathways can be estimated relatively well, but faulty inferences may emerge when pathways converge or diverge; similar levels of biodiversity at particular time points may suggest that sites have well-matched abiotic and biotic conditions, but in fact the sites experience different trajectories of recovery (e.g. as a result of priority effects or different land-use history). Multiple distinct pathways may similarly result in faulty inferences. Inferences can be especially misleading if pathways are directly opposing.

diverge (Walker et al., 2010; Fig. 2D). Highly diverse communities have many more potential interactions, which can also widen the range of recovery trajectories (Cramer et al., 2008), meaning that tropical forests—where chronosequences have been widely used—may be one of the least suitable systems for using them. Sites of a similar age may be in different successional stages because of differences in initial conditions or random colonisation events. Fragmentation can reduce the utility of chronosequences, with young sites likely to diverge in community composition if they vary in distance from propagule sources (e.g. primary vegetation), particularly if some species have limited dispersal ability (Anderson, 2007; Capers et al., 2005). In summary, each additional layer of complexity in a study system reduces the predictive ability of chronosequences (Walker et al., 2010).

Some of the key assumptions made in chronosequence studies can be tested if sampling is repeated at each site for sufficiently long to 'join the dots' between the initial ages of the sites (e.g. Clarkson, 1997)—coined the 'chronochain' technique. The chronochain approach can be used to test the assumption that all sites sampled are following the same recovery trajectory.

2.2 Extensions of Control–Impact: Chronosequences, After-Control–Impact and Reference-After

Control–Impact designs can be extended to estimate how effects of land-use change unfold over time, by comparing sites that differ in how long ago the change occurred, assuming space-for-time substitution; this is known as a *chronosequence*. In landscapes containing habitat patches of very different ages, chronosequences present the possibility of inferring long-term biodiversity dynamics from a single field season. However, the basic premise of a chronosequence study—that all sites are following the same biodiversity trajectory—can often be incorrect and is seldom tested (see Box 2 for further discussion of chronosequences and their limitations).

Whereas archetypal chronosequences require only a single field season, other modifications of Control–Impact designs sample repeatedly to assess temporal dynamics. Sampling may be repeated in both Control and Impact sites (which we term After-Control–Impact; Fig. 1B), or only in Impact sites (a design we term Reference-After; Fig. 1C). The data from these designs are less susceptible to stochastic effects that can influence a single sampling campaign. However, because sites were not sampled in advance of the land-use change, neither of these designs can reveal the immediate biodiversity response. After-Control–Impact designs are robust against environmental trends and fluctuations that affect all sites equally, because each sampling campaign compares Impact sites with Controls. Inferences from Reference-After designs are more variable, depending on whether the Reference site was sampled in a 'good' or a 'bad' year. This difference between these two designs highlights a crucial general point: many study designs involving repeated sampling assume what might be termed *'time-for-time substitution'*—that the conditions in which sites are sampled are comparable among years. Depending on the taxonomic group being sampled, this assumption can be very hard to meet (see Table 1). Designs in which all sites are sampled at the same time (e.g. Control–Impact, After-Control–Impact) rely less heavily on time-for-time substitution.

2.3 Before–After and After

Before–After studies (Fig. 1D), in which biodiversity at a site is assessed before and after a disturbance, are conceptually the simplest and most direct way to assess biotic consequences of human activity (Smith, 2002; Underwood, 1991). Compared with designs that substitute space for time, Before–After studies have the advantage that the land use before conversion is known, rather than assumed or inferred. These studies can be done opportunistically, where ongoing monitoring is interrupted by an unexpected disturbance (e.g. a severe fire; Pons et al., 2003). To undertake Before–After sampling in a more planned way, however, researchers need to either: (i) know about the disturbance in advance (Buczkowski and Richmond, 2012; e.g. construction projects; Zwart et al., 2015); (ii) resample sites from previous studies (Martin et al., 2015a; Ohwaki et al., 2008) (often a difficult task, as original sites or methods will often be unclear) or (iii) impose experimental treatments (e.g. silvicultural treatments; Wardell-Johnson and Williams, 2000). These constraints limit sample sizes in time and space (although there may be pseudo-replicated subsamples of sites; Smith, 2002; Underwood, 1991). In addition, the lack of contemporaneously sampled control sites makes inferences very fragile to violations of time-for-time substitution (Table 1): land-use change impacts are confounded by coincidental changes (Underwood, 1989), including extreme weather events and natural turnover in diversity (Magurran et al., 2010; Smith, 2002; Underwood, 1991).

Studies in which one or more sites are surveyed repeatedly at known times after a land-use change can provide information on rates of change over time; we term these After studies (Fig. 1E). The After designs have all the limitations of Before–After studies, and additionally cannot inform on how diversity differs from the preimpact condition. Some studies link together time series from several sites of different ages into a chronochain (see Box 2) that provides an integrated assessment of biodiversity change. For example, using this approach Norden et al. (2015) showed that species density of secondary forests is difficult to predict from only time since last disturbance.

2.4 Before–After Control–Impact

Before–After-Control–Impact studies (Fig. 1F) assess biodiversity in (ideally replicated) disturbed sites and (ideally replicated and well matched) unimpacted control sites, both before and after the time of the disturbance. Before–After-Control–Impact methods can estimate impacts accurately: they account for potentially confounding long-term trends (which can mislead studies without control sites, e.g. Before–After; Fig. 1) and the temporal dynamics at each site (which Control–Impact comparisons miss; Fig. 1). Repeated temporal sampling allows Before–After-Control–Impact studies to detect transient impacts of disturbance and, if they continue to sample for long enough, assess biotic lag and rates of recovery.

However, Before–After–Control–Impact studies are relatively rare (Bennett and Adams, 2004; Bernes et al., 2015), because the approach is logistically demanding, expensive and does not offer rapid results that will meet the timescales of funding programmes. The logistical and cost challenges mean that Before–After–Control–Impact studies tend to have little replication, and are heavily biased towards North America and western Europe, easily identified taxa having stable population dynamics (e.g. birds and trees), and land-use transitions that are likely to be known in advance (e.g. those relating to industrial forestry). Even under such favourable circumstances, Before–After–Control–Impact studies are rarer than Control–Impact. For instance, a systematic map of set-aside management impacts on biodiversity in temperate and boreal forests found that only 29% of relevant papers were of a Before–After–Control–Impact design, while 65% were Control–Impact designs (Bernes et al., 2015).

2.5 Time Series Designs

The most common design of study for quantifying temporal change in biodiversity change—and one with many applications (e.g. Vellend et al., 2013; Dornelas et al., 2014)—is time series (Fig. 1G), in which biodiversity is measured at one or more sites for multiple years. However, while time series can provide estimates of net change, this approach is not well suited to understanding how disturbances and other pressures influence biodiversity, unless data on pressures are available for the sampling locations and dates, ideally according to a design such as one of those considered above. Without such data, the average trend across a set of time series will only correctly estimate the average trend across the entire region of interest if the sites sampled are fully representative, in terms of the disturbances or other pressure changes that they show (Section 3.3). Few data sets if any meet this demanding criterion (Gonzalez et al., 2016; Vellend et al., 2017). Long-term standardised monitoring schemes with broad spatial coverage—like the United Kingdom Butterfly Monitoring Scheme—contain richer information about populations' dynamics, so the impact of a disturbance can sometimes be distinguished from natural population variability (Magurran et al., 2010; Oliver et al., 2013; Stewart-Oaten et al., 1986; Underwood, 1989, 1992). Even so, attributing population changes to a specific impact can be difficult if there are other processes and drivers that can influence biodiversity (Pechmann and Wilbur, 1994), and/or significant time lags in responses (Box 1).

3. SAMPLING CONSIDERATIONS

The study designs outlined earlier (and summarised in Fig. 1) are all susceptible to sampling-related issues, including the comparability and replication of spatial and temporal samples. We outline key aspects of sampling design that need to be considered to obtain meaningful comparisons of biodiversity change due to a disturbance event, highlighting the designs for which each aspect is particularly important.

3.1 'Time-for-Time Substitution': Comparability Between Sampling Events

If samples are not comparable over time, then coincidental effects (e.g. extreme weather events) can confound inferences about the disturbance impact, especially if control sites are not sampled concurrently with impact sites (i.e. in Before–After and After designs and time series, and to a lesser extent Reference-After designs, in which only one campaign includes contemporaneous sampling). The lack of concurrently sampled controls means that these designs rely heavily on 'time-for-time substitution', i.e., the assumption that the difference between biodiversity samples is driven by the passage of time rather than unaccounted-for differences between sampling campaigns. Table 1 lists some possible sources of such differences.

3.2 'Space-for-Time' Substitution: Comparability vs Independence

Studies that did not sample before the land–use change (i.e. Control–Impact, After-Control–Impact and Reference-After designs) rely strongly on the assumption that the biodiversity difference among sites in different land uses was caused by land use rather than other factors. However, Control and Impact sites may have differed in initial conditions such as elevation and slope (Wills et al., 2017), or in subsequent changes such as hunting pressure or habitat fragmentation. Comparability of sites can be improved by matching them for environmental characteristics (e.g. topography, soil type and land-use history) through matched pairs or blocked designs. Often, sites are selected based on spatial proximity, with control sites placed close enough to treatment sites that preimpact conditions are likely to have been similar, but far enough away to reduce the risk that they have been affected by the impact (Stewart-Oaten et al., 1986). For instance, Kremen and M'Gonigle (2015) placed control sites within 3 km of the matched treatment

Table 1 Potential Sources of Differences Between Samples Taken in Different Campaigns

Factor	Explanation	Taxon and References
Weather	Day-to-day variation in weather can influence abundance patterns, activity rates and detection in ecological samples, particularly for ectotherms; weather therefore should be matched among samples	Butterflies: Waltz and Covington (2004) Bees and flies: Kremen and M'Gonigle (2015) Slugs: Kappes (2006)
Season	Sampling season is most commonly considered in studies of birds, which might—depending on the research focus—sample in a particular season or across all seasons. However, sampling season can also matter for other groups including mammals and invertebrates. Importantly, sampling can take place in the same month in each year but be in different seasons, because of among-year variation (e.g. early vs late spring); the different biota sampled as a result may be incorrectly interpreted as showing radical turnover from year to year, rather than normal seasonal succession. Sampling over more rather than fewer months is likely to reduce the problems caused by seasonal variation, and replicated sampling on a monthly basis can minimise them	Gastropods: Willig et al. (2007) Mammals and birds: Bicknell et al. (2015)

Continued

Table 1 Potential Sources of Differences Between Samples Taken in Different Campaigns—cont'd

Factor	Explanation	Taxon and References
Activity period of taxon (e.g. flight season for butterflies)	Unless the focal taxa are always present, detectable and identifiable, sampling needs to coincide with when they can be recorded. For example, migratory birds are only present in the relevant season, and many angiosperms are most easily identified when in flower	Spiders: Fuller et al. (2013) Birds: Haché and Villard, (2010)
Recorder	Different researchers undertaking sampling and identification could have an impact on results, particularly for taxa that are difficult to identify	Bryophytes and lichens: Rudolphi et al. (2014) Butterflies: Isaac et al. (2011)
Sampling methodology	Sampling methodologies can change over time as new technologies emerge or as research questions change. Additionally, reports in the older literature often do not fully describe how samples were collected, meaning that any resurvey is unlikely to be fully comparable	Common Birds Census (CBC; https://www.bto.org/about-birds/birdtrends/2011/methods/common-birds-census) and the Breeding Bird Survey (BBS; https://www.bto.org/volunteer-surveys/bbs)

All of the above have more scope to cause problems if studies lack concurrently sampled controls.

site (i.e. 'within the same landscape context'), but more than 1 km from other sites to maintain independence of sampling sites.

In reality, robust matching of sites may be impossible if site characteristics affecting conversion to a particular land-use class also affect biodiversity. For instance, areas of high agricultural suitability are exposed to more intense human pressures globally (using the Human Footprint Index as a measure of human pressure; Venter et al., 2016a,b). It is also often difficult to ensure the independence of sites when matching by geographic proximity. Among-treatment differences will be muted by 'spillover' from unimpacted to impacted sites that are too close, and by loss of species from the entire landscape. Conversely, the difference will be inflated if conversion of some patches causes spillover from impacted to unimpacted patches. These spillover effects set up a tension between the wish to match sites well (pushing control and impact sites together) and the wish for independence (pushing them further apart). However, the choice of distance between control and impact sites in practice depends not only on ecological issues of abiotic conditions and species movements but also on accessibility, funding, time and permissions to use the land. In addition, researchers will seldom know fully the scale of impact of a disturbance event, although knowledge of the species' biology can help to choose appropriate spacing between plots. For instance Fox et al. (2003) separated all plots in their study of small mammals by at least 100 m in order to reduce the likelihood of movement between plots. If it is not possible to minimise the risk of 'impacted' control sites influencing results, the assumption that a control site has not been impacted by the disturbance event should be tested (Underwood, 1994). For example, Torres et al. (2011) recommend having an intermediate zone between control and impact sites to test whether impacts are observed at different distances from the control sites. This is rarely done, however, and proximity to any undisturbed habitat (not just chosen control sites) can influence recovery trajectories in treatment sites.

3.3 'Space-for-Space' Substitution: Representativeness of Sites

'Space-for-space' substitution is a particularly strong assumption in time series studies that do not capture pressure data. In such cases, for the average observed trend to be applied to the whole area of interest, the sites sampled have to be strictly representative (Gonzalez et al., 2016)—which is seldom the case (Vellend et al., 2017)—or at least part of a stratified design (e.g. The UK Butterfly Monitoring Scheme and the UK Countryside Survey).

Designs that directly relate biodiversity to land use, whether using spatial comparisons, temporal comparisons or both, make only the less stringent assumption that their sites are representative in terms of how their biodiversity responds to a given pressure.

3.4 Replication and Comparability

Adequate replication can mitigate some of the problems of time-for-time and space-for-time substitution, but only if the replication scheme avoids biases. For instance, if temporal comparability of samples is low, multiple, frequent samples will be needed so that the average will provide an informative measure of how biodiversity changes with a disturbance event—but this will not overcome biases caused by a systematic difference in sampling methods before vs after the disturbance. Replication boosts the signal-to-noise ratio in ecological data, but does not by itself affect the proportion of the signal that is caused by bias. Likewise, problems associated with low replication cannot necessarily be mitigated by high comparability of samples. Having only a single control site is problematic regardless of how well matched it is to the impact sites, because even sites with similar current abiotic conditions may have different intrinsic dynamics (Norden et al., 2015), potentially confounding disturbance effects with natural variability among sites (Underwood, 1992). If feasible, it would be better to sample more sites that capture environmental variation in the landscape; for example, Fox et al. (2003) and Monamy and Fox (2010) arranged several treatment and control plots so as to capture geographic variation in an elevational decline from dry heath towards swamp land.

Replicated sampling before a disturbance can help disentangle natural variation from the impact of disturbance (Bernstein and Zalinski, 1983; Stewart-Oaten et al., 1986). In addition, longer time series after a disturbance are more likely to capture fully the biotic effects of 'press' disturbances, probably because of biotic lags (e.g. extinction debts; Gonzalez et al., 2016), and allow estimation of natural turnover and changes in population variability (Magurran et al., 2010). However, long-term studies are expensive, time consuming and can be logistically problematic if changes in landowners and/ or permission for researchers to access sites occurs. For example, Holmes et al. (2012) had to curtail their planned long-term study of how bird communities recover after selective logging at 12 years when a new land owner decided to harvest from the research blocks. Changes in property ownership also affected a study of regeneration after fire and grazing abandonment in

Kenya, resulting in a sample size reduction from five sites to four (Gregory et al., 2010).

3.5 Scale

The spatial scale of sampling can influence the magnitude of population changes and processes that are detected (Underwood, 1989), with continuous and large fluctuations in abundance detected at smaller scales (Connell and Sousa, 1983) and small fluctuations in species abundance detectable at larger scales (unless the spatial extent of the disturbance is particularly large; Underwood, 1989). Temporal scales can also influence the detectability of certain responses; for instance, sampling at finer temporal scales may be necessary to adequately capture and disentangle impact and recovery of species with high reproductive rates (Connell and Sousa, 1983; Pons, 2015; Underwood, 1989) and to resolve dynamic effects of 'pulse' impacts (Underwood, 1989).

4. MANIPULATIVE VS CORRELATIONAL APPROACHES

Manipulative experiments are those where a land-use change or disturbance is imposed upon a system, whereas correlational approaches rely on sites showing preexisting gradients or contrasts of pressures. As mentioned in previous sections, capturing data before a land-use change requires either knowledge of when the land-use change is due to occur or a fortuitous choice of sampling sites, thus manipulative experiments often give rise to Before–After–Control–Impact or Before–After designs. Control–Impact designs and their extensions on the other hand generally rely on quasi- or mensurative experiments. Both approaches have advantages and disadvantages related to the fact that land-use change is not a single pressure; rather, it is a combination of environmental changes that can interact to influence biodiversity. For instance, at the site scale, agricultural expansion into forest can result in habitat loss, changes in microclimate and habitat quality, and increased edge effects, while at the landscape scale leading to changes in fragmentation, patch area and isolation (Ewers and Didham, 2006; Fahrig, 2003). Any and all of these pressures could impact biodiversity in interactive ways: for example, local habitat quality may have stronger effects on biodiversity in only some landscape contexts (Deans and Chalcraft, 2017). If not accounted for, interactions among pressures can affect our inferences and the ability to generalise these inferences. For example, Fahrig (2003) suggested that by failing to study the independent effects of fragmentation and habitat loss, the evidence appears contentious, but in fact generalisation is possible only for separate features of fragmentation. In order to separate out components of land-use change, manipulative experiments are usually required. While this could aid our ability to elucidate general patterns (Fahrig, 2003), manipulative experiments often have smaller spatial and temporal scales and replication, reducing the generalisability of results. Mensurative studies have the opposite problem; they allow large spatial and temporal scales and replication, but have limited ability to disentangle different drivers of biodiversity change associated with land use. There are exceptions, however. Phillips et al. (2017) used an extensive compilation of Control–Impact studies (Hudson et al., 2017) to demonstrate the additive opposing effects of patch area and edge effects on local species density.

Regardless of the design chosen, however, any field-based experiment on land-use change (either manipulative or mensurative) will be subject to interactive effects with other change drivers, such as invasive species (Didham et al., 2007) or climate change (Oliver and Morecroft, 2014). For instance, a meta-analysis showed that land-use impacts on biodiversity were related to both current temperature and average change in precipitation over the last 100 years (Mantyka-pringle et al., 2012).

5. CHALLENGES FOR SYNTHESES

5.1 Literature Bias: Realms, Regions and Research Fields

Logistical, funding and time constraints cause biases in where and how ecologists carry out their research, and what they sample. In the context of individual studies these biases are often unimportant, but they can undermine syntheses and policy decisions based on them if the choices of individual researchers result in wider systematic bias in the ecological literature (De Palma et al., 2016; Gonzalez et al., 2016; Martin et al., 2012; McRae et al., 2017). Some study methodologies are likely to be more strongly affected by biases than others; here, we discuss these biases and the potential problems they may cause for syntheses of biodiversity change.

Control–Impact designs are the most straightforward, spatially expansive and tend to be the least geographically biased. Even so, published Control–Impact studies are best represented in North America and Europe (Martin et al., 2012). Time series designs also suffer these biases (e.g. McRae et al., 2017), as well as often underrepresenting regions where drivers of biodiversity change are most intense (Gonzalez et al., 2016). Geographic bias can limit the generality of findings (Gonzalez et al., 2016; Vellend et al., 2017) as ecological responses to disturbances can vary across regions (e.g. De Palma et al., 2016). Geographic and taxonomic biases may be even more pronounced in Before–After–Control–Impact and Before–After studies, given the relative complexity of these methods (see Box 3; Bernes et al., 2015).

Temporal biases are also prevalent in ecological research. Most studies only last for 2 or 3 years (Gaston and Blackburn, 2000; Tilman, 1989;

BOX 3 Biases in logging studies in temperate boreal forests

Temperate boreal forests are favourable areas for more complicated study designs, such as Before–After–Control–Impact: temperate areas are generally more extensively studied (e.g. Martin et al., 2012), and—unlike many disturbances—the timing of logging events is often known in advance. However, even in circumstances favourable for Before–After–Control–Impact designs, Bernes et al.'s (2015) systematic map found that such designs were still underrepresented: only 29% of studies were Before–After–Control–Impact, 67% were Control–Impact and 3% Before–After. In addition, Control–Impact datasets were more globally spread, spanning more continents than both Before–After and Before–After–Control–Impact studies, and more countries than Before–After and Before–After–Control–Impact studies combined. In general, all study designs showed geographic bias, with most studies coming from North America (64% of all studies) followed by Europe (29%); only 6% of all studies were from Asia, Australia and South America combined.

We further interrogated the dataset compiled by Bernes et al. (2015) to assess whether geographic biases varied among study designs (considering only Before–After–Control–Impact, Before–After and Control–Impact designs), by matching study locations to global layers of accessibility to the nearest city (Nelson, 2008), human population density (Venter et al., 2016a,b) and a measure of exposure to a range of human impacts (the Human Footprint Index, Venter et al., 2016a,b; note that human population is one of the impacts considered). Analysis of variance revealed Before–After designs did not differ significantly from Control–Impact designs in terms of accessibility to the nearest city, Human Footprint Index or human population density. While Before–After–Control–Impact designs did not differ from Control–Impact designs in terms of accessibility to the nearest city, they tended to have a lower Human Footprint Index than Control–Impact studies (estimate $= -0.185$, $t = -2.32$, $P < 0.05$) and a lower human population density in the surrounding $1 km^2$ grid square (estimate $= -0.291$, $t = -1.98$, $P < 0.05$).

Weatherhead, 1986) making the detection of lagged effects of land-use change difficult (Kuussaari et al., 2009). Studies that specifically aim to monitor biodiversity change over time tend to be of a longer duration (e.g. a median of 9 years for freshwater studies; Jackson and Füreder, 2006). However, even though Before–After–Control–Impact studies aim to assess the dynamics of biodiversity, they commonly last only 2–5 years in terrestrial systems (Bicknell et al., 2015; França et al., 2016), which may not be enough time to account for time lags such as extinction debts (Kuussaari et al., 2009). Such short studies risk having the worst of both worlds: too short a duration for effect sizes to be large and too little replication for estimates to be precise. Although Control–Impact studies sample in only a single year, they may do so many years after the land-use change of interest, potentially giving enough time for extinction debts to be repaid. For instance, in the database of the PREDICTS (Projecting Responses of Ecological Diversity in Changing Terrestrial Systems) project (Hudson et al., 2017), agricultural sites of known age (~34% of sites) are on average 39 years old (sd = 31).

The choice of what type of ecosystem to work in reflects a mix of researchers' personal interests (e.g. protected areas are overrepresented in the literature; Martin et al., 2012), logistical concerns (e.g. canopy fauna is underrepresented) and funders' changing priorities (e.g. the soil biota is now receiving much more attention than previously). However, does this cause different study methods to be more common in particular systems? Before–After–Control–Impact methods are generally used to assess biodiversity change where the timing of disturbances is known in advance (e.g. tree harvesting in forestry). However, even in these situations Before–After–Control–Impact studies are relatively rare—for example, only 2% of studies on the ecological impacts of forestry in Australia reviewed by Bennett and Adams (2004) used this method. In contrast, a recent systematic review of the impacts of conservation management in forests found a much higher proportion (29%) of studies used a Before–After–Control–Impact design (Bernes et al., 2015). Methods that use space-for-time substitution in contrast can be used wherever there is a suitable reference ecosystem, greatly increasing the types of ecosystems and disturbance that they can be used to investigate.

The utility of systematic reviews and syntheses is compromised when the literature itself is biased. Whereas publication bias in terms of the underreporting of nonsignificant or heterodox results (Koricheva, 2003) is often considered, biases in study location, focus, duration and design seldom are, despite their potential to cause similar problems. Opinions vary on whether syntheses should be exclusive (restricting attention to the highest quality studies; e.g. Whittaker, 2010) or inclusive (with less stringent eligibility criteria but exploration of heterogeneity; e.g. Lajeuness, 2010). If the highest quality studies are more strongly biased in terms of the regions, kinds of landscape and taxa they consider, focusing exclusively on them could come at a very high cost. Thus, focussing on well-designed experimental Before–After–Control–Impact designs may provide the most robust evidence for causality (as they are less susceptible to confounding variables), but will likely have limited potential to provide general inferences because of geographic, ecological or taxonomic unrepresentativeness. A synthesis that incorporated a wider range of evidence would in contrast have the potential to detect more general correlational patterns, but would provide little insight into causality. In either case, the context-specificity of studies causes potential biases in meta-analyses if disproportionately many of the studies come from particular contexts.

Likely biases must either be acknowledged or preferably accounted for in syntheses (e.g. Gibson et al., 2011; Gonzalez et al., 2016; McRae et al., 2017; Newbold et al., 2015).

5.2 Different Study Designs Can Give Different Results

Given the relative paucity of Before–After and Before–After-Control–Impact studies compared with Control–Impact designs, some syntheses have treated samples taken before a disturbance as equivalent to those taken from control site (e.g. Martin et al., 2015b). This approach has been used in meta-analyses (e.g. Moreno-Mateos et al., 2017; Pleieninger et al., 2014) and to analyse multiyear Before–After-Control–Impact designs (Kremen and M'Gonigle, 2015; in this approach, any temporal change in control sites that could confound inferences is partly, though not entirely, averaged away). The validity of this analytical treatment depends upon different study designs providing similar results in spite of their varying assumptions and limitations. To our knowledge only two studies have directly compared space-for-time substitution and temporal designs for estimating land-use impacts on terrestrial biodiversity. First, Chai et al. (2012) compared estimates of recovery in experimentally disturbed tropical forest in Jamaica and found that methods using space-for-time substitution marginally overestimated recovery of plant diversity when compared to Before–After (e.g. when considering stem density, the spatial comparisons found 78% similarity between control and impact plots, while the temporal comparisons found 68% similarity between before and after plots). Second, França et al. (2016) reported that a Control–Impact design dramatically underestimated the impact of logging on dung beetle diversity in forest in Pará, Brazil (a Before–After-Control–Impact approach revealed double the species richness loss indicated by the Control–Impact design). However, neither of these studies can be relied upon to provide a general guide. All of Chai et al.'s (2012) plots were subject to hurricane disturbance following the experimental treatments, which may have affected control plots more severely (Chai et al., 2012) and which can homogenise both forest structure (Flynn et al., 2010) and biodiversity (Uriarte et al., 2009). The control sites used by França et al. (2016) may have differed from impact sites in ways other than logging—control sites were grouped together, about 6.5 km away from impact sites. More such studies are therefore required to infer generality.

As few studies formally compare Control–Impact and Before–After-Control–Impact methods, there is little empirical evidence detailing the circumstances under which spatial comparisons can be as informative as—or even better than—temporal data, but ecological theory provides some suggestions. Space-for-time substitution may be most effective where community turnover is rapid, resulting in shorter biotic lags in response to perturbation (Banet and Trexler, 2013; Kuussaari et al., 2009) and in regions with low beta diversity, which have relatively small species pools (Banet and Trexler, 2013; Walker et al., 2010). They may also be the only feasible approach in logistically challenging settings, e.g., where repeated surveys would not be possible or could not go on long enough avoid the problems of 'time-for-time substitution'. However, Control–Impact approaches may be less helpful when spatial variation in a disturbance or driver far exceeds temporal variation (e.g. the extent of climatic change; Blois et al., 2013). For example, when plant communities exist in toposequences (adjacent soils with different characteristics due to differing topographies or vegetation histories), their position in the landscape becomes more of an influence on plant assemblages than any temporal factors (Walker et al., 2010). In addition, Control–Impact designs may underestimate biodiversity loss in landscapes where pristine habitats are rare, as the diversity of control sites may already be affected by direct pressures or pressures in the surrounding landscape, such as fragmentation, intensive management of surrounding land uses and invasive species.

6. METHODS FOR SYNTHESES

Study designs are not all equally informative for syntheses: they differ in quality, scope (in terms of geography, taxonomy and research field) and assumptions (Fig. 1). However, developing an accurate global picture of biodiversity responses to land-use change requires syntheses to make use of all available data, to recognise the fundamental differences in what a given study design can and cannot provide evidence for and to use modern statistical tools to account for study design when estimating effect sizes (Table 2).

One possible approach for synthesising data from multiple study designs is to use nonlinear models to estimate the functional response of biodiversity over time for each sampled site (Fig. 3). Parameters of interest (e.g. those characterising recovery after a disturbance; see Table 3) could then be extracted and analysed in a second step to explore how additional factors (e.g. the land-use transition or prior land use) influence the parameter values; this second stage of analysis would need to account for change in control sites. This modelling strategy would allow multiple study designs to inform different parameters (for instance, After studies may only be able to inform the parameter describing the rate of recovery to its prior state; see Table 3).

An alternative approach is to fit all data sets into a Before–After–Control–Impact framework by treating all other studies as having incomplete Before–After–Control–Impact designs, allowing for missing data in the response variable. In such a framework, different studies highlight the variation among treatments (Control–Impact), periods (Before–After), years or any combination of these. If multiple land-use transitions are of interest, disturbance type could also be included to account for the expected variation in response across different land-use changes. This approach uses all of the available data without assuming control sites to be equivalent to sites sampled before a disturbance and allows tests of interactive effects on biodiversity of treatments (Control vs Impact), periods (Before vs After) and years relating to questions of interest. For instance, the interaction between treatment and time period assesses whether the disturbance event had a significant effect on diversity in the impact site, after accounting for any coincidental events that may have influenced both control and impact sites equally. It also permits the incorporation of pressures other than land-use change (if some of the studies have considered multiple pressures), provided that the status of such pressures is known for all sites. Such integration methods will be feasible only for aspects of diversity that can be calculated from a single sample, such as species

Table 2 Assumptions of Different Study Designs

Design	Time-for-time	Space-for-time	Space-for-space
Control–Impact	NR Does not include temporal sampling	× Inherent assumption, although careful matching of sites for biotic and abiotic conditions can help	✓ Does not assume that sites are representative of whole region of interest
After-Control–Impact	✓ Contemporaneous sampling of Control and Impact sites largely removes problems of temporal comparability	× Same as Control–Impact above	✓ Same as Control–Impact above
Reference–After	× Inferences are sensitive to choice of reference year	× Same as Control–Impact above	✓ Same as Control–Impact above
Before–After	× No contemporaneous sampling of control and impact sites so temporal matching is crucial	NR Includes temporal sampling	✓ Same as Control–Impact above
After	× Same as Before–After above	NR Same as Before–After above	✓ Same as Control–Impact above
Before–After-Control–Impact	✓ Same as After-Control–Impact above	NR Same as Before–After above	✓ Same as Control–Impact above
Time series	× Same as Before–After above	NR Same as Before–After above	× Uniquely sensitive because of lack of pressure data: sites must be representative of the region of interest

× indicates that the assumption is potentially problematic; ✓ indicates that the assumption is unlikely to strongly influence results in a well-designed study; NR indicates that the assumption is not relevant for the given study type.

richness, overall abundance or functional diversity. Synthesis will be more difficult for measures that require a baseline for comparison, such as compositional similarity to the predisturbance assemblage; for such questions, only Before–After and Before–After-Control–Impact designs will be directly

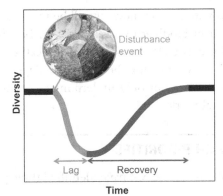

Fig. 3 The effect of different disturbances on biodiversity. *Coloured line* segments indicate different parameters that could be estimated.

Table 3 Relevance of Each Study Design to Different Aspects of Biodiversity Response to Disturbance

Parameter/Effect of Interest	Suitable Study Designs
Resistance to change after disturbance and lag time	Before–After and Before–After-Control–Impact designs with sampling shortly after the disturbance event
Recovery after a disturbance	Any design with temporal sampling after a disturbance event should provide evidence of the rate of recovery. Most Before–After-Control–Impact studies are not long enough to report on the full recovery curve; other designs can help to provide a clearer overall picture
Overall impact (difference between initial diversity and new equilibrium level)	All designs with a counterfactual (either control or impact) are suitable provided sampling is long enough after the disturbance to test (or assume) that sites have reached equilibrium

Each study design provides some information on how biodiversity responds to disturbance. Before–After-Control–Impact studies may most accurately quantify the impact of disturbance, but even After studies (arguably the most information-poor study design) can show the temporal trajectory of change, albeit without any indication of the starting or end point of this trajectory.

useful without ancillary assumptions such as equating control sites with predisturbance sites. Not all data sets are informative for all questions.

The two methods of integration presented earlier attempt to find general patterns, and as with any synthesis must balance generality with realism. Ecological responses are highly context-dependent, depending on the taxa, sampling methodology and region, as well as the type, frequency and magnitude of disturbances. As with any synthesis, it will be important to capture these sources of variation in biodiversity response (e.g. Newbold et al., 2015): both the criteria for data collation and the analytical framework should be carefully designed to account for these context dependencies to provide general inferences that are not only underpinned by large datasets but are also useful in the real world.

7. RESEARCH PRIORITIES

As a result of our review, we have identified three priorities for future research:

1. *How does study design shape conclusions about human impacts on local biodiversity?* Very few studies have assessed whether Control–Impact and Before–After-Control–Impact designs provide similar inferences regarding biodiversity response to land use, and there is as yet no indication about whether the results emerging from these approaches can be generalised. Given the prevalence of Control–Impact studies in the literature, it is important that we answer the question of when and how these designs are likely to provide meaningful inferences as opposed to offering systematically inaccurate inferences with the potential to mislead. The trend towards syntheses of large compilations of data provides the potential to model human impacts on biodiversity with ever greater statistical power, but with great power comes great responsibility to ensure estimates are accurate as well as precise.

2. *A synthetic analysis of land-use impacts on temporal dynamics of biodiversity using multiple study designs.* Previous syntheses that have explored how biodiversity responds to land-use change have generally only considered a single study design or used a simplistic assumption that counterfactuals (whether they are samples before a disturbance or at an undisturbed control site) are all equivalent. No synthesis has yet been performed that incorporates the temporal dynamics of biodiversity response to land-use changes while accepting the fundamental differences among study designs, but such a synthesis could greatly enhance our understanding of human impacts on local biodiversity.

3. *More long-duration Before–After-Control–Impact studies—of more taxa, in more regions and in more ecological settings—are needed in the longer term.* Before–After-Control–Impact data are relevant for the widest range of questions, and these designs (particularly manipulative experiments) have the greatest potential for rigorous tests of different components of land-use change. Even though such studies will always be in a

minority, they are disproportionately important in syntheses because they can effectively be used to validate—and where necessary correct—inferences from other, simpler designs.

8. CONCLUSIONS

1. Many different study designs can provide insight into how biodiversity responds to land-use changes and related pressures, at least when the research question is carefully tailored to consider the inherent limitations and vulnerabilities of each design.
2. However, study designs vary in their assumptions, scope and results. Those that assume space-for-time substitution cannot capture initial biodiversity dynamics after a disturbance, but are widespread in the literature allowing powerful syntheses. On the other hand, Before–After-Control–Impact designs may be the most accurate when it comes to assessing effects of a disturbance and, because they are often experimental, can attempt to disentangle different component pressures of land-use change, but offer less comprehensive insights; low replication and biases in the literature make syntheses based solely on such data difficult.
3. Assessing the temporal dynamics of how biodiversity responds is necessary for more accurate synthesis and more realistic projections of change in the future. Control–Impact and related study designs are therefore much more useful for inference if the ages of the land uses are known.
4. For a comprehensive assessment of how biodiversity responds to land-use change, syntheses should aim to make use of all available data, while accounting for the fundamental differences between study designs.

ACKNOWLEDGEMENTS

This work was supported by NERC (NE/M014533/1 to A.P.). P.A.M. was supported by the MAVA foundation (Project PFZH/118). This chapter is a contribution from the Imperial College Grand Challenges in Ecosystem and the Environment Institute. PREDICTS is endorsed by the Group on Earth Observations Biodiversity Observation Network (GEO BON).

GLOSSARY

Global biodiversity The diversity of life at global scales, e.g., the number of species present on earth. This large-scale biodiversity is driven by both local diversity (alpha diversity—the diversity of life at local scales) and turnover among communities (beta diversity—the differences in biodiversity from one local area to the next).

Local biodiversity The diversity of life at local scales, e.g., the number of species at a small plot. Also referred to as alpha diversity.

Manipulative experiments An experiment where treatments are assigned and imposed by the researcher, while holding other variables constant.

Mensurative experiments; quasi-experiments Experiments where the researcher makes use of existing variation across space or time.

Control site Depending on the nature of the study, control sites may either be sites where the pressure was never imposed (i.e. undisturbed sites) or sites where the pressure persists (e.g. unrestored sites). Some restoration studies include control sites of both kinds.

Impact site Depending on the nature of the study, impact sites may either be sites where a pressure has been imposed (e.g. land-use change) or the pressure has been released (e.g. restored sites).

Control–Impact Study design that makes use of spatial variation in the treatment of interest to infer temporal responses, comparing control sites and impact sites or comparing sites with different types of impact.

After-Control–Impact Control–Impact studies where sampling at each site is repeated over more than one sampling season. Samples are not taken before the treatment of interest has been imposed.

Reference-After Control–Impact design where sampling at the impact sites (but not the control sites) is repeated over more than one sampling season.

Chronosequence Study design where sites being compared differ in how long ago the change occurred, in order to infer the temporal dynamics of change.

Chronochain A chronosequence where sampling is repeated at each site.

Before–After A study design where impact sites are monitored both before and after the treatment of interest (e.g. land-use change), but control sites are absent.

After A study design where impact sites, but not control sites, are sampled after the treatment of interest is imposed (e.g. land-use change) and sampling is repeated over more than one sampling season.

Before–After-Control–Impact A study design where both control and impact sites are sampled both before and after the treatment of interest is imposed (e.g. land-use change). Often, sampling only takes place in two time periods (once before and once after the change), but here we use an extended definition to include designs that sample over multiple time points.

REFERENCES

Anderson, K.J., 2007. Temporal patterns in rates of community change during succession. Am. Nat. 169, 780–793.

Balmford, A., 1996. Extinction filters and current resilience: the significance of past selection pressures for conservation biology. Trends Ecol. Evol. 11, 193–196.

Banet, A.I., Trexler, J.C., 2013. Space-for-time substitution works in Everglades ecological forecasting models. PLoS One 8, e81025.

Bennett, L.T., Adams, M.A., 2004. Assessment of ecological effects due to forest harvesting: approaches and statistical issues. J. Appl. Ecol. 41, 585–598.

Bernes, C., Jonsson, B.G., Junninen, K., Lõhmus, A., Macdonald, E., Müller, J., Sandström, J., 2015. What is the impact of active management on biodiversity in boreal and temperate forests set aside for conservation or restoration? A systematic map. Environ. Evid. 4, 25.

Bernstein, B.B., Zalinski, J., 1983. An optimum sampling design and power tests for environmental biologists. J. Environ. Manage. 16, 35–43.

Bicknell, J.E., Struebig, M.J., Davies, Z.G., 2015. Reconciling timber extraction with biodiversity conservation in tropical forests using reduced-impact logging. J. Appl. Ecol. 52, 379–388.

Block, W.M., Franklin, A.B., Ward, J.P., Ganey, J.L., White, G.C., 2001. Design and implementation of monitoring studies to evaluate the success of ecological restoration on wildlife. Restor. Ecol. 9, 293–303.

Blois, J.L., Williams, J.W., Fitzpatrick, M.C., Jackson, S.T., Ferrier, S., 2013. Space can substitute for time in predicting climate-change effects on biodiversity. Proc. Natl. Acad. Sci. U. S. A. 110, 9374–9379.

Blüthgen, N., Simons, N.K., Jung, K., Prati, D., Renner, S.C., Boch, S., Fischer, M., Hölzel, N., Klaus, V.H., Kleinebecker, T., Tschapka, M., Weisser, W.W., Gossner, M.M., 2016. Land use imperils plant and animal community stability through changes in asynchrony rather than diversity. Nat. Commun. 7, 10697.

Brummitt, N.A., Bachman, S.P., Griffiths-Lee, J., Lutz, M., Moat, J.F., Farjon, A., Donaldson, J.S., Hilton-Taylor, C., Meagher, T.R., Albuquerque, S., Aletrari, E., Andrews, A.K., Atchison, G., Baloch, E., Barlozzini, B., Brunazzi, A., Carretero, J., Celesti, M., Chadburn, H., Cianfoni, E., Cockel, C., Coldwell, V., Concetti, B., Contu, S., Crook, V., Dyson, P., Gardiner, L., Ghanim, N., Greene, H., Groom, A., Harker, R., Hopkins, D., Khela, S., Lakeman-Fraser, P., Lindon, H., Lockwood, H., Loftus, C., Lombrici, D., Lopez-Poveda, L., Lyon, J., Malcolm-Tompkins, P., McGregor, K., Moreno, L., Murray, L., Nazar, K., Power, E., Tuijtelaars, M.Q., Salter, R., Segrott, R., Thacker, H., Thomas, L.J., Tingvoll, S., Watkinson, G., Wojtaszekova, K., Lughadha, E.M.N., 2015. Green plants in the red: a baseline global assessment for the IUCN Sampled Red List Index for Plants. PLoS One 10, e0135152.

Buczkowski, G., Richmond, D.S., 2012. The effect of urbanization on ant abundance and diversity: a temporal examination of factors affecting biodiversity. PLoS One 7, e41729.

Capers, R.S., Chazdon, R.L., Brenes, A.R., Alvarado, B.V., 2005. Successional dynamics of woody seedling communities in wet tropical secondary forests. J. Ecol. 93, 1071–1084.

Cardinale, B.J., Duffy, J.E., Gonzalez, A., Hooper, D.U., Perrings, C., Venail, P., Narwani, A., Mace, G.M., Tilman, D., Wardle, D.A., Kinzig, A.P., Daily, G.C., Loreau, M., Grace, J.B., Larigauderie, A., Srivastava, D.S., Naeem, S., 2012. Biodiversity loss and its impact on humanity. Nature 486, 59–67.

Chai, S.-L., Healey, J.R., Tanner, E.V.J., 2012. Evaluation of forest recovery over time and space using permanent plots monitored over 30 years in a Jamaican montane rain forest. PLoS One 7, e48859.

Chazdon, R.L., Letcher, S.G., van Breugel, M., Martínez-Ramos, M., Bongers, F., Finegan, B., 2007. Rates of change in tree communities of secondary Neotropical forests following major disturbances. Philos. Trans. R. Soc. Lond. B Biol. Sci. 362, 273–289.

Clarkson, B.R., 1997. Vegetation recovery following fire in two Waikato peatlands at Whangamarino and Moanatuatua, New Zealand. N. Z. J. Bot. 35, 167–179.

Connell, J.H., Sousa, W.P., 1983. On the evidence needed to judge ecological stability or persistence. Am. Nat. 121, 789–824.

Cramer, V.A., Hobbs, R.J., Standish, R.J., 2008. What's new about old fields? Land abandonment and ecosystem assembly. Trends Ecol. Evol. 23, 104–112.

De Palma, A., Abrahamczyk, S., Aizen, M.A., Albrecht, M., Basset, Y., Bates, A., Blake, R.J., Boutin, C., Bugter, R., Connop, S., Cruz-López, L., Cunningham, S.A., Darvill, B., Diekötter, T., Dorn, S., Downing, N., Entling, M.H., Farwig, N., Felicioli, A., Fonte, S.J., Fowler, R., Franzén, M., Goulson, D., Grass, I., Hanley, M.E., Hendrix, S.D., Herrmann, F., Herzog, F., Holzschuh, A., Jauker, B., Kessler, M., Knight, M.E., Kruess, A., Lavelle, P., Le Féon,

V., Lentini, P., Malone, L.A., Marshall, J., Pachón, E.M., McFrederick, Q.S., Morales, C.L., Mudri-Stojnic, S., Nates-Parra, G., Nilsson, S.G., Öckinger, E., Osgathorpe, L., Parra-H, A., Peres, C.A., Persson, A.S., Petanidou, T., Poveda, K., Power, E.F., Quaranta, M., Quintero, C., Rader, R., Richards, M.H., Roulston, T., Rousseau, L., Sadler, J.P., Samnegård, U., Schellhorn, N.A., Schüepp, C., Schweiger, O., Smith-Pardo, A.H., Steffan-Dewenter, I., Stout, J.C., Tonietto, R.K., Tscharntke, T., Tylianakis, J.M., Verboven, H.A.F., Vergara, C.H., Verhulst, J., Westphal, C., Yoon, H.J., Purvis, A., 2016. Predicting bee community responses to land-use changes: effects of geographic and taxonomic biases. Sci. Rep. 6, 31153.

Deans, R.A., Chalcraft, D.R., 2017. Matrix context and patch quality jointly determine diversity in a landscape-scale experiment. Oikos 126, 874–887.

Díaz, S., Fargione, J., Chapin III, F.S., Tilman, D., 2006. Biodiversity loss threatens human well-being. PLoS Biol. 4, e277.

Díaz, S., Demissew, S., Carabias, J., Joly, C., Lonsdale, M., Ash, N., Larigauderie, A., Adhikari, J.R., Arico, S., Báldi, A., Bartuska, A., Baste, I.A., Bilgin, A., Brondizio, E., Chan, K.M., Figueroa, V.E., Duraiappah, A., Fischer, M., Hill, R., Koetz, T., Leadley, P., Lyver, P., Mace, G.M., Martin-Lopez, B., Okumura, M., Pacheco, D., Pascual, U., Pérez, E.S., Reyers, B., Roth, E., Saito, O., Scholes, R.J., Sharma, N., Tallis, H., Thaman, R., Watson, R., Yahara, T., Hamid, Z.A., Akosim, C., Al-Hafedh, Y., Allahverdiyev, R., Amankwah, E., Asah, S.T., Asfaw, Z., Bartus, G., Brooks, L.A., Caillaux, J., Dalle, G., Darnaedi, D., Driver, A., Erpul, G., Escobar-Eyzaguirre, P., Failler, P., Fouda, A.M.M., Fu, B., Gundimeda, H., Hashimoto, S., Homer, F., Lavorel, S., Lichtenstein, G., Mala, W.A., Mandivenyi, W., Matczak, P., Mbizvo, C., Mehrdadi, M., Metzger, J.P., Mikissa, J.B., Moller, H., Mooney, H.A., Mumby, P., Nagendra, H., Nesshover, C., Oteng-Yeboah, A.A., Pataki, G., Roué, M., Rubis, J., Schultz, M., Smith, P., Sumaila, R., Takeuchi, K., Thomas, S., Verma, M., Yeo-Chang, Y., Zlatanova, D., 2015. The IPBES conceptual framework—connecting nature and people. Curr. Opin. Environ. Sustain. 14, 1–16.

Didham, R.K., Tylianakis, J.M., Gemmell, N.J., Rand, T.A., Ewers, R.M., 2007. Interactive effects of habitat modification and species invasion on native species decline. Trends Ecol. Evol. 22, 489–496.

Dornelas, M., Gotelli, N.J., McGill, B., Shimadzu, H., Moyes, F., Sievers, C., Magurran, A.E., 2014. Assemblage time series reveal biodiversity change but not systematic loss. Science 344, 296–299.

Dullinger, S., Essl, F., Rabitsch, W., Erb, K.-H., Gingrich, S., Haberl, H., Hülber, K., Jarošík, V., Krausmann, F., Kühn, I., Pergl, J., Pyšek, P., Hulme, P.E., 2013. Europe's other debt crisis caused by the long legacy of future extinctions. Proc. Natl. Acad. Sci. U. S. A. 110, 7342–7347.

Dumbrell, A.J., Hill, J.K., 2005. Impacts of selective logging on canopy and ground assemblages of tropical forest butterflies: implications for sampling. Biol. Conserv. 125, 123–131.

Essl, F., Dullinger, S., Rabitsch, W., Hulme, P.E., Pyšek, P., Wilson, J.R.U., Richardson, D.M., 2015. Delayed biodiversity change: no time to waste. Trends Ecol. Evol. 30, 375–378.

Estrada, A., Garber, P.A., Rylands, A.B., Roos, C., Fernandez-Duque, E., Di Fiore, A., Nekaris, A.-I., Nijman, V., Heymann, E.W., Lambert, J.E., Rovero, F., Barelli, C., Setchell, J.M., Gillespie, T.R., Mittermeier, R.A., Arregoitia, L.V., de Guinea, M., Gouveia, S., Dobrovolski, R., Shanee, S., Shanee, N., Boyle, S.A., Fuentes, A., MacKinnon, K.C., Amato, K.R., Meyer, A.L.S., Wich, S., Sussman, W., Pan, R.,

Kone, I., Li, B., 2017. Impending extinction crisis of the world's primates: why primates matter. Sci. Adv. 3, e1600946.

Ewers, R.M., Didham, R.K., 2006. Confounding factors in the detection of species responses to habitat fragmentation. Biol. Rev. 81, 117–142.

Ewers, R.M., Didham, R.K., Fahrig, L., Ferraz, G., Hector, A., Holt, R.D., Kapos, V., Reynolds, G., Sinun, W., Snaddon, J.L., Turner, E.C., 2011. A large-scale forest fragmentation experiment: the Stability of Altered Forest Ecosystems Project. Philos. Trans. R. Soc. Lond. B Biol. Sci. 366, 3292–3302. rstb.royalsocietypublishing.org.

Fahrig, L., 2003. Effects of habitat fragmentation on biodiversity. Annu. Rev. Ecol. Evol. Syst. 34, 487–515.

Flinn, K.M., Vellend, M., 2005. Recovery of forest plant communities in post-agricultural landscapes. Front. Ecol. Environ. 3, 243–250.

Flynn, D.F.B., Uriarte, M., Crk, T., Pascarella, J.B., Zimmerman, J.K., Aide, T.M., Caraballo Ortiz, M.A., 2010. Hurricane disturbance alters secondary forest recovery in Puerto Rico. Biotropica 42, 149–157.

Fox, B.J., Taylor, J.E., Thompson, P.T., 2003. Experimental manipulation of habitat structure: a retrogression of the small mammal succession. J. Anim. Ecol. 72, 927–940.

França, F., Louzada, J., Korasaki, V., Griffiths, H., Silveira, J.M., Barlow, J., 2016. Do space-for-time assessments underestimate the impacts of logging on tropical biodiversity? An Amazonian case study using dung beetles. J. Appl. Ecol. 53, 1098–1105.

Fukami, T., Wardle, D.A., 2005. Long-term ecological dynamics: reciprocal insights from natural and anthropogenic gradients. Proc. Biol. sci. 272, 2105–2115.

Fuller, L., Oxbrough, A., Gittings, T., Irwin, S., Kelly, T.C., O'Halloran, J., 2013. The response of ground-dwelling spiders (Araneae) and hoverflies (Diptera: Syrphidae) to afforestation assessed using within-site tracking. Forestry 87, 301–312.

Gaston, K.J., Blackburn, T.M., 2000. Pattern and Process in Macroecology. Wiley.

Gibson, L., Lee, T.M., Koh, L.P., Brook, B.W., Gardner, T.A., Barlow, J., Peres, C.A., Bradshaw, C.J.A., Laurance, W.F., Lovejoy, T.E., Sodhi, N.S., 2011. Primary forests are irreplaceable for sustaining tropical biodiversity. Nature 478, 378–381.

Gibson, L., Lynam, A.J., Bradshaw, C.J.A., He, F., Bickford, D.P., Woodruff, D.S., Bumrungsri, S., Laurance, W.F., 2013. Near-complete extinction of native small mammal fauna 25 years after forest fragmentation. Science 341, 1508–1510.

Gonzalez, A., Cardinale, B.J., Allington, G.R.H., Byrnes, J., Arthur Endsley, K., Brown, D.G., Hooper, D.U., Isbell, F., O'Connor, M.I., Loreau, M., 2016. Estimating local biodiversity change: a critique of papers claiming no net loss of local diversity. Ecology 97, 1949–1960.

Gregory, N.C., Sensenig, R.L., Wilcove, D.S., 2010. Effects of controlled fire and livestock grazing on bird communities in East African savannas. Conserv. Biol. 24, 1606–1616.

Griggs, D., Stafford-Smith, M., Gaffney, O., Rockström, J., Öhman, M.C., Shyamsundar, P., Steffen, W., Glaser, G., Kanie, N., Noble, I., 2013. Policy: sustainable development goals for people and planet. Nature 495, 305–307.

Haché, S., Villard, M.-A., 2010. Age-specific response of a migratory bird to an experimental alteration of its habitat. J. Anim. Ecol. 79, 897–905.

Holmes, S.B., Pitt, D.G., McIlwrick, K.A., Hoepting, M.K., 2012. Response of bird communities to single-tree selection system harvesting in northern hardwoods: 10–12 years post-harvest. For. Ecol. Manage. 271, 132–139.

Hooper, D.U., Adair, E.C., Cardinale, B.J., Byrnes, J.E.K., Hungate, B.A., Matulich, K.L., Gonzalez, A., Duffy, J.E., Gamfeldt, L., O'Connor, M.I., 2012. A global synthesis reveals biodiversity loss as a major driver of ecosystem change. Nature 486, 105–108.

Hudson, L.N., Newbold, T., Contu, S., Hill, S.L.L., Lysenko, I., De Palma, A., Phillips, H.R.P., Alhusseini, T.I., Bedford, F.E., Bennett, D.J., Booth, H.,

Burton, V.J., Chng, C.W.T., Choimes, A., Correia, D.L.P., Day, J., Echeverría-Londoño, S., Emerson, S.R., Gao, D., Garon, M., Harrison, M.L.K., Ingram, D.J., Jung, M., Kemp, V., Kirkpatrick, L., Martin, C.D., Pan, Y., Pask-Hale, G.D., Pynegar, E.L., Robinson, A.N., Sanchez-Ortiz, K., Senior, R.A., Simmons, B.I., White, H.J., Zhang, H., Aben, J., Abrahamczyk, S., Adum, G.B., Aguilar-Barquero, V., Aizen, M.A., Albertos, B., Alcala, E.L., del Mar Alguacil, M., Alignier, A., Ancrenaz, M., Andersen, A.N., Arbeláez-Cortés, E., Armbrecht, I., Arroyo-Rodríguez, V., Aumann, T., Axmacher, J.C., Azhar, B., Azpiroz, A.B., Baeten, L., Bakayoko, A., Báldi, A., Banks, J.E., Baral, S.K., Barlow, J., Barratt, B.I.P., Barrico, L., Bartolommei, P., Barton, D.M., Basset, Y., Batáry, P., Bates, A.J., Baur, B., Bayne, E.M., Beja, P., Benedick, S., Berg, Å., Bernard, H., Berry, N.J., Bhatt, D., Bicknell, J.E., Bihn, J.H., Blake, R.J., Bobo, K.S., Bóçon, R., Boekhout, T., Böhning-Gaese, K., Bonham, K.J., Borges, P.A.V., Borges, S.H., Boutin, C., Bouyer, J., Bragagnolo, C., Brandt, J.S., Brearley, F.Q., Brito, I., Bros, V., Brunet, J., Buczkowski, G., Buddle, C.M., Bugter, R., Buscardo, E., Buse, J., Cabra-García, J., Cáceres, N.C., Cagle, N.L., Calviño-Cancela, M., Cameron, S.A., Cancello, E.M., Caparrós, R., Cardoso, P., Carpenter, D., Carrijo, T.F., Carvalho, A.L., Cassano, C.R., Castro, H., Castro-Luna, A.A., Rolando, C.B., Cerezo, A., Chapman, K.A., Chauvat, M., Christensen, M., Clarke, F.M., Cleary, D.F.R., Colombo, G., Connop, S.P., Craig, M.D., Cruz-López, L., Cunningham, S.A., D'Aniello, B., D'Cruze, N., da Silva, P.G., Dallimer, M., Danquah, E., Darvill, B., Dauber, J., Davis, A.L.V., Dawson, J., de Sassi, C., de Thoisy, B., Deheuvels, O., Dejean, A., Devineau, J.-L., Diekötter, T., Dolia, J.V., Domínguez, E., Dominguez-Haydar, Y., Dorn, S., Draper, I., Dreber, N., Dumont, B., Dures, S.G., Dynesius, M., Edenius, L., Eggleton, P., Eigenbrod, F., Elek, Z., Entling, M.H., Esler, K.J., de Lima, R.F., Faruk, A., Farwig, N., Fayle, T.M., Felicioli, A., Felton, A.M., Fensham, R.J., Fernandez, I.C., Ferreira, C.C., Ficetola, G.F., Fiera, C., Filgueiras, B.K.C., Fırıncıoğlu, H.K., Flaspohler, D., Floren, A., Fonte, S.J., Fournier, A., Fowler, R.E., Franzén, M., Fraser, L.H., Fredriksson, G.M., Freire, G.B., Frizzo, T.L.M., Fukuda, D., Furlani, D., Gaigher, R., Ganzhorn, J.U., García, K.P., Garcia-R, J.C., Garden, J.G., Garilleti, R., Ge, B.-M., Gendreau-Berthiaume, B., Gerard, P.J., Gheler-Costa, C., Gilbert, B., Giordani, P., Giordano, S., Golodets, C., Gomes, L.G.L., Gould, R.K., Goulson, D., Gove, A.D., Granjon, L., Grass, I., Gray, C.L., Grogan, J., Gu, W., Guardiola, M., Gunawardene, N.R., Gutierrez, A.G., Gutiérrez-Lamus, D.L., Haarmeyer, D.H., Hanley, M.E., Hanson, T., Hashim, N.R., Hassan, S.N., Hatfield, R.G., Hawes, J.E., Hayward, M.W., Hébert, C., Helden, A.J., Henden, J.-A., Henschel, P., Hernández, L., Herrera, J.P., Herrmann, F., Herzog, F., Higuera-Diaz, D., Hilje, B., Höfer, H., Hoffmann, A., Horgan, F.G., Hornung, E., Horváth, R., Hylander, K., Isaacs-Cubides, P., Ishida, H., Ishitani, M., Jacobs, C.T., Jaramillo, V.J., Jauker, B., Hernández, F.J., Johnson, M.F., Jolli, V., Jonsell, M., Juliani, S.N., Jung, T.S., Kapoor, V., Kappes, H., Kati, V., Katovai, E., Kellner, K., Kessler, M., Kirby, K.R., Kittle, A.M., Knight, M.E., Knop, E., Kohler, F., Koivula, M., Kolb, A., Kone, M., Kőrösi, Á., Krauss, J., Kumar, A., Kumar, R., Kurz, D.J., Kutt, A.S., Lachat, T., Lantschner, V., Lara, F., Lasky, J.R., Latta, S.C., Laurance, W.F., Lavelle, P., Le Féon, V., LeBuhn, G., Légaré, J.-P., Lehouck, V., Lencinas, M.V., Lentini, P.E., Letcher, S.G., Li, Q., Litchwark, S.A., Littlewood, N.A., Liu, Y., Lo-Man-Hung, N., López-Quintero, C.A., Louhaichi, M., Lövei, G.L., Lucas-Borja, M.E., Luja, V.H., Luskin, M.S., MacSwiney, G.M.C., Maeto, K., Magura, T., Mallari, N.A., Malone, L.A., Malonza, P.K., Malumbres-Olarte, J., Mandujano, S., Måren, I.E., Marin-Spiotta, E., Marsh, C.J., Marshall, E.J.P., Martínez, E., Martínez, P.G., Moreno, M.D.,

Mayfield, M.M., Mazimpaka, V., McCarthy, J.L., McCarthy, K.P., McFrederick, Q.S., McNamara, S., Medina, N.G., Medina, R., Mena, J.L., Mico, E., Mikusinski, G., Milder, J.C., Miller, J.R., Miranda-Esquivel, D.R., Moir, M.L., Morales, C.L., Muchane, M.N., Muchane, M., Mudri-Stojnic, S., Munira, A.N., Muoñz-Alonso, A., Munyekenye, B.F., Naidoo, R., Naithani, A., Nakagawa, M., Nakamura, A., Nakashima, Y., Naoe, S., Nates-Parra, G., Navarrete Gutierrez, D.A., Navarro-Iriarte, L., Ndang'ang'a, P.K., Neuschulz, E.L., Ngai, J.T., Nicolas, V., Nilsson, S.G., Noreika, N., Norfolk, O., Noriega, J.A., Norton, D.A., Nöske, N.M., Nowakowski, A.J., Numa, C., O'Dea, N., O'Farrell, P.J., Oduro, W., Oertli, S., Ofori-Boateng, C., Oke, C.O., Oostra, V., Osgathorpe, L.M., Otavo, S.E., Page, N.V., Paritsis, J., Parra-H, A., Parry, L., Pe'er, G., Pearman, P.B., Pelegrin, N., Pélissier, R., Peres, C.A., Peri, P.L., Persson, A.S., Petanidou, T., Peters, M.K., Pethiyagoda, R.S., Phalan, B., Philips, T.K., Pillsbury, F.C., Pincheira-Ulbrich, J., Pineda, E., Pino, J., Pizarro-Araya, J., Plumptre, A.J., Poggio, S.L., Politi, N., Pons, P., Poveda, K., Power, E.F., Presley, S.J., Proença, V., Quaranta, M., Quintero, C., Rader, R., Ramesh, B.R., Ramirez-Pinilla, M.P., Ranganathan, J., Rasmussen, C., Redpath-Downing, N.A., Reid, J.L., Reis, Y.T., Rey Benayas, J.M., Rey-Velasco, J.C., Reynolds, C., Ribeiro, D.B., Richards, M.H., Richardson, B.A., Richardson, M.J., Ríos, R.M., Robinson, R., Robles, C.A., Römbke, J., Romero-Duque, L.P., Rös, M., Rosselli, L., Rossiter, S.J., Roth, D.S., Roulston, T.H., Rousseau, L., Rubio, A.V., Ruel, J.-C., Sadler, J.P., Sáfián, S., Saldaña-Vázquez, R.A., Sam, K., Samnegård, U., Santana, J., Santos, X., Savage, J., Schellhorn, N.A., Schilthuizen, M., Schmiedel, U., Schmitt, C.B., Schon, N.L., Schüepp, C., Schumann, K., Schweiger, O., Scott, D.M., Scott, K.A., Sedlock, J.L., Seefeldt, S.S., Shahabuddin, G., Shannon, G., Sheil, D., Sheldon, F.H., Shochat, E., Siebert, S.J., Silva, F.A.B., Simonetti, J.A., Slade, E.M., Smith, J., Smith-Pardo, A.H., Sodhi, N.S., Somarriba, E.J., Sosa, R.A., Soto, Q.G., St-Laurent, M.-H., Starzomski, B.M., Stefanescu, C., Steffan-Dewenter, I., Stouffer, P.C., Stout, J.C., Strauch, A.M., Struebig, M.J., Su, Z., Suarez-Rubio, M., Sugiura, S., Summerville, K.S., Sung, Y.-H., Sutrisno, H., Svenning, J.-C., Teder, T., Threlfall, C.G., Tiitsaar, A., Todd, J.H., Tonietto, R.K., Torre, I., Tóthmérész, B., Tscharntke, T., Turner, E.C., Tylianakis, J.M., Uehara-Prado, M., Urbina-Cardona, N., Vallan, D., Vanbergen, A.J., Vasconcelos, H.L., Vassilev, K., Verboven, H.A.F., Verdasca, M.J., Verdú, J.R., Vergara, C.H., Vergara, P.M., Verhulst, J., Virgilio, M., Van, V.L., Waite, E.M., Walker, T.R., Wang, H.-F., Wang, Y., Watling, J.I., Weller, B., Wells, K., Westphal, C., Wiafe, E.D., Williams, C.D., Willig, M.R., Woinarski, J.C.Z., Wolf, J.H.D., Wolters, V., Woodcock, B.A., Wu, J., Wunderle, J.M., Yamaura, Y., Yoshikura, S., Yu, D.W., Zaitsev, A.S., Zeidler, J., Zou, F., Collen, B., Ewers, R.M., Mace, G.M., Purves, D.W., Scharlemann, J.P.W., Purvis, A., 2017. The database of the PREDICTS (Projecting Responses of Ecological Diversity in Changing Terrestrial Systems) project. Ecol. Evol. 7, 145–188.

Hurtt, G., Chini, L., Sahajpal, R., Frolking, S., et al., 2017. Harmonization of global land-use change and management for the period 850–2100. Geosci. Model Dev. (in preparation). Accessed from, http://luh.umd.edu. in June 2017.

Hylander, K., Ehrlén, J., 2013. The mechanisms causing extinction debts. Trends Ecol. Evol. 28, 341–346.

Isaac, N.J.B., Cruickshanks, K.L., Weddle, A.M., Rowcliffe, J.M., Brereton, T.M., Dennis, R.L.H., Shuker, D.M., Thomas, C.D., 2011. Distance sampling and the challenge of monitoring butterfly populations. Methods Ecol. Evol. 2, 585–594.

Jackson, J.K., Füreder, L., 2006. Long-term studies of freshwater macroinvertebrates: a review of the frequency, duration and ecological significance. Freshwater Biol. 51, 591–603.

Jackson, S.T., Sax, D.F., 2010. Balancing biodiversity in a changing environment: extinction debt, immigration credit and species turnover. Trends Ecol. Evol. 25, 153–160.

Jasinski, E., Morton, D., DeFries, R., Shimabukuro, Y., Anderson, L., Hansen, M., 2005. Physical landscape correlates of the expansion of mechanized agriculture in Mato Grosso, Brazil. Earth Interact. 9, 1–18.

Johnson, E.A., Miyanishi, K., 2008. Testing the assumptions of chronosequences in succession. Ecol. Lett. 11, 419–431.

Kappes, H., 2006. Relations between forest management and slug assemblages (Gastropoda) of deciduous regrowth forests. For. Ecol. Manage. 237, 450–457.

Koricheva, J., 2003. Non-significant results in ecology: a burden or a blessing in disguise? Oikos 102, 397–401.

Kremen, C., M'Gonigle, L.K., 2015. Small-scale restoration in intensive agricultural landscapes supports more specialized and less mobile pollinator species. J. Appl. Ecol. 52, 602–610.

Kuussaari, M., Bommarco, R., Heikkinen, R.K., Helm, A., Krauss, J., Lindborg, R., Öckinger, E., Pärtel, M., Pino, J., Rodà, F., Stefanescu, C., Teder, T., Zobel, M., Steffan-Dewenter, I., 2009. Extinction debt: a challenge for biodiversity conservation. Trends Ecol. Evol. 24, 564–571.

Lajeuness, M.J., 2010. Achieving synthesis with meta-analysis by combining all available studies. Ecology 91, 2561–2564.

Lawton, J.H., Bignell, D.E., Bolton, B., Bloemers, G.F., Eggleton, P., Hammond, P.M., Hodda, M., Holt, R.D., Larsen, T.B., Mawdsley, N.A., Stork, N.E., Srivastava, D.S., Watt, A.D., 1998. Biodiversity inventories, indicator taxa and effects of habitat modification in tropical forest. Nature 391, 72–76.

Lebrija-Trejos, E., Meave, J.A., Poorter, L., Pérez-García, E.A., Bongers, F., 2010. Pathways, mechanisms and predictability of vegetation change during tropical dry forest succession. Perspect. Plant Ecol. Evol. Syst. 12, 267–275.

Magurran, A.E., Baillie, S.R., Buckland, S.T., Dick, J.M., Elston, D.A., Scott, E.M., Smith, R.I., Somerfield, P.J., Watt, A.D., 2010. Long-term datasets in biodiversity research and monitoring: assessing change in ecological communities through time. Trends Ecol. Evol. 25, 574–582.

Mantyka-pringle, C.S., Martin, T.G., Rhodes, J.R., 2012. Interactions between climate and habitat loss effects on biodiversity: a systematic review and meta-analysis. Glob. Chang. Biol. 18, 1239–1252.

Martin, L.J., Blossey, B., Ellis, E., 2012. Mapping where ecologists work: biases in the global distribution of terrestrial ecological observations. Front. Ecol. Environ. 10, 195–201.

Martin, P.A., Newton, A.C., Cantarello, E., Evans, P., 2015a. Stand dieback and collapse in a temperate forest and its impact on forest structure and biodiversity. For. Ecol. Manage. 358, 130–138.

Martin, P.A., Newton, A.C., Pfeifer, M., Khoo, M., Bullock, J.M., 2015b. Impacts of tropical selective logging on carbon storage and tree species richness: a meta-analysis. For. Ecol. Manage. 356, 224–233.

McGarigal, K., Cushman, S.A., 2002. Comparative evaluation of experimental approaches to the study of habitat fragmentation effects. Ecol. Appl. 12, 335–345.

McGill, B.J., Dornelas, M., Gotelli, N.J., Magurran, A.E., 2015. Fifteen forms of biodiversity trend in the Anthropocene. Trends Ecol. Evol. 30, 104–113.

McRae, L., Deinet, S., Freeman, R., 2017. The diversity-weighted Living Planet Index: controlling for taxonomic bias in a global biodiversity indicator. PLoS One 12, e0169156.

Monamy, V., Fox, B.J., 2010. Responses of two species of heathland rodents to habitat manipulation: vegetation density thresholds and the habitat accommodation model. Austral Ecol. 35, 334–347.

Moreno-Mateos, D., Barbier, E.B., Jones, P.C., Jones, H.P., Aronson, J., López-López, J.A., McCrackin, M.L., Meli, P., Montoya, D., Benayas, J.M.R., 2017. Anthropogenic ecosystem disturbance and the recovery debt. Nat. Commun. 8, 14163.

Murphy, G.E.P., Romanuk, T.N., 2014. A meta-analysis of declines in local species richness from human disturbances. Ecol. Evol. 4, 91–103.

Nelson, A., 2008. Estimated Travel Time to the Nearest City of 50,000 or More People in Year 2000. http://bioval.jrc.ec.europa.eu/products/gam/index.htm. Accessed October 2016.

Newbold, T., Hudson, L.N., Hill, S.L.L., Contu, S., Lysenko, I., Senior, R.A., Börger, L., Bennett, D.J., Choimes, A., Collen, B., Day, J., De Palma, A., Díaz, S., Echeverria-Londoño, S., Edgar, M.J., Feldman, A., Garon, M., Harrison, M.L.K., Alhusseini, T., Ingram, D.J., Itescu, Y., Kattge, J., Kemp, V., Kirkpatrick, L., Kleyer, M., Correia, D.L.P., Martin, C.D., Meiri, S., Novosolov, M., Pan, Y., Phillips, H.R.P., Purves, D.W., Robinson, A., Simpson, J., Tuck, S.L., Weiher, E., White, H.J., Ewers, R.M., Mace, G.M., Scharlemann, J.P.W., Purvis, A., 2015. Global effects of land use on local terrestrial biodiversity. Nature 520, 45–50.

Norden, N., Angarita, H.A., Bongers, F., Martínez-Ramos, M., Granzow-de la Cerda, I., van Breugel, M., Lebrija-Trejos, E., Meave, J.A., Vandermeer, J., Williamson, G.B., Finegan, B., Mesquita, R., Chazdon, R.L., 2015. Successional dynamics in Neotropical forests are as uncertain as they are predictable. Proc. Natl. Acad. Sci. U. S. A. 112, 8013–8018.

Ohwaki, A., Tanabe, S.-I., Nakamura, K., 2008. Effects of anthropogenic disturbances on the butterfly assemblage in an urban green area: the changes from 1990 to 2005 in Kanazawa Castle Park, Japan. Ecol. Res. 23, 697–708.

Oliver, T.H., Morecroft, M.D., 2014. Interactions between climate change and land use change on biodiversity: attribution problems, risks, and opportunities. WIREs Clim. Change 5, 317–335.

Oliver, T.H., Brereton, T., Roy, D.B., 2013. Population resilience to an extreme drought is influenced by habitat area and fragmentation in the local landscape. Ecography 36, 579–586.

Pechmann, J.H.K., Wilbur, H.M., 1994. Putting declining amphibian populations in perspective: natural fluctuations and human impacts. Herpetologia 50, 65–84.

Phillips, H.R.P., Halley, J.M., Urbina-Cordona, J.N., Purvis, A., 2017. The effect of fragment area on site-level biodiversity. Ecography https://doi.org/10.1111/ecog.02956.

Pickett, S.T.A., 1989. Space-for-time substitution as an alternative to long-term studies. In: Likens, G.E. (Ed.), Long-Term Studies in Ecology. Springer, New York, NY, pp. 110–135.

Pleieninger, T., Hui, C., Gaertner, M., Huntsinger, L., 2014. The impact of land abandonment on species richness and abundance in the Mediterranean Basin: a meta-analysis. PLoS One 9, e98355.

Pons, P., 2015. Delayed effects of fire and logging on cicada nymph abundance. J. Insect Conserv. 19, 601–606.

Pons, P., Rakotobearison, G., Wendenburg, C., 2003. Immediate effects of a fire on birds and vegetation at Ankarafantsika Strict Nature Reserve, NW Madagascar. Ostrich 74, 146–148.

Rudolphi, J., Jönsson, M.T., Gustafsson, L., 2014. Biological legacies buffer local species extinction after logging. J. Appl. Ecol. 51, 53–62.

Smith, E.P., 2002. BACI design. In: El-Shaarawi, A.H., Piegorsch, W.W. (Eds.), In: Encyclopedia of Environmetrics, vol. 1. John Wiley & Sons, Chichester, UK, pp. 141–148.

Stewart-Oaten, A., Murdoch, W.W., Parker, K.R., 1986. Environmental impact assessment: 'Pseudoreplication' in time? Ecology 67, 929–940.

Tilman, D., 1989. Ecological experimentation: strengths and conceptual problems. In: Likens, G.E. (Ed.), Long-Term Studies in Ecology. Springer, New York, NY, pp. 136–157.

Tilman, D., May, R.M., Lehman, C.L., Nowak, M.A., 1994. Habitat destruction and the extinction debt. Nature 371, 65–66.

Torres, A., Palacín, C., Seoane, J., Alonso, J.C., 2011. Assessing the effects of a highway on a threatened species using Before–During–After and Before–During–After–Control–Impact designs. Biol. Conserv. 144, 2223–2232.

Underwood, A.J., 1989. The analysis of stress in natural populations. Biol. J. Linn. Soc. 37, 51–78.

Underwood, A.J., 1991. Beyond BACI: experimental designs for detecting human environmental impacts on temporal variations in natural populations. Mar. Freshw. Res. 42, 569–587.

Underwood, A.J., 1992. Beyond BACI: the detection of environmental impacts on populations in the real, but variable, world. J. Exp. Mar. Biol. Ecol. 161, 145–178.

Underwood, A.J., 1994. On beyond BACI: sampling designs that might reliably detect environmental disturbances. Ecol. Appl. 4, 3–15.

Uriarte, M., Canham, C.D., Thompson, J., Zimmerman, J.K., Murphy, L., Sabat, A.M., Fetcher, N., Haines, B.L., 2009. Natural disturbance and human land use as determinants of tropical forest dynamics: results from a forest simulator. Ecol. Monogr. 79, 423–443.

Vellend, M., Verheyen, K., Jacquemyn, H., Kolb, A., Van Calster, H., Peterken, G., Hermy, M., 2006. Extinction debt of forest plants persists for more than a century following habitat fragmentation. Ecology 87, 542–548.

Vellend, M., Baeten, L., Myers-Smith, I.H., Elmendorf, S.C., Beauséjour, R., Brown, C.D., De Frenne, P., Verheyen, K., Wipf, S., 2013. Global meta-analysis reveals no net change in local-scale plant biodiversity over time. Proc. Natl. Acad. Sci. U. S. A. 110, 19456–19459.

Vellend, M., Dornelas, M., Baeten, L., Beauséjour, R., Brown, C.D., De Frenne, P., Elmendorf, S.C., Gotelli, N.J., Moyes, F., Myers-Smith, I.H., Magurran, A.E., McGill, B.J., Shimadzu, H., Sievers, C., 2017. Estimates of local biodiversity change over time stand up to scrutiny. Ecology 98, 583–590.

Venter, O., Sanderson, E.W., Magrach, A., Allan, J.R., Beher, J., Jones, K.R., Possingham, H.P., Laurance, W.F., Wood, P., Fekete, B.M., Levy, M.A., Watson, J.E.M., 2016a. Sixteen years of change in the global terrestrial human footprint and implications for biodiversity conservation. Nat. Commun. 7, 12558.

Venter, O., Sanderson, E.W., Magrach, A., Allan, J.R., Beher, J., Jones, K.R., Possingham, H.P., Laurance, W.F., Wood, P., Fekete, B.M., Levy, M.A., Watson, J.E.M., 2016b. Data From: Global Terrestrial Human Footprint Maps for 1993 and 2009. Dryad Digital Repository, https://doi.org/10.5061/dryad.052q5.2.

Walker, L.R., Wardle, D.A., Bardgett, R.D., Clarkson, B.D., 2010. The use of chronosequences in studies of ecological succession and soil development. J. Ecol. 98, 725–736.

Waltz, A.E.M., Covington, W.W., 2004. Ecological restoration treatments increase butterfly richness and abundance: mechanisms of response. Restor. Ecol. 12, 85–96.

Wardell-Johnson, G., Williams, M., 2000. Edges and gaps in mature karri forest, southwestern Australia: logging effects on bird species abundance and diversity. For. Ecol. Manage. 131, 1–21.

Wearn, O.R., Reuman, D.C., Ewers, R.M., 2012. Extinction debt and windows of conservation opportunity in the Brazilian Amazon. Science 337, 228–232.

Weatherhead, P.J., 1986. How unusual are unusual events? Am. Nat. 128, 150–154.

Whittaker, R.J., 2010. Meta-analyses and mega-mistakes: calling time on meta-analysis of the species richness-productivity relationship. Ecology 91, 2522–2533.

Willig, M.R., Bloch, C.P., Brokaw, N., Higgins, C., Thompson, J., Zimmermann, C.R., 2007. Cross-scale responses of biodiversity to hurricane and anthropogenic disturbance in a tropical forest. Ecosystems 10, 824–838.

Wills, J., Herbohn, J., Moreno, M.O.M., Avela, M.S., Firn, J., 2017. Next-generation tropical forests: reforestation type affects recruitment of species and functional diversity in a human-dominated landscape. J. Appl. Ecol. 54, 772–783.

Zwart, M.C., Robson, P., Rankin, S., Whittingham, M.J., McGowan, P.J.K., 2015. Using environmental impact assessment and post-construction monitoring data to inform wind energy developments. Ecosphere 6, 1–11.

FURTHER READING

Brooks, T.M., Pimm, S.L., Oyugi, J.O., 1999. Time lag between deforestation and bird extinction in tropical forest fragments. Conserv. Biol. 13, 1140–1150.

Butchart, S.H.M., Walpole, M., Collen, B., van Strien, A., Scharlemann, J.P.W., Almond, R.E.A., Baillie, J.E.M., Bomhard, B., Brown, C., Bruno, J., Carpenter, K.E., Carr, G.M., Chanson, J., Chenery, A.M., Csirke, J., Davidson, N.C., Dentener, F., Foster, M., Galli, A., Galloway, J.N., Genovesi, P., Gregory, R.D., Hockings, M., Kapos, V., Lamarque, J.-F., Leverington, F., Loh, J., McGeoch, M.A., McRae, L., Minasyan, A., Hernández Morcillo, M., Oldfield, T.E.E., Pauly, D., Quader, S., Revenga, C., Sauer, J.R., Skolnik, B., Spear, D., Stanwell-Smith, D., Stuart, S.N., Symes, A., Tierney, M., Tyrrell, T.D., Vié, J.-C., Watson, R., 2010. Global biodiversity: indicators of recent declines. Science 328, 1164–1168.

Lefort, P., Grove, S., 2009. Early responses of birds to clearfelling and its alternatives in lowland wet eucalypt forest in Tasmania, Australia. For. Ecol. Manage. 258, 460–471.

Martin, P.A., Newton, A.C., Bullock, J.M., 2013. Carbon pools recover more quickly than plant biodiversity in tropical secondary forests. Proc. Biol. Sci. 280, 20132236.

McKinney, M.L., 1997. Extinction, vulnerability and selectivity: combining ecological and paleontological veiws. Annu. Rev. Ecol. Syst. 28, 495–516.

Metzger, J.P., Martensen, A.C., Dixo, M., Bernacci, L.C., Ribeiro, M.C., Teixeira, A.M.G., Pardini, R., 2009. Time-lag in biological responses to landscape changes in a highly dynamic Atlantic forest region. Biol. Conserv. 142, 1166–1177.

Soga, M., Koike, S., 2013. Mapping the potential extinction debt of butterflies in a modern city: implications for conservation priorities in urban landscapes. Anim. Conserv. 16, 1–11.

Tittensor, D.P., Walpole, M., Hill, S.L.L., Boyce, D.G., Britten, G.L., Burgess, N.D., Butchart, S.H.M., Leadley, P.W., Regan, E.C., Alkemade, R., Baumung, R., Bellard, C., Bouwman, L., Bowles-Newark, N.J., Chenery, A.M., Cheung, W.W.L., Christensen, V., Cooper, H.D., Crowther, A.R., Dixon, M.J.R., Galli, A., Gaveau, V., Gregory, R.D., Gutierrez, N.L., Hirsch, T.L., Hoft, R., Januchowski-Hartley, S.R., Karmann, M., Krug, C.B., Leverington, F.J., Loh, J., Lojenga, R.K., Malsch, K., Marques, A., Morgan, D.H.W., Mumby, P.J., Newbold, T., Noonan-Mooney, K., Pagad, S.N., Parks, B.C., Pereira, H.M., Robertson, T., Rondinini, C., Santini, L., Scharlemann, J.P.W., Schindler, S., Sumaila, U.R., Teh, L.S.L., van Kolck, J., Visconti, P., Ye, Y., 2014. A mid-term analysis of progress toward international biodiversity targets. Science 346, 241–245.

CHAPTER FIVE

Modelling and Projecting the Response of Local Terrestrial Biodiversity Worldwide to Land Use and Related Pressures: The PREDICTS Project

Andy Purvis*,†,1, Tim Newbold‡, Adriana De Palma*, Sara Contu*,
Samantha L.L. Hill*,§, Katia Sanchez-Ortiz*, Helen R.P. Phillips¶,
Lawrence N. Hudson*, Igor Lysenko†, Luca Börger‖,
Jörn P.W. Scharlemann#,§

*Natural History Museum, London, United Kingdom
†Grand Challenges in Ecosystems and the Environment, Imperial College London, Ascot, United Kingdom
‡Centre for Biodiversity and Environment Research, University College London, London, United Kingdom
§UN Environment World Conservation Monitoring Centre, Cambridge, United Kingdom
¶German Centre for Integrative Biodiversity Research (iDiv) Halle-Jena-Leipzig, Leipzig, Germany
‖College of Science, Swansea University, Swansea, United Kingdom
#School of Life Sciences, University of Sussex, Brighton, United Kingdom
1Corresponding author: e-mail address: andy.purvis@nhm.ac.uk

Contents

Advances in Ecological Research, Volume 58
ISSN 0065-2504
https://doi.org/10.1016/bs.aecr.2017.12.003

Abstract

The PREDICTS project (Projecting Responses of Ecological Diversity In Changing Terrestrial Systems) has collated ecological survey data from hundreds of published biodiversity comparisons of sites facing different land-use and related pressures, and used the resulting taxonomically and geographically broad database (abundance and occurrence data for over 50,000 species and over 30,000 sites in nearly 100 countries) to develop global biodiversity models, indicators, and projections. After outlining the science and science-policy gaps that motivated PREDICTS, this review discusses the key design decisions that helped it to achieve its objectives. In particular, we discuss basing models on a large, taxonomically, and geographically representative database, so that they may be applicable to biodiversity more broadly; space-for-time substitution, which allows estimation of pressure-state models without the need for representative time-series data; and collation of raw data rather than statistical results, greatly expanding the range of response variables that can be modelled. The heterogeneity of data in the PREDICTS database has presented a range of modelling challenges: we discuss these with a focus on our implementation of the Biodiversity Intactness Index, an indicator with considerable policy potential but which had not previously been estimated from primary biodiversity data. We then summarise the findings from analyses of how land use and related pressures affect local (α) diversity and spatial turnover (β diversity), and how these effects are mediated by ecological attributes of species. We discuss the relevance of our findings for policy, before ending with some directions of ongoing and possible future research.

1. INTRODUCTION: PREDICTS' SCIENTIFIC AND SCIENCE-POLICY OBJECTIVES

Multiple drivers and pressures are causing biodiversity change. Human conversion of natural habitats and land–use intensification are two of the major pressures on biological diversity (Maxwell et al., 2016). With a growing human population and concomitant increases in consumption, the extent of anthropogenic landscapes on Earth has expanded substantially over past millennia and is projected to continue to do so (Hurtt et al., 2011). Understanding the past and future impacts of land use on biodiversity is therefore important, particularly as biodiversity underpins human wellbeing on Earth by providing essential ecosystem services, such as food, fibres, clean air, water, and climate regulation.

Models and scenarios of biodiversity and ecosystem services can help inform policy and contribute to scientific understanding. Scenarios—representations of possible futures of drivers (e.g. human population, land use) or policy interventions (e.g. protected areas)—can help with both problem identification (exploratory scenarios) and policy evaluation (intervention scenarios) (Ferrier et al., 2016). Scenarios can then be translated into projected consequences for biodiversity and ecosystem services by using models—quantitative or qualitative descriptions of key components of a system and relationships between components (Ferrier et al., 2016). Models can help clarify how biodiversity is affected by both indirect drivers (e.g. trade) and direct drivers (e.g. land-use change). Models can help us understand the mechanisms that have led to current patterns of biodiversity, understand the processes underlying the response of biodiversity to global change to project future changes, synthesise a wide range of disparate sources of information, and help evaluate the effectiveness of mitigation and adaptation policies (Ferrier et al., 2016).

For the past three decades, models and scenarios have been used extensively by the Intergovernmental Panel on Climate Change (IPCC) to encourage science-policy dialogue and communicate climate change issues (e.g. IPCC Assessment Reports 1–5, Houghton et al., 1990; Stocker et al., 2013). The 'hockey stick' graph, depicting a reconstruction of annual surface temperatures over the past six centuries (Mann et al., 1998) and extended to show scenarios of future climate change (e.g. Figure SPM.6 in IPCC, 2013), has become an extremely successful tool in communicating the importance of climatic change (e.g. An Inconvenient Truth: Guggenheim, 2006).

The use of models and scenarios for biodiversity and ecosystem services assessments is more recent than for climate change, and an (inverted) 'hockey stick' graph to communicate biodiversity loss had not been produced when we started the Projecting Responses of Ecological Diversity In Changing Terrestrial Systems (PREDICTS) project in March 2012. The first global assessment that used scenarios and models of biodiversity extensively was the United Nations Environment Programme's Global Environment Outlook 3 (GEO3: UNEP, 2002). The GLOBIO2 method (Nellemann et al., 2001), a distance–decay function relating animal species' abundance or survival to distance from infrastructure (e.g. roads, power lines, pipelines, urban areas) based on a literature review of over 200 published studies, was used to globally map the impact of infrastructure expansion under a set of scenarios to 2030 (UNEP, 2002). For future assessments (e.g. UNEP, 2007), the GLOBIO2 modelling framework was extended to include dose–response relationships for four other pressures (i.e. land use, nitrogen deposition, fragmentation, and climate change) in addition to infrastructure, and a new output metric was developed—'mean species abundance' (MSA)—the mean abundance (or species richness) of the original species relative to their abundance in pristine ecosystems (Alkemade et al., 2009). The MSA only considers declines of native species (or species richness), ignoring any species thriving in novel ecosystems (e.g. croplands, urban areas) or the arrival of invasive species (Alkemade et al., 2009).

Other global assessments using biodiversity models include the Millennium Ecosystem Assessment (2005, abbreviated MA) and the Convention on Biological Diversity's (CBD) Global Biodiversity Outlooks (GBO: Secretariat of the Convention on Biological Diversity, 2010, 2014). To assess past trends and current state of biodiversity, the MA used trends in populations of well-studied species (mostly vertebrates) and compared current to palaeontological extinction rates (Mace et al., 2005). To assess future impacts, the MA used species–area relationships combined with four scenarios describing declines of 13%–20% in habitat availability, to infer a projected loss of 12%–16% of terrestrial plant species from 1970 to 2050 globally (Sala et al., 2005). The GBOs (Secretariat of the Convention on Biological Diversity, 2010, 2014) used models to assess achievement of the CBD biodiversity targets, showing that the 2010 target was missed and that few of the Aichi targets for 2020 will be met (Butchart et al., 2010; Tittensor et al., 2014).

Climate has often been summarised and communicated using a single measure—average temperature—but biodiversity (i.e. biological diversity) means different things to different people. The CBD defined biodiversity as 'the variability among living organisms from … terrestrial, marine and other aquatic ecosystems and the ecological complexes of which they are part; this includes diversity within species, between species and of ecosystems' (Article 2 of United Nations, 1992), emphasising the multiple scales from genetic to landscape diversity. In another influential definition, Noss (1990) recognised three attributes of biodiversity that range along these scales: compositional (genes, species, populations, ecosystems), structural (genetic structure to landscape patterns), and functional (genetic to landscape processes) biodiversity. Hence, adequately representing the complexity of biodiversity requires multiple measures (McGill et al., 2015; Purvis and Hector, 2000).

One of the simplest species-level measures of compositional biodiversity is species richness, i.e., the number of different kinds of organisms. To model global species richness is difficult, however, as not all species have been described with only a rough estimate of the total number of species on Earth available (e.g. Mora et al., 2011), while local species richness is not additive across space or time. Abundance—the number of individuals—is mathematically easier to model and assess, and aggregated population trends of vertebrate abundance have been used to assess the rate of change in the status of biodiversity and used in policy documents like the Living Planet Reports (Collen et al., 2009; Loh et al., 1998, 2005). The Red List Index, based on repeated assessments of groups of species by the IUCN Red List, has been used as a measure of changes in extinction risk (Butchart et al., 2007) and to assess progress towards the CBD biodiversity targets (Convention on Biological Diversity, 2017). Most broad-scale measures, estimates, and indicators of terrestrial species richness and abundance have focused on large vertebrates and plants in temperate regions, leaving large knowledge gaps especially for invertebrates and tropical biomes. To overcome some of these shortcomings, Scholes and Biggs (2005) developed the Biodiversity Intactness Index (BII: Box 1) to provide a scientifically robust overall measure of the state of biodiversity for any given area, synthesising land use, ecosystem extent, species richness, and population abundance data across multiple taxa (see Box 1). Because of data limitations, BII was not estimated globally, and Scholes and Biggs (2005) based their estimates for selected biomes in Africa on expert assessment rather than primary data.

The PREDICTS project began in 2012 aiming to address several science and science-policy gaps. Foremost, scientifically robust biodiversity measures were needed for a broader set of taxonomic groups and geographically more representative than previously developed measures; measures that assess multiple aspects of biodiversity but that allow arresting visualisations of rigorous scientific findings such as the 'hockey-stick' graph; and measures that can feed into policy making by addressing current policy goals and targets such as the CBD Aichi biodiversity targets and UN Sustainable Development Goals. This review explains and discussed the project's key design decisions, discusses how we have tackled the challenges of modelling a large and highly heterogeneous database (including a detailed account of our implementation of BII), and reviews the results that have so far emerged.

BOX 1 Scholes and Biggs' (2005) Biodiversity Intactness Index

The Biodiversity Intactness Index (BII) was developed with the intention of providing a scientifically sound, transparent, sensitive, disaggregable indicator of broad-sense (as opposed to taxon-specific) biodiversity that could be compared with a baseline, applied at any spatial scale and readily interpreted (Scholes and Biggs, 2005). BII is defined as 'the average abundance of a large and diverse set of organisms in a given geographical area, relative to their reference populations' (Scholes and Biggs, 2005). The inclusion of a diverse set of taxonomic groups reflects the importance for ecosystem health of a wide range of functional types and recognises that higher taxa can differ systematically in their response to human pressures (e.g. Baillie et al., 2004; Lawton et al., 1998). The choice of abundance as the basis of the index greatly facilitates reporting at any spatial scale: BII estimates the average condition of local biodiversity across the region of interest, rather than needing to upscale in order to estimate the total biodiversity within a larger region (a more challenging task, though suitable methods are becoming available, e.g. Azaele et al., 2015). Scholes and Biggs (2005) suggested that the preindustrial state of an ecosystem would be the ideal reference condition, but that this could be approximated in practice by contemporary data from minimally impacted sites—large protected areas—given the dearth of true historical baseline data.

If I_{ijk} is the population of species group i in ecosystem j under land use k, relative to a preindustrial population in the same ecosystem type, then Scholes and Biggs (2005) estimate BII according to:

$$BII = 100 \times \sum_i \sum_j \sum_k R_{ij} A_{jk} I_{ijk} / \sum_i \sum_j \sum_k R_{ij} A_{jk}$$

where R_{ij} is the species-richness of taxon i in ecosystem j and A_{jk} is the area of ecosystem j under land use k. Taxa are weighted in proportion to their species-richness, making BII a diversity-weighted index of abundance (relative to the reference state): its coverage of functional diversity depends on the range of taxa included. Scholes and Biggs (2005) used vertebrates and angiosperms but not, due to poor data, invertebrates. Alien species are explicitly excluded from the calculation. To compute BII for seven countries in sub-Saharan Africa, Scholes and Biggs (2005) compiled estimates of overall abundance for plants, mammals, birds, reptiles, and amphibians in each of five land-use classes (moderate use, degraded, cultivated, plantation, and urban) relative to the baseline (protected areas). As not enough empirical data could be marshalled, estimates came from a survey of expert opinion rather than primary biodiversity data. Scholes and Biggs (2005) allowed higher taxa to vary in their responses to land use, but implicitly treated land use as capturing all relevant human pressures (so, for example, the intensity of hunting might differ systematically among land uses, but the framework does not attempt to disentangle effects of hunting from effects of

BOX 1 Scholes and Biggs' (2005) Biodiversity Intactness Index—cont'd

land use per se). By combining their relative abundance estimates with broad-scale data on areas under each land use, they estimated BII to average 84% across seven sub-Saharan African countries, and that it had declined by 0.8% in the 1990s. Biggs et al. (2008), using the same expert estimates, projected that BII would decline much more by 2100 than it has to date under each of three scenarios of land-use change. A key strength of BII is that it can be updated using data on land-use change, which can be much easier to obtain than new biodiversity data from extensive monitoring.

Faith et al. (2008) criticised the design of BII, noting that—because it is a measure of relative abundance—it can rise even if species are lost, provided that the remaining species increase sufficiently in abundance. This point highlights that no indicator is ideal: for instance, although measures of biodiversity exist that decline whenever any one of richness, abundance, and evenness decline (Buckland et al., 2005), they tend to be less intuitive and harder to communicate than BII. Rouget et al. (2006) showed that smaller-scale data on land use, better reflecting land degradation, produced markedly lower estimates, highlighting the sensitivity of the final estimate to the land-use data used to drive spatial projections. Vackar et al. (2012) review BII alongside a range of other indicators of how biodiversity responds to human pressures.

BII has become prominent in recent years because of its inclusion within the 'Planetary Boundaries' framework (Mace et al., 2014; Steffen et al., 2015), an influential way of framing questions about the effects of human pressures on global sustainability (Rockstrom et al., 2009; Steffen et al., 2015). The framework aims to delimit a 'safe operating space' for humanity, by identifying—for each of nine components of the earth system—boundaries beyond which the system may deteriorate sharply. One of the two core boundaries proposed in the most recent framework is the integrity of the biosphere (the other being climate change: Steffen et al., 2015); as this is eroded, the biota's ability to meet humanity's needs must become compromised, though the form of the relationship is not known (Fig. 1C). BII is one of two measures proposed as indicators of biosphere integrity (Mace et al., 2014; Steffen et al., 2015), reflecting the functional diversity within local ecosystems needed to ensure they continue to provide functions and services over the short term (Fig. 1). The threshold value proposed, using the precautionary principle, is a reduction of 10% relative to the baseline condition, though there is much uncertainty about the precise placement (Fig. 1C, and see Steffen et al., 2015). The previous regional estimates of BII (Scholes and Biggs, 2005) suggest that this boundary has been exceeded, but BII had not been estimated based on primary biodiversity data.

Continued

BOX 1 Scholes and Biggs' (2005) Biodiversity Intactness Index— cont'd

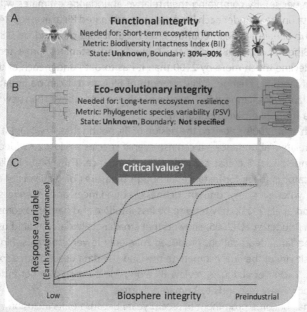

A Functional integrity
Needed for: Short-term ecosystem function
Metric: Biodiversity Intactness Index (BII)
State: **Unknown**, Boundary: **30%–90%**

B Eco-evolutionary integrity
Needed for: Long-term ecosystem resilience
Metric: Phylogenetic species variability (PSV)
State: **Unknown**, Boundary: **Not specified**

C Critical value?

Response variable (Earth system performance)

Low   Biosphere integrity   Preindustrial

Fig. 1 Planetary boundaries for biosphere integrity proposed by Mace et al. (2014) and Steffen et al. (2015). (A) BII is proposed as the most suitable indicator for the functional integrity needed for delivery of ecosystem functions and services in the short term. (B) Phylogenetic species variability (PSV) is suggested as a measure of the eco-evolutionary integrity needed for longer-term adaptation to rapidly changing environments. (C) Possible relationships between biosphere integrity and performance of the earth system; the uncertainty of this relationship means that the critical values for BII and PSV cannot be specified precisely, and that the precautionary principle should apply.

PREDICTS has implemented a version of BII that is based on primary biodiversity data rather than expert judgement (Newbold et al., 2016a) and that addresses previous criticisms (Faith et al., 2008; Rouget et al., 2006). We discuss our implementation in Box 2.

2. KEY DESIGN DECISIONS AND METHODS

Some key decisions were taken at or before the start of the project that, despite bringing their own limitations, have facilitated attainment of many of the objectives. We discuss the most important of these here and give more detail on how these decisions shaped the methods we used to collate and analyse data.

2.1 Aiming to Collate a Taxonomically and Geographically Representative, Large Database

Because species' responses to human impacts depend on their functional response traits, which are likely to often be phylogenetically patterned, responses often differ significantly among clades (McKinney, 1997; Purvis, 2008). Consequently, the biotic effects of land use inferred from a global database are likely to reflect its taxonomic biases. Global biodiversity knowledge greatly overrepresents vertebrates and underrepresents invertebrates and fungi despite their importance in driving many ecosystem functions (Dobson, 2005). Geographic bias is also evident, overrepresenting economically developed, logistically accessible regions, which tend to be temperate, and underrepresenting remote, poorer, largely tropical areas (Hortal et al., 2015; Meyer et al., 2016). Responses can vary among regions as well as among clades (e.g. De Palma et al., 2016; Gibson et al., 2011; Phillips et al., 2017b), so analyses of unrepresentative databases can yield misleading results.

Because our aim was to represent animal, plant, and fungal diversity, as we collated data we tracked cumulative numbers of species in the database belong to each of 59 higher taxa (e.g. vertebrate classes and insect orders) and compared them with estimated numbers of formally described species (Chapman, 2009). To track geographic representativeness, we compared numbers of sites in each major terrestrial biome with the biome's contribution to global net primary production. Taxa or regions that stood out as being underrepresented were then targeted in literature searches (Hudson et al., 2014) leading to reasonable representativeness (shown in Figures 2 and 4 of Hudson et al., 2017). Note that this attempt at representativeness deliberately differs from the approach used in many systematic reviews that try instead to represent the published literature. Extrapolating model results beyond the taxonomic, ecological, or geographic scope of the data is always risky (Petchey et al., 2015); the broad coverage of the PREDICTS database reduces the need for such extrapolation, while its representativeness increases the likelihood that statistical results will be relevant to animal, plant, and fungal biodiversity as a whole.

To facilitate collation of a large and taxonomically representative database, we did not require each study we included to contain a site that could be viewed as 'pristine' for the purpose of modelling (i.e. used as a baseline), in contrast to the GLOBIO model (Alkemade et al., 2009). Truly pristine areas are rare or absent over much of the world (Watson et al., 2016), particularly in the well-funded regions that are overrepresented in the conservation literature (Wilson et al., 2016), so relaxing the requirement for them also improved geographic coverage. Pristine sites in each study would make the estimation of the BII much simpler (as they would directly inform about which species were present in the original assemblage, as in GLOBIO: Alkemade et al., 2009). The less direct two-step approach that we have instead used to estimate BII is detailed in Box 2, after a discussion of issues relating to statistical modelling with the PRE-DICTS database.

BOX 2 PREDICTS' Approach to Estimating the Biodiversity Intactness Index

Scholes and Biggs (2005) defined BII as 'the average abundance of a large and diverse set of organisms in a given geographical area, relative to their reference populations' excluding species not originally present, but the paucity and patchiness of truly pristine sites precludes reliable identification of which species were present originally. Consequently, a two-step modelling approach, followed by spatial projection, is needed to estimate BII from PREDICTS data. The first step is to model how site-level total abundance varies as a function of site-level human pressure: Newbold et al. (2016a, following Newbold et al., 2015) considered land use, land-use intensity, human population density, and proximity to the nearest road, allowing the continuous variables to determine ln-transformed abundance nonlinearly and to interact with land-use class, rather than being kept separate as in, e.g., the GLOBIO model (Alkemade et al., 2009). Site-level response variables, such as total abundance, can—when the whole database is used—support models having several pressure variables; and this number will increase as the database grows. This model infers only the net effect of pressures on overall abundance, so incoming species can compensate for decline or loss of some of the originally present species. The second modelling step is therefore to estimate the compositional similarity between a site in the reference land-use class (typically, Primary vegetation) and an immediately adjacent site in each other land use (this is the 'initial similarity' of Soininen et al., 2007, and its use avoids conflating distance–decay with human impact). The most appropriate similarity measure to use is the fraction of overall abundance that is made up of species also present in the reference land-use class, which is calculated using an asymmetric form of the Jaccard index (Newbold et al., 2016a). Only studies having sites in the reference class can contribute to the compositional similarity models, and

BOX 2 PREDICTS' Approach to Estimating the Biodiversity Intactness Index—cont'd

the pairwise comparison has so far limited these models to a simpler form (i.e. land use class as the only fixed effect of anthropogenic pressure). Nonetheless, BII can be estimated and projected by combining a model of overall abundance with a model of how land use affects the fraction of that abundance that is composed of originally present species (Newbold et al., 2016a). Fig. 4 maps global BII estimated in this way, with an inset zooming in on Mount Taranaki, a near-circular protected area in New Zealand whose high BII values stand out against those of the surrounding pasture.

Use of species- rather than abundance-based response variables—site-level species richness and proportion of species shared between sites—permits estimation of analogues of BII that focus on the intactness of taxonomic diversity rather than abundance, addressing Faith et al.'s (2008) main criticism of BII. The inclusion of use intensity, human population density, and proximity to roads in the statistical models goes some way to address Rouget et al.'s (2006) criticism, as the biodiversity estimated for a patch of land depends not only on the land-use class but also other forms of degradation. However, the resulting BII estimates are still dependent on the accuracy of the land-use and other data used when making spatial projections. High-resolution land-use data are becoming available (Hoskins et al., 2016) based on statistical downscaling of coarser global datasets; and many other pressures are now available as global maps than was the case when PREDICTS began in 2012, even though many gaps remain (Joppa et al., 2016). As more pressure data become available at finer temporal resolution—ideally annually—estimation of BII at each time step is relatively straightforward conceptually, even if the sizes of the rasters being used makes it computationally demanding.

Because our implementation of BII can reflect only the effects of the pressures we model, because our models do not (yet) incorporate lagged responses, and because the sites in the reference class may still bear the imprint of human pressures, the resulting estimates of BII are likely to be optimistic (Newbold et al., 2016a). However, demonstrating this bias is difficult, given the lack of geographically representative long-term multiclade data that could be used to test the model's estimates. Such testing will become easier with the development of annual estimates of BII at fine spatial scales.

At the time of writing, the PREDICTS database contains 3,857,790 data records (species × site combinations) coming from 767 data sets, each having multiple sites (median = 18 sites) giving a total of 32,076 sites in 98 countries; over 50,000 taxa are represented. Hudson et al. (2017) shows the geographic, taxonomic, and ecological coverage of the database. Over 65% of the rows are counts (i.e. numbers of individuals of a taxon at a site), and these total 65,573,366 counted individuals. Other measures related to abundance (e.g. percent cover) make up around a further 13% of the rows with the remainder being either occurrence (i.e. presence/absence; 15% of rows) or prevalence (6.6% of rows). The first release of the PREDICTS database, making 3.2 million records freely available, was at the end of 2016 (Hudson et al., 2016, 2017).

2.2 Focus on Site-Level Diversity Data

Data collation focused on site-level biodiversity data, i.e., data (usually species abundances) collected within a sampling frame—which might be, e.g., a series of transects, a single transect, an array of sample points, or a quadrat—rather than, for instance, inventories of species within a larger patch or region. The median size of a sampling frame (as measured by its maximum linear extent: Hudson et al., 2014) is 60 m, and most are between 22 and 400 m (Hudson et al., 2016). This concentration on local biodiversity reflects the fact that the rates of delivery of many ecosystem functions and services depend on the local, rather than global, state of biodiversity (Cardinale et al., 2012; Hooper et al., 2012; Isbell et al., 2011). This focus on the average status of biodiversity across local ecosystems carries through to projections from PREDICTS' statistical models: it is easy to map the average expected values of response variables in spatial and temporal projections, without any need for upscaling. PREDICTS estimates, for instance, the average number of species lost from local assemblages, but says nothing whatsoever about numbers of species going globally or even regionally extinct, or the rates at which their global status is declining.

2.3 The Space-for-Time Gambit

Fig. 2 illustrates the key problem facing any attempt to estimate how biodiversity is changing over time across a region. Between Time 1 and Time 2, some of the natural habitat (dark shading) is converted to agricultural use, and there is also a general degradation in the quality of even the more natural habitat (i.e. it becomes paler). The symbols indicate some sets of sites that might be monitored. Some sites remain as natural habitat throughout (blue triangles), some are converted to agricultural land between Time 1 and Time 2 (orange stars), and some had been converted before Time 1 (red squares). As the graph shows, the estimated biodiversity trajectory is very different

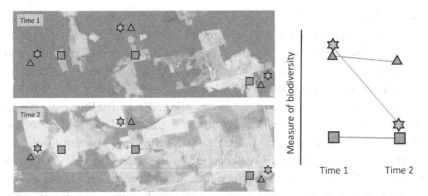

Fig. 2 Landscape shown at Time 1 and Time 2 highlighting three possible sets of monitoring sites. *Blue triangles* are natural habitat at Time 1 and remain so at Time 2. *Red squares* had been converted to a low-biodiversity use before Time 1. *Orange stars* are converted from natural habitat to low-biodiversity land use between Time 1 and Time 2. None of the sets is representative of the whole landscape, so none of the trajectories shown in the graph correctly reflects how biodiversity has been affected by land-use change. PREDICTS attempts to produce a more accurate estimate by combining a statistical model of how land-use change affects site-level diversity with information on how much land has changed use between Time 1 and Time 2.

between these three classes of site: blue triangles and red squares see no land conversion so will underestimate the true rate of biodiversity decline, whereas the orange star sites are all converted and so overestimate the rate of decline (graph in Fig. 2). For temporal data alone to correctly capture the average trajectory across the region of interest, the sites would have to be representative of the entire region, in terms of the changes in pressure that they experience. Few monitoring schemes are designed to be truly representative of the whole region they cover (Buckland et al., 2005, who also highlight some exceptions), and global compilations of data are unlikely to even closely approach representativeness (Gonzalez et al., 2016; Vellend et al., 2017).

PREDICTS has so far taken a different approach to estimate the rate of decline. Rather than collate and analyse time-series data (as done by, e.g. Dornelas et al., 2014, Vellend et al., 2013), we have targeted spatial comparisons between sites with different land-use and related pressures (Control-Impact designs: De Palma et al., 2018, Chapter 7, this volume). These data then form the basis for statistical estimation of how site-level biodiversity is influenced by pressures, under the assumption of 'space-for-time substitution', i.e., that the spatial comparison between sites in different land uses will detect the same biodiversity difference as would ensue if a site changed from one land use to the other. Thus, in Fig. 2, if the sites were surveyed at Time 1, the blue triangles and orange stars might be classed as primary vegetation and the red squares as cropland; other pressures for which PREDICTS has so far scored each site include land-use intensity (on a coarse scale described by Hudson et al., 2014), human population density (CIESIN et al., 2011), proximity to the nearest road (e.g. from gROADS: CIESIN et al., 2013), and whether a site is inside or outside a protected area (IUCN and UNEP, 2014). The statistical model estimates the change in site-level biodiversity entailed by the land-use change, as well as the effects of each other pressure, and can be used to test the significance of putative pressures. Then, combining the model coefficients with maps of the significant pressures for Time 1 allows estimation of the average level of the response variable across the landscape at that time; likewise, such estimates can be produced for any time for which maps of pressure data, estimates or projections are available. PREDICTS is therefore conceptually much more similar to the GLOBIO model (Alkemade et al., 2009) than to approaches such as the Living Planet Index (Collen et al., 2009), Red List Index (Butchart et al., 2007), or Bio-Time (Dornelas et al., 2014).

PREDICTS' explicit linking of biodiversity to pressures facilitates hindcasting and future projections (as in Newbold et al., 2015)—an advantage over approaches that synthesise observed trends in possibly unrepresentative data without a pressure-state model (Vackar et al., 2012). However, this model-based approach also makes major assumptions (discussed further by De Palma et al., 2018, Chapter 7, this volume). Notably, the assumption of space-for-time substitution means that the biodiversity differences among sites within the same survey are assumed to be attributable to differences in land use and the other pressures in the model. Diffuse pressures acting across the whole landscape are implicitly assumed to be unimportant in shaping present and future diversity (Vackar et al., 2012): a statistical model based on sites at Time 1 could not inform about the effects of, for instance, climatic change between Time 1 and Time 2. These assumptions are more reasonable when the spatial comparisons can be made over sufficiently short distances that site-level assemblages are drawn from the same source pool, with the result that PREDICTS is best able to model effects of pressures that have a fine spatial grain. It is therefore much better suited to quantifying effects of land use, fragmentation, roads, and human population than coarse-grained pressures such as atmospheric nitrogen deposition and climate change. (The GLOBIO model, which includes both fine- and coarse-grained pressures, does so by using different modelling approaches for different pressures and combining the results: Alkemade et al., 2009). Nonetheless, even when considering fine-grained pressures, there are several reasons why the space-for-time assumption is likely to underestimate the effects of land-use on biodiversity, these are discussed alongside the relevant findings in Section 4. A further assumption of the model-based approach is that the pressure data used to make spatial and temporal projections are consistent with those used in modelling; a later subsection outlines how our choice of land-use classes was made with this point in mind (see Section 2.5).

2.4 Collating Raw Data Rather Than Results-Based Meta-Analysis

No single measure can ever capture all of biodiversity's dimensions—the numbers, evenness, and difference of genes, species, and ecosystems at local, regional, and global scales—or all the ways that anthropogenic pressures have affected biodiversity (McGill et al., 2015; Purvis and Hector, 2000). Most of the many thousands of studies that have compared biodiversity at multiple sites report only one or a few of the possible measures (Naeem et al., 2016). As a result, meta-analysis of such studies—i.e., statistical combination of effects sizes across studies (Koricheva and Gurevitch, 2014)— have usually been constrained to analyse only the most-commonly reported measures, such as species richness (e.g. Gibson et al., 2011; Murphy and Romanuk, 2014; Tuck et al., 2014). This constraint is unfortunate given these measures' limitations as indicators of biodiversity change (Buckland et al., 2005; Dornelas et al., 2014; Hillebrand et al., 2017; Schipper et al., 2016). Comparisons among different response variables are sometimes possible (e.g. Gibson et al., 2011), but it is not possible to estimate the effect size for measures other than those presented in the source.

As Fig. 3 illustrates, PREDICTS has taken a different approach, collating the species-by-site abundance or occurrence matrices from each source study, meaning that many different taxon-based measures of site-level diversity (α_t) and among-site spatial turnover (β_t) can be estimated and used in statistical modelling, greatly expanding the range of uses to which the collated data can be put. Because the taxon names from each source have been recorded and curated (Hudson et al., 2014), it is also possible to calculate measures of α_t and β_t diversity for taxonomic subsets (e.g. bees: De Palma et al., 2016). There has been an increasing interest in recent years in characterising species in terms of their ecological characteristics or functional traits, as an alternative to considering just species occurrence and abundance (McGill et al., 2006). Relevant ecological characteristics are those that determine how an organism interacts with other organisms and with its environment. Such characteristics can relate both to how species

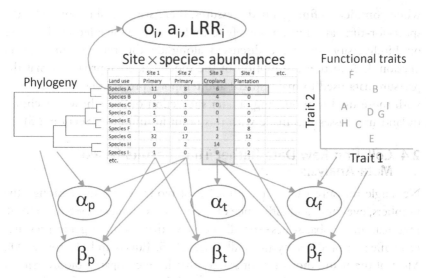

Fig. 3 Rather than collate statistical results from published studies, PREDICTS has collated the underlying species × site matrices of abundances (*central table*) or, less commonly, occurrences (i.e. presence–absence data) allowing estimation of a much wider range of response variables. Species' site-level data (*shaded row* of data table) can be used to fit models of how species i's occurrence (o_i) and/or abundance (a_i) are affected by land use and other site-level pressures, or of factors determining species' log response ratios (LRR_i). Each site's species-level data (*shaded column* of data table) can used to estimate a range of taxon-based measures of α diversity (α_t), such as within-sample richness, rarefaction-based richness, overall abundance or evenness (reviewed by Magurran, 2004), while comparisons among sites yield a range of taxonomic measures of spatial β diversity (β_t). Combining the site × species data with a phylogeny yields phylogenetic measures of diversity at a site (α_p) or among sites (β_p) (reviewed by Tucker et al., 2017); whereas combination with species-level functional trait data yields analogous measures of functional diversity, α_f and β_p (reviewed by Schleuter et al., 2010) as well as community-weighted mean trait values.

respond to environmental changes ('functional response traits', e.g. fecundity; not be confused with functional responses in the sense of Holling, 1959) and how species contribute to ecosystem functions ('functional effect traits', e.g. pollination efficiency) (Hooper et al., 2005). Often the same traits are both response and effect traits (e.g. position on the leaf economic spectrum: Suding et al., 2008). Combining species-level functional trait data with the species × site abundance data allows estimation of trait-based diversity at and among sites (α_f and β_f, respectively); while instead combining the biodiversity data with a phylogeny likewise allows estimation of phylogeny-based diversity measures (α_p and β_p), which are increasingly being used in community ecology and conservation research (Tucker et al., 2017). It is therefore possible to, for instance, model taxonomic, functional, and phylogenetic diversity in parallel (De Palma et al., 2017); and such parallel analyses can give more insight into the underpinning mechanisms than considering any single diversity measure (Naeem et al., 2016; Phillips et al., 2017a).

The availability of the raw species-by-site data also permits analyses of species-level (as opposed to site-level) response variables, modelling species' occurrence, and abundance (o_i and a_i in Fig. 3) as a function of, e.g., environmental and trait variables (e.g. De Palma et al., 2015; Newbold et al., 2014).

2.5 Land-Use Classes Chosen to Facilitate Global Spatial and Temporal Projections

Given our limited resources, we had to choose a single classification of land uses, out of the many available, into which we could score each site having biodiversity data. The objective of making global spatial and temporal projections under alternative scenarios led us to Hurtt et al.'s (2011) harmonised land-use data, which mapped fractional coverage of primary forest, primary nonforest, secondary vegetation, cropland, pasture, and urban within 0.5 degree grid cells, from 1500 to 2005 (based on Goldewijk et al., 2011) and from 2005 to 2100 under each of the four Representative Concentration Pathways embodying different approaches for climate mitigation (van Vuuren et al., 2011). Because the Hurtt et al. (2011) classes neglected some distinctions likely to be important for site-level biodiversity, we therefore split secondary vegetation into three successional classes (young, intermediate, and mature) and added a Plantation Forest category (Hudson et al., 2014); we further split each class into three use intensities (minimal, light, and intense). The resulting classification allowed us to test whether the extra classes and use intensities were sufficiently important in explanatory models to try to include in projections. Using the Hurtt et al. (2011) classes as a basis greatly facilitated projections, notwithstanding the complications our additions caused. Choosing a single land-use scheme greatly sped up the scoring of sites, but has the negative consequence that switching to a different scheme would be very time-consuming. For example, the harmonised land-use data derived from the Shared Socioeconomic Pathways (O'Neill et al., 2014) split agricultural land more finely than Hurtt et al. (2011), and the wholesale curation of all of PREDICTS' agricultural sites into the new classification took over 6 person-months.

3. MODELLING CONSIDERATIONS

3.1 General Challenges

The main challenge in analysing the PREDICTS data is its hierarchical structure. Data in the PREDICTS database have come from very many source publications, and each sampled different taxonomic groups, using different methods and levels of sampling effort, in sites of different sizes, and in different parts of the world. Each of these sources of variability affects the sampled measures of biodiversity. For example, one would expect many more species to be reported from tropical forest plots comprehensively assessed for plant species than from sites in a survey of bat species in the United Kingdom. The heterogeneity in the underlying data means that one cannot compare directly the biodiversity values from different datasets in the PREDICTS database. Furthermore, the variability caused by sampling methods, taxonomy, and geography is essentially unrelated to the main effects of interest, namely, the effects of land use and other human pressures on biodiversity.

The challenges of data structure can be overcome by using hierarchically structured statistical models. All of the analyses of PREDICTS data to date have used generalised linear mixed-effects models (GLMMs: Bolker et al., 2009), usually implemented through the lme4 package in R (e.g. Bates et al., 2014). In a GLMM, the identity of the study from which the data originally came can be fitted as a random intercept, which accounts for the massive variability in sampled biodiversity across studies (Newbold et al., 2015). Models that include random slopes—which additionally allow the relationship between biodiversity and pressures to vary among studies—have also been considered when possible (e.g. De Palma et al., 2017; Echeverría-Londoño et al., 2016) but not all data sets support such complexity (Bates et al., 2015). Robustness of coefficient estimates can be tested by cross-validation; e.g., Newbold et al. (2015) refitted their models omitting each biome's set of studies in turn (a better approach for hierarchical data than random cross-validation: Roberts et al., 2017).

Even within studies, sites often show nonrandom spatial structure. For example, each land use will often have a clumped distribution within the landscape; and sampled sites will often be arranged in distinct spatial blocks (Hudson et al., 2014). Such spatial structure creates a problem for any analysis if there is some environmental gradient that causes variation in biodiversity among the blocks independent of land-use effects. It can be accounted for by fitting a random intercept of spatial block identity, nested within a random intercept of study identity (Newbold et al., 2015).

Another kind of variability among the studies in the database is the level of detail in the metadata. Many studies do not report the spatial extent of the area sampled, for example, or the size of the patch of habitat within which the site is located, markedly reducing the sample size for some analyses (e.g. Phillips et al., 2017a). Similarly, studies vary widely in the geographic precision of site coordinates and the temporal precision in sampling dates, making it harder to match diversity data to remotely sensed data from the same time and space. Studies also differ in how much detail they provide about land use and intensity at each site. Given this, we developed detailed classification criteria (given in full in Table S1 of Hudson et al., 2014); a formal repeatability analysis indicated substantial interrater agreement in land-use class and fair to moderate agreement in use intensity (Hudson et al., 2014).

3.2 Modelling Species-Level Response Variables: o_i, a_i, and LRR_i

The most fundamental measures of biodiversity are based on whether or not species occur in a location, and how abundant they are if they do. However, species abundance data present several difficulties for analysis. They typically have a positive mean–variance relationship and therefore a nonnormal error distribution, but the variance is commonly higher than would be expected from a Poisson error structure (i.e. overdispersion) and there are often more zeros than expected even from quasi-Poisson or negative binomial distributions (i.e. zero inflation) (Zuur et al., 2009). The zeros are likely to reflect both real differences in the presence of a species (caused by either ecological exclusion or chance) and imperfect detection ('false zeros'—Martin et al., 2005), which can be accommodated in zero-inflated models. Furthermore, not all of the abundance data in the PREDICTS database are integers, as many datasets report densities or relative abundance rather than absolute counts; this currently precludes the use of discrete and zero-inflated error structures (e.g. zero-inflated quasi-Poisson) in a number of modelling frameworks (e.g. lme4 in R).

Faced with these difficulties, we have used a two-part modelling framework (De Palma et al., 2015; Newbold et al., 2014). Firstly, the presence or absence of species is modelled as a binary outcome, using a binomial distribution. Then, the ln-transformed abundances of the present species are modelled using a Gaussian error distribution; the noninteger nature of many of the abundances preclude the use of Poisson errors, which would otherwise be preferable (O'Hara and Kotze, 2010). This two-step approach is similar to fitting a hurdle model (Loeys et al., 2012; Zuur et al., 2009), except that the nonzero data are taken from a continuous distribution rather than a truncated discrete distribution.

A second possible approach to modelling species' abundance data in PREDICTS is to model the sensitivity of each species separately to given land use transitions, e.g., natural/semi-natural vegetation vs human-dominated land uses. As a given species rarely appears in large numbers of studies in the PREDICTS database, it is more appropriate to treat study identity as a fixed rather than random effect (Fox et al., 2015). In this case, generalised linear (rather than mixed-effects) models can be used, where Poisson and quasi-Poisson distributions can be applied to noninteger data (Zuur et al., 2009). Such a model estimates the log-response ratio (Hedges et al., 1999) reflecting a species' sensitivity to the particular land-use transition. With data such as these, a different problem with zeros can arise: complete separation. This occurs when there are species that have nonzero abundances in only one of the land-use categories (e.g. they are sometimes found in natural/semi-natural vegetation but never in the human–dominated land uses). In generalised linear models, such complete separation results in extreme and uncertain estimates. A commonly used approach is to add a small nonzero value to the zeros, but this can produce anticonservative standard errors and overly conservative estimates in GLMMs. Bayesian generalised linear models with weakly informative priors instead provide more realistic estimates with wider standard errors (bayesglm function, arm package: Gelman et al., 2008).

3.3 Modelling Site-Level Response Variables: α_t, α_p, and α_f

Species abundance data at a site can be summarised by various measures of overall species richness, by the total abundance of all organisms, and by diversity indices that combine the occurrence of species and their relative abundances (Magurran, 2004). The simplest measure of richness, within-sample species richness, is a count and so is best modelled with a suitable discrete error distribution (O'Hara and Kotze, 2010); rarefaction-based estimates of species richness can also be modelled as counts after rounding (e.g. Newbold et al., 2015). Mixed-effects models of species richness from the PREDICTS database often show overdispersion, which can be accommodated within lme4 using a site-level random effect (Harrison, 2014). Overall abundance is not always an integer, because some of our datasets neither present counts nor give relative abundances in a way that can be converted back to counts, so we have modelled overall abundance with Gaussian errors after ln-transformation (e.g. Newbold et al., 2015).

Not all of the datasets within the PREDICTS database were intending to sample assemblage diversity; some focused on abundance or occurrence of a predetermined list of species, with 18 studies considering only a single species (Hudson et al., 2017). We have excluded single-species studies, and sometimes (depending on the research question) the 'closed-list' studies, from our models of site-level diversity (De Palma et al., 2016; Echeverría-Londoño et al., 2016; Gray et al., 2016; Jung et al., 2017; Newbold et al., 2015; Phillips et al., 2017b).

For site-level measures using functional trait data, the information on traits must be aggregated in some way to derive some overall measure of the trait structure of an assemblage of species. The simplest measure is the mean value of a trait across all species in an assemblage, weighted where possible by species relative abundances, i.e., community-aggregated trait values (Shipley et al., 2006) or community-weighted mean trait values (Violle et al., 2007). In addition to measures based on single traits at a time, there are also many different measures of functional diversity (α_f) that aggregate information on several traits of the species in an assemblage (Schleuter et al., 2010). Several PREDICTS analyses have modelled either community-weighted mean trait values (Gray et al., 2016; Newbold et al., 2015; Phillips et al., 2017b) or community-level measures of α_f (De Palma et al., 2017); this last paper is also the first to use PREDICTS to model a community-level measure of phylogenetic diversity (α_p).

PREDICTS aims to produce statistical models that improve understanding of how human pressures affect ecological assemblages, but also aims to use models for spatial and temporal projections of expected values of response variables given spatial and temporal variation in human pressures. These two objectives can conflict. Maximally explanatory models are likely to contain pressure variables that, although important in shaping site-scale diversity, are not available at a larger scale or for a different time. Models used in projections therefore need to include only those pressures for which data are available for the requisite area and time. This consideration led us at the outset of PREDICTS to develop a land-use classification (Hudson et al., 2014) based on that used in Hurtt et al.'s (2011) harmonised data for historical and projected future land-use changes from 1500 to 2100.

3.4 Modelling Among-Site Response Variables: β_t, β_p, and β_f

Measures of α diversity can fail to detect important changes in ecological assemblages (Dornelas et al., 2014; Hillebrand et al., 2017; Schipper et al., 2016). For example, anthropogenic and natural ecosystems might have similar numbers of species because losses of native species can be offset by the arrival of newcomers (Banks-Leite et al., 2012; Sobral et al., 2016). Analyses of how disturbance changes species composition may therefore provide additional insight.

Analyses of β diversity must be able to separate anthropogenic from natural turnover; otherwise, the component of β diversity attributable to natural turnover inflates estimates of human impact (Scholes et al., 2010). PREDICTS analyses have taken two approaches to this problem. One (e.g. Echeverría-Londoño et al., 2016; Newbold et al., 2015) is to rescale estimates of compositional similarity relative to the average similarity between sites in the reference condition (e.g. primary vegetation). The other, useable with larger data sets (Newbold et al., 2016a,b), is to model compositional similarity as a function of geographic and/or environmental distance (to capture the distance–decay relationship: Soininen et al., 2007), as well as land use; the distance–decay relationship among sites in the reference condition (e.g. primary vegetation) then provides a baseline against which additional effects of human pressures can be seen.

Compositional similarity can be calculated (using any of several measures: e.g. Newbold et al., 2016b) between every pair of sites within each study, but the estimates are highly nonindependent. Pseudoreplication can be avoided by statistically analysing only a subset of the comparisons, chosen randomly with the constraint that no site is used in more than one comparison; this selection process can be repeated many times and the model results averaged across the selections (see Newbold et al., 2016a,b for details). However, this random selection process is inefficient when the focus is estimating the compositional similarity between other land uses and the reference condition, which can causes sample-size problems when analysing smaller data sets (e.g. within clades or biomes), especially if some land uses (notably Urban) are seldom sampled in the same studies as the reference land use (e.g. minimally used Primary vegetation). We are currently evaluating an alternative method for modelling compositional similarity—multiple regression on distance matrices (MRM) (Lichstein, 2007)—which utilises all pairwise comparisons between sites within studies. MRM is an extension of partial Mantel analysis; the test involves performing a multiple regression of a response matrix on any number of explanatory matrices (Lichstein, 2007) (e.g. site × site matrices for environmental and geographic distance). The test can accommodate different types of data (e.g. continuous, categorical, count data) and different types of relationships between variables, such as nonlinear or nonparametric (Lichstein, 2007).

Distance–decay relationships are commonly modelled after logarithmic transformation of both variables, allowing estimation of initial similarity (i.e. the level of similarity when ln-distance is zero) as well as the decay exponent (i.e. the slope of the log–log regression) (Soininen et al., 2007), and we have also used this approach (Newbold et al., 2016a,b) as well as others (Echeverría-Londoño et al., 2016; Phillips et al., 2017b). However, most similarity indices range only from 0 to 1 (minus infinity to zero once ln-transformed), meaning that this modelling approach—which does not constrain estimates of initial similarity to lie within this range—risks predicting values that are not meaningful (Millar et al., 2011). Such problems can be avoided by modelling compositional similarity using binomial errors with a logit link (Warton and Hui, 2011). We have found that this last approach, especially when comparisons are weighted by logarithm of the total abundance in the reference site, yields much better model diagnostics than our original approach when applied to smaller data sets, though both perform reasonably with our larger data sets.

The choice among measures of compositional similarity is shaped by the purpose of the analysis. Measures are available that reflect proportions of species in common, proportions of individuals in shared species, or similarity in species' relative abundances (Hillebrand et al., 2017; Magurran, 2004). They can also be symmetric or asymmetric; asymmetric measures are particularly useful when the analytical focus is the fraction of site-level diversity that was also present in the reference land use, as is the case when estimating the BII (see Box 2 and Newbold et al., 2016a).

Although methods have been developed to measure turnover in the functional trait composition of ecological assemblages between locations, β_f (de Bello et al., 2009), these have not so far been applied to PREDICTS data.

4. SUMMARY OF FINDINGS

By applying the PREDICTS data in the different types of model described in the previous section, a number of different results and patterns have emerged. Here, we summarise the main findings of these studies and attempt to synthesise some general patterns.

4.1 Local (α) Diversity

Most studies using the PREDICTS data so far have focused either on the local occurrence or abundance of species, or on site-level species diversity or total organism abundance. These studies have shown, more or less consistently, that species diversity, and to a slightly lesser extent abundance, are lower on average in land uses dominated by human activities: arable croplands, pastures (livestock grazing areas), wood plantations (e.g. fruit trees, coffee plantations, oil-palm plantations), and human settlements (De Palma et al., 2016; Echeverría-Londoño et al., 2016; Gray et al., 2016; Jung et al., 2017; Newbold et al., 2013, 2014, 2015; Phillips et al., 2017b). This result is not surprising, and is consistent with other more regionally or taxonomically focused analyses (e.g. Gerstner et al., 2014; Gibson et al., 2011), but the PREDICTS data have enabled the effect to be quantified for the first time globally and for a broad set of taxonomic groups.

What is perhaps surprising is the relatively small magnitude of the effect. Even the most heavily disturbed classes of habitat are estimated to support, on average, around half as many species and half as many individual organisms present in natural habitat (Newbold et al., 2015). There are several reasons why analysing spatial comparisons of biodiversity (such as those in the PREDICTS database) will lead to conservative estimates of the effects of land use. First, even the most natural habitats sampled are not completely free of present or past human influence (Watson et al., 2016): these habitats can, according to the criteria used for classifying land use, experience low levels of human disturbance such as hunting of particular species (Hudson et al., 2014); primary habitat can exist as small patches surrounded by anthropogenic land uses; and some species that would have occurred historically within natural habitat have disappeared as a result of past human influences (Balmford, 1996), such as the extirpation of megafauna in Europe and North America. Second, a species may be sampled within nonnatural habitats even if it cannot form a self-sustaining population there (i.e. such populations might constitute 'sink' populations reliant on dispersal from nearby natural habitat: Gilroy and Edwards, 2017; Pulliam, 1988), making the overall condition of the landscape more important and muting the contrast among land uses. A related issue is that sampling in landscapes where the land-use change has been recent may not capture the full biotic impacts, because of lagged responses (reviewed by De Palma et al., 2018, Chapter 7, this volume; Essl et al., 2015). Lastly, when researchers study sites of very different vegetational complexity (e.g. primary forest vs pasture), the surveys typically use methods geared towards the simpler habitat (Echeverría-Londoño et al., 2016): canopy surveys are not done in meadows. Despite these issues, it is clear that even habitats intensively modified by humans can often support a level of biodiversity that is substantial relative to the level in natural habitats.

The same type of land use does not have the same effects on biodiversity in all settings. Some of this variability is captured by measuring not only the broad class of land use (e.g. cropland, pasture) but also land-use intensity. At a global scale, areas used more intensively by humans generally have lower species diversity and abundance than less intensively used areas (Newbold et al., 2015). At a much smaller scale, Jung et al. (2017) showed that low-intensity agricultural systems in East Africa often had higher biodiversity than average for agricultural areas in Africa, probably because these systems were characterised by a higher level of forest and vegetation cover than average. Variation in biodiversity among intensity levels is particularly wide for human settlements, which range from rural settlements or extensive green spaces to city centres (Newbold et al., 2015). Biodiversity levels have also been shown to vary markedly among different types of wood plantation, at least in tropical forests, with oil-palm plantations and to a lesser extent cocoa plantations being shown to have much lower biodiversity than other plantation types such as coffee or timber plantations (Phillips et al., 2017b). Different facets of biodiversity can respond differently to land-use pressures leading to spatial mismatches: for example, bee assemblages in urban areas in Europe retain relatively high α_t but relatively low α_p and α_f (i.e. local taxonomic remains high but phylogenetic and functional diversity do not: De Palma et al., 2017).

Wider aspects of the landscape may also play a role in shaping the local communities. For example, Phillips et al. (2017a) showed that site-level diversity was higher within fragments of a greater area. By analysing multiple response variables, they show that this area-scaling is driven by larger fragments supporting a higher density of individuals. The relationship between fragment area and diversity was consistent across land uses, surprising given that larger areas of human–dominated land uses often have lower diversity (Fahrig et al., 2015).

Although global analyses of the PREDICTS data have shown that responses of local biodiversity to land use do not vary substantially among broad biogeographical regions (i.e. biomes) or among major taxonomic groups (vertebrates, invertebrates, and plants) (Newbold et al., 2015), more focused analyses have found some differences. For example, Phillips et al. (2017b) showed that biodiversity in tropical forests responds differently to land use among continents (Central America, South America, Africa, and Asia), probably because the proportion of different land use types varies between continents. Similarly, De Palma et al. (2016) showed that responses of bee species to land use varied markedly across geographical regions, and that in this case the total abundance of bees was not always lower in human-dominated land uses than in natural habitat. Moving to taxonomic differences, tropical forest invertebrates, reptiles, amphibians, and mammals each respond differently to different types of land use (Newbold et al., 2014); and bumblebees respond differently from other bees (De Palma et al., 2016). Analyses not using PREDICTS data have also suggested taxonomic differences in responses of local biodiversity to land use in tropical forests (e.g. Gibson et al., 2011; Lawton et al., 1998). More systematic analyses of global taxonomic differences remain to be done.

4.2 Spatial Turnover in Assemblage Composition (β Diversity)

Small net differences in overall diversity can mask considerable turnover in the composition of ecological assemblages. Indeed, analysis of the PREDICTS data has shown that anthropogenic land uses have a significantly different species composition to natural habitats (Echeverría-Londoño et al., 2016; Newbold et al., 2015, 2016b; Phillips et al., 2017b). Even vegetation recovering after a previous disturbance—which may have similar numbers of species and organisms to natural habitat (Newbold et al., 2015)—contains a distinct set of species (Echeverría-Londoño et al., 2016; Newbold et al., 2015, 2016b; Phillips et al., 2017b). At least in the tropics, compositional differences are driven purely by turnover, and are not just caused by the lower diversity in anthropogenic habitats compared with natural habitat (Newbold et al., 2016b). Previous smaller-scale analyses of the effects of land use on beta diversity have produced mixed results (e.g. Gabriel et al., 2006; Mayor et al., 2015; Norfolk et al., 2015; Tylianakis et al., 2005, 2006; Vellend et al., 2007). Nevertheless, the global results using the PREDICTS data are reasonably consistent with analyses of collated time-series data that suggest little or no net change in α diversity (Newbold et al., 2015 infers a slight decline on average, whereas Dornelas et al., 2014 and Vellend et al., 2013 infer no net change) but substantial turnover in assemblage composition.

4.3 Ecological Characteristics and Land Use

Although there is much scope for more work in this area, a few analyses have already combined the PREDICTS data with information on species' traits to investigate how these characteristics influence responses to land use, and what this might mean for the structure of ecological assemblages. Results so far show that effects of land use are generally strongly nonrandom with respect to ecological characteristics. At the level of whole assemblages, human-dominated land uses are characterised by a reduced average height of plants compared with natural habitats, but seem to differ little in terms of the average body mass of animals (Newbold et al., 2015). The decrease in average plant height in nonnatural land uses is consistent with studies at more local scale (Mayfield et al., 2013). In tropical forests, assemblages in human-dominated land uses are composed of species with a larger average range size than assemblages in natural habitats (Phillips et al., 2017b), as has been shown before in studies at smaller spatial scales (Cleary and Mooers, 2006; Phalan et al., 2011).

Analyses of species-level responses have revealed some more detailed patterns. For example, among tropical forest bird species (a group with particularly good data on ecological characteristics), long-lived, large-sized, nonmigratory, habitat specialist species respond more to land use than do other species (Newbold et al., 2013). Across both birds and mammals in tropical forests, narrow-ranged habitat specialists respond more to human land use and reduced forest cover than do widespread habitat generalists (Newbold et al., 2014). These results for vertebrates are consistent with other regional studies (Tscharntke et al., 2008; Vetter et al., 2011). Among bees, solitary, univoltine, nest-excavating (as opposed to using existing cavities) species with short flying seasons and more specialised diets are relatively less likely than other species to be present in human-dominated land uses than in natural habitat (De Palma et al., 2015).

4.4 Policy-Relevant Results

There have been three main ways in which the PREDICTS data have been applied to answer questions of direct relevance to policy: to assess the current global status of local terrestrial biodiversity; to predict potential futures for biodiversity under different scenarios of changes in land use; and to assess whether protected areas help to maintain local biodiversity.

Understanding the current status of biodiversity is important in order to monitor progress towards meeting international biodiversity targets

(Tittensor et al., 2014). The BII (see Boxes 1 and 2) has been proposed recently as a useful indicator, serving as a proxy for the functional diversity on which important ecosystem processes depend (Mace et al., 2014; Scholes and Biggs, 2005; Steffen et al., 2015), and is now recognised by the Intergovernmental Science-Policy Platform on Biodiversity and Ecosystem Services (IPBES, 2017) and Group on Earth Observations Biodiversity Observation Network (GEO BON, 2015), among others. It has been suggested that a 10% loss of BII risks crossing a 'planetary boundary' for biodiversity, beyond which the ability of biodiversity to support key ecosystem functions becomes increasingly uncertain (Steffen et al., 2015). Applying models of land-use impacts on biodiversity based on the PREDICTS data suggested that loss of BII has exceeded 10% across nearly two-thirds of the Earth's land surface (Newbold et al., 2016a, Fig. 4), even allowing for the likely bias (discussed earlier) towards underestimating human impacts. Whether a single planetary boundary for biodiversity exists, and if so whether it is possible to quantify, remains contentious (Brook et al., 2013; Mace et al., 2014). Nevertheless, such widespread losses of biodiversity are likely to imply a reduced resilience of ecosystem functions to further environmental changes (Oliver, 2016).

Hindcasting of how average local species richness has changed since 1500 provided a 'hockey stick' graph, with most of the loss to date having taken place in the 20th century (Newbold et al., 2015). Projections of how biodiversity might change under different scenarios of human socio-economic development are also very relevant for policy discussions, providing an opportunity to identify pathways that are more beneficial for biodiversity. Global projections of the PREDICTS models under the four Representative Concentrations Pathways (RCP) scenarios (van Vuuren et al., 2011) showed clear differences in biodiversity impacts (Newbold et al., 2015). The scenario with the highest level of projected climate change, as well as rapid human population growth and land-use change, led to the most negative impacts on biodiversity (Newbold et al., 2015). Interestingly though, the scenario with the lowest predicted level of climate change was associated with the second-most negative impacts on biodiversity, probably because this scenario assumed that climate mitigation would be achieved largely through the expansion of biofuels, which support relatively low levels of biodiversity (Newbold et al., 2015). The same ordering of the RCP scenarios in terms of their biodiversity impacts was seen in a national-scale study for Colombia, although the magnitudes of the differences among scenarios varied by subnational region (Echeverría-Londoño et al., 2016).

Fig. 4 Map showing the Biodiversity Intactness Index as estimated by Newbold et al. (2016a) using high-resolution land use estimates from Hoskins et al. (2016). The *inset* highlights the region surrounding Mount Taranaki in New Zealand, a protected area whose circular outline can clearly be seen.

BII value

<60%
60%–70%
70%–75%
75%–80%
80%–85%
85%–90%
90%–95%
95%–97.5%
97.5%–100%
>100%

The RCP scenarios are mostly concerned with modelling climate change, and as such do not explore policy options directed at reducing biodiversity loss. In contrast, the Rio + 20 scenarios, which were used in the Convention on Biological Diversity's *Global Biodiversity Outlook 4* report (Secretariat of the Convention on Biological Diversity, 2014), were designed to explore policy options targeted at biodiversity conservation. Projections of the PREDICTS models onto these scenarios suggested that policies targeted at reducing biodiversity loss would achieve their desired goal, and that the scenario assuming improved agricultural technology, and thus more efficient agriculture, would lead to greater reductions in local biodiversity loss than the scenario assuming human lifestyle changes, such as dietary shifts (Leadley et al., 2014).

Protected areas have been a key part of biodiversity conservation efforts in the past decades. The target to conserve 17% of global terrestrial area with protected areas is one of the few current international biodiversity targets that is likely to be met by 2020 (Tittensor et al., 2014). However, whether protected areas are effective at conserving biodiversity remains uncertain, with previous studies producing contradictory results (Geldmann et al., 2013). A study of the effectiveness of protected areas at conserving local biodiversity, based on PREDICTS data, showed that protected areas contain on average higher local species richness and abundance, and a greater representation of narrow-ranged species (Gray et al., 2016). This is consistent with previous global-scale analyses (Coetzee et al., 2014; Geldmann et al., 2013). An important advance made by this analysis, though, was to consider the interaction between land use and protection status. Many protected areas contain land that is used by humans, for agriculture or settlement. The PREDICTS analysis showed for the first time globally that human-used areas within protected areas have higher levels of biodiversity than comparable areas outside protected areas—in other words, the benefits of protected areas do not arise only by reducing the rate of land-use conversion (Gray et al., 2016).

5. SYNTHESIS AND PROSPECTS

PREDICTS has demonstrated the value and power of collating even a small fraction of the 'long tail' of mostly 'dark' (i.e. not directly accessible) data (Heidorn, 2008) into a reasonably representative global database of terrestrial biodiversity responding to land use and related pressures (Box 3). Although most of the constituent data sets are small, together they present

BOX 3 Tips from a Large Project

PREDICTS became much bigger than originally envisaged and budgeted for: we did not anticipate how many researchers would be willing to share their unpublished data files, and we expanded the research team (especially through over 50 project students). Here, we briefly discuss some practical steps that helped the project to become so large and to overcome some of the problems that its large size caused, as they may be helpful for others involved with or planning large projects.

Incentives for data contributors. Around one-third of the researchers we approached shared their data with us, even though (from autumn 2012) our request included that we would make the data publicly available. A likely reason for this unexpectedly high response was our offer of coauthorship on an open-access paper to make the collated database available, and on which they would have the opportunity to comment (Hudson et al., 2017); and of collaboration on any papers where their data made up a substantial fraction of the dataset being analysed. Some researchers explained their reasons for sharing their data: as well as authorship, these included wishing to have the data publicly available, the endorsement GEO BON had given to PREDICTS, having finished their own analyses, and our research questions being complementary to theirs rather than competing. Up until the publication of the database, we also kept our contacts updated through a quarterly newsletter.

Using a customer relationship management (CRM) system. The volume of correspondence—over 1800 contacts relating to over 1600 source papers—would not have been tractable without CRM software. We used Insightly (https://www.insightly.com), which let us track, for each potential source paper, who had contacted whom, whether data were shared, permissions for the use and public sharing of shared data, whether any restrictions were placed on its use, whether any information needed to be withheld (e.g. locations of sites containing particular rare species), and whether data had already been uploaded into the database. Insightly's API allowed some integration with the database, and its user-defined fields gave flexibility. Even so, the number of manual steps means that, despite much checking and curation, some mistakes were inevitable and some contacts were understandably unhappy. To overcome this problem would have required a mechanism for automatically forwarding to the system all the correspondence with our contacts, and a more detailed handover meeting with team members at the end of their involvement with PREDICTS to ensure that ongoing correspondence was picked up.

Inflexible data-capture tools. Even for a conceptually simple data structure (site × species matrix of occurrences or abundances), researchers can store the data in many different ways. We used a rigidly structured Excel spreadsheet to capture the site- and species-level data, which worked well to enforce a degree

Continued

BOX 3 Tips from a Large Project—cont'd

of comparability among datasets and allowed some validation before uploading to the database.

Well-maintained training documents. Up-to-date training documents covering all aspects of PREDICTS' procedures were shared online (PREDICTS, 2017), and the key sections kept under open review. Google Groups (more recently, Slack) allowed team members to ask and answer questions efficiently and see answers to previous questions, further helping consistency.

Successive improvements to data structures and definitions. Shortcomings of our original data structures soon came to light. In particular, we realised the importance of collecting additional information about our sites (e.g. the physical size of the sampling unit), and a formal assessment of repeatability of assigning sites to land uses (described in Hudson et al., 2014) highlighted ambiguity in some of the then-current definitions. Because our project team had the necessary skills, we were able to develop new versions of our data capture tools, change our data ingestion and validation procedures, and recurate already-processed data sets as required.

One final point is that we greatly underestimated the skill and effort needed to manage biodiversity data on this scale, which led to bottlenecks and unexpectedly heavy workloads. With hindsight, we should have doubled the resource allocated to this heading.

a much richer picture of human impacts on terrestrial biodiversity than could be obtained through traditional meta-analyses of statistical results. Collating the long tail of dark data is rapidly becoming easier: text-mining approaches are making relevant data easier to find and extract (Peters et al., 2014); open-data policies and data repositories mean more data are accessible directly; and metadata in global repositories can now accommodate ecological samples (De Pooter et al., 2017).

Despite PREDICTS' models being entirely phenomenological, they can do more than just quantify human impacts (though that is valuable in itself: Ferrier et al., 2016; Steffen et al., 2015; Tittensor et al., 2014). Analysing multiple diversity measures can discriminate among alternative mechanisms: for example, the area-scaling of species density appears to result from a higher density of individuals in larger habitat patches (Phillips et al., 2017a), but such abundance effects cannot explain differences in species richness among land uses (Newbold et al., 2015). Integrating the database with species traits allows testing of a range of hypotheses about which traits confer susceptibility or resilience (e.g. De Palma et al., 2015; Newbold et al., 2014). Comparative modelling of species' susceptibility to land-use change, inferring susceptibility even of species missing from the PREDICTS database, would be a natural extension of these models, allowing projections to be made for individual species as well as composite measures of biodiversity.

Our implementation of the BII (Newbold et al., 2016a; Scholes and Biggs, 2005) is already a useful indicator of the average state of broad-sense biodiversity in local terrestrial assemblages, but—as with many indicators—can be improved with further refinement. There are several main strands of ongoing work. First, we are testing the space-for-time assumption by collating datasets that capture the dynamics of biotic responses to land-use change (De Palma et al., 2018, Chapter 7, this volume). Second, we have begun using annual fine-scale pressure data to drive annual estimates of how BII has changed in the recent past, increasing its usefulness as an indicator; a remaining challenge is the lack of a global high-resolution map of plantation forest. Third, looking forward, we are modelling and projecting BII and other measures using historical and future estimates of land use and other pressures from the Shared Socioeconomic Pathways (O'Neill et al., 2014). Fourth, islands present an unusual opportunity for separating native from nonnative species across a wide range of taxa, allowing a direct test of our two-step approach for BII estimation. Fifth, we are exploring the possibility of combining PREDICTS' models of the effects of land use with GLOBIO's models of the effects of some other human pressures, in order to produce spatial and temporal projections that reflect a wider range of human impacts. Lastly, we are exploring related abundance-based measures that are more sensitive than BII to changes in species' relative abundances (e.g. Hillebrand et al., 2017), as possible complements to BII itself.

Looking further ahead, a limitation of all current global biodiversity models is that they are impact models, i.e., they assess the impacts on biodiversity of scenarios of pressures such as land use and climate change often derived from integrated assessment models (Harfoot et al., 2014). Recognising that biodiversity, and the ecosystem services it provides, ultimately underpin human wellbeing and life on Earth, we suggest that future models should be inverted, modelling socio-economic systems as being 'impacted' by, or dependent on, biodiversity and ecosystem services.

ACKNOWLEDGEMENTS

We thank all the many researchers worldwide who shared with us their hard-won primary data, and all the PREDICTS team members who collated biodiversity, trait, pressure, and phylogenetic information that have fed into analyses. Georgina Mace, Rob Ewers, Drew Purves, Ben Collen, and Neil Burgess all helped with the development of PREDICTS, and Juan Gallego Zamorano and David Peek helped with preparing this manuscript. We also thank Michelle Jackson and anonymous referee for comments that improved the paper. PREDICTS has been supported by NERC (NE/J011193/2 and NE/M014533/1), BBSRC (BB/F017324), UNEP-WCMC, Microsoft Research, the University of Sussex, the Natural History Museum, PBL Netherlands, a Hans Rausing Scholarship, and the National Council of Science & Technology of Mexico (CONACyT). This paper is a contribution from the Imperial College Grand Challenges in Ecology and the Environment Initiative. PREDICTS is endorsed by the GEO BON.

REFERENCES

Alkemade, R., van Oorschot, M., Miles, L., Nellemann, C., Bakkenes, M., ten Brink, B., 2009. GLOBIO3: a framework to investigate options for reducing global terrestrial biodiversity loss. Ecosystems 12, 374–390.

Azaele, S., Maritan, A., Cornell, S.J., Suweis, S., Banavar, J.R., et al., 2015. Towards a unified descriptive theory for spatial ecology: predicting biodiversity patterns across spatial scales. Methods Ecol. Evol. 6, 324–332.

Baillie, J.E.M., Hilton-Taylor, C., Stuart, S.N. (Eds.), 2004. 2004 IUCN Red List of Threatened Species. A Global Species Assessment. IUCN, Gland, Switzerland and Cambridge, UK.

Balmford, A., 1996. Extinction filters and current resilience: the significance of past selection pressures for conservation biology. Trends Ecol. Evol. 11, 193–196.

Banks-Leite, C., Ewers, R.M., Metzger, J.P., 2012. Unraveling the drivers of community dissimilarity and species extinction in fragmented landscapes. Ecology 93, 2560–2569.

Bates, D., Maechler, M., Bolker, B., Walker, S., 2014. lme4: Linear Mixed-Effects Models Using Eigen and S4. R Package Version 1.1-7, http://cran.r-project.org/package=lme4.

Bates, D., Kliegl, R., Vasishth, S. & Baayen, H. 2015. Parsimonious mixed models. arXiv, 1506.04967 [stat.ME].

Biggs, R., Simons, H., Bakkenes, M., Scholes, R.J., Eickhout, B., et al., 2008. Scenarios of biodiversity loss in southern Africa in the 21st century. Glob. Environ. Chang. 18, 296–309.

Bolker, B.M., Brooks, M.E., Clark, C.J., Geange, S.W., Poulsen, J.R., et al., 2009. Generalized linear mixed models: a practical guide for ecology and evolution. Trends Ecol. Evol. 24, 127–135.

Brook, B.W., Ellis, E.C., Perring, M.P., Mackay, A.W., Blomqvist, L., 2013. Does the terrestrial biosphere have planetary tipping points? Trends Ecol. Evol. 28, 396–401.

Buckland, S.T., Magurran, A.E., Green, R.E., Fewster, R.M., 2005. Monitoring change in biodiversity through composite indices. Philos. Trans. R. Soc. Lond. B Biol. Sci. 360, 243–254.

Butchart, S.H.M., Akcakaya, H.R., Chanson, J., Baillie, J.E.M., Collen, B., et al., 2007. Improvements to the Red List Index. PLos One 2, e140.

Butchart, S.H.M., Walpole, M., Collen, B., van Strien, A., Scharlemann, J.P.W., et al., 2010. Global biodiversity: indicators of recent declines. Science 328, 1164–1168.

Cardinale, B.J., Duffy, J.E., Gonzalez, A., Hooper, D.U., Perrings, C., et al., 2012. Biodiversity loss and its impact on humanity. Nature 486, 59–67.

Chapman, A.D., 2009. Numbers of Living Species in Australia and the World, second ed. Department of the Environment, Water, Heritage and the Arts, Australian Government, Canberra, Australia. https://www.environment.gov.au/node/13876.

CIESIN, IFPRI, CIAT, 2011. Global Rural-Urban Mapping Project, Version 1 (GRUMPv1): Population Density Grid. NASA SEDAC. http://sedac.ciesin.columbia. edu/data/dataset/grump-v1-population-density.

CIESIN, ITOS, University of Georgia, 2013. Global Roads Open Access Data Set, Version 1 (gROADSv1). https://doi.org/10.7927/H4VD6WDCT NASA SEDAC.

Cleary, D.F.R., Mooers, A.O., 2006. Burning and logging differentially affect endemic vs. widely distributed butterfly species in Borneo. Divers. Distrib. 12, 409–416.

Coetzee, B.W.T., Gaston, K.J., Chown, S.L., 2014. Local scale comparisons of biodiversity as a test for global protected area ecological performance: a meta-analysis. PLos One 9, e105824.

Collen, B., Loh, J., Whitmee, S., Mcrae, L., Amin, R., Baillie, J.E.M., 2009. Monitoring change in vertebrate abundance: the living planet index. Conserv. Biol. 23, 317–327.

Convention on Biological Diversity, 2017. Aichi Biodiversity Targets. (Online). Secretariat of the Convention on Biological Diversity, Montreal, Canada. Accessed 20/7/2017. https://www.cbd.int/sp/targets/.

de Bello, F., Thuiller, W., Leps, J., Choler, P., Clement, J.C., et al., 2009. Partitioning of functional diversity reveals the scale and extent of trait convergence and divergence. J. Veg. Sci. 20, 475–486.

De Palma, A., Kuhlmann, M., Roberts, S.P.M., Potts, S.G., Borger, L., et al., 2015. Ecological traits affect the sensitivity of bees to land-use pressures in European agricultural landscapes. J. Appl. Ecol. 52, 1567–1577.

De Palma, A., Abrahamczyk, S., Aizen, M.A., Albrecht, M., Basset, Y., et al., 2016. Predicting bee community responses to land-use changes: effects of geographic and taxonomic biases. Sci. Rep. 6, 31153.

De Palma, A., Kuhlmann, M., Bugter, R., Ferrier, S., Hoskins, A.J., et al., 2017. Dimensions of biodiversity loss: spatial mismatch in land-use impacts on species, functional and phylogenetic diversity of European bees. Divers. Distrib. 23, 1435–1446.

De Palma, A., Sanchez-Ortiz, K., Martin, P.A., Chadwick, A., Gilbert, G., et al., 2018. Challenges with inferring how land-use affects terrestrial biodiversity: study design, time, space and synthesis. Adv. Ecol. Res. 58, 163–199.

De Pooter, D., Appeltans, W., Bailly, N., Bristol, S., Deneudt, K., et al., 2017. Toward a new data standard for combined marine biological and environmental datasets—expanding OBIS beyond species occurrences. Biodivers. Data J. 5, e10989.

Dobson, A., 2005. Monitoring global rates of biodiversity change: challenges that arise in meeting the Convention on Biological Diversity (CBD) 2010 goals. Philos. Trans. R. Soc. Lond. B Biol. Sci. 360, 229–241.

Dornelas, M., Gotelli, N.J., McGill, B., Shimadzu, H., Moyes, F., et al., 2014. Assemblage time series reveal biodiversity change but not systematic loss. Science 344, 296–299.

Echeverría-Londoño, S., Newbold, T., Hudson, L.N., Contu, S., Hill, S.L.L., et al., 2016. Modelling and projecting the response of local assemblage composition to land use change across Colombia. Divers. Distrib. 22, 1099–1111.

Essl, F., Dullinger, S., Rabitsch, W., Hulme, P.E., Pysek, P., et al., 2015. Delayed biodiversity change: no time to waste. Trends Ecol. Evol. 30, 375–378.

Fahrig, L., Girard, J., Duro, D., Pasher, J., Smith, A., et al., 2015. Farmlands with smaller crop fields have higher within-field biodiversity. Agr. Ecosyst. Environ. 200, 219–234.

Faith, D.P., Ferrier, S., Williams, K.J., 2008. Getting biodiversity intactness indices right: ensuring that 'biodiversity' reflects 'diversity'. Glob. Chang. Biol. 14, 207–217.

Ferrier, S., Ninan, K.N., Leadley, P., Alkemade, R., Acosta, L.A. et al., (Eds.), 2016. Summary for Policymakers of the Methodological Assessment Report on Scenarios and Models of Biodiversity and Ecosystem Services of the Intergovernmental Science-Policy Platform on Biodiversity and Ecosystem Services. Secretariat of the Intergovernmental Science-Policy Platform on Biodiversity and Ecosystem Services, Bonn, Germany.

Fox, G.A., Negrete-Yankelevich, S., Sosa, V.S., 2015. Ecological Statistics: Contemporary Theory and Application. Oxford University Press, Oxford.

Gabriel, D., Roschewitz, I., Tscharntke, T., Thies, C., 2006. Beta diversity at different spatial scales: plant communities in organic and conventional agriculture. Ecol. Appl. 16, 2011–2021.

Geldmann, J., Barnes, M., Coad, L., Craigie, I.D., Hockings, M., Burgess, N.D., 2013. Effectiveness of terrestrial protected areas in reducing habitat loss and population declines. Biol. Conserv. 161, 230–238.

Gelman, A., Jakulin, A., Pittau, M.G., Su, Y.S., 2008. A weakly informative default prior distribution for logistic and other regression models. Ann. Appl. Stat. 2, 1360–1383.

GEO BON, 2015. Global Biodiversity Change Indicators. Version 1.2. Group of Earth Observations Biodiversity Observation Network Secretariat, Leipzig.

Gerstner, K., Dormann, C.F., Stein, A., Manceur, A.M., Seppelt, R., 2014. Effects of land use on plant diversity—a global meta-analysis. J. Appl. Ecol. 51, 1690–1700.

Gibson, L., Lee, T.M., Koh, L.P., Brook, B.W., Gardner, T.A., et al., 2011. Primary forests are irreplaceable for sustaining tropical biodiversity. Nature 478, 378–381.

Gilroy, J.J., Edwards, D.P., 2017. Source-sink dynamics: a neglected problem for landscape-scale biodiversity conservation in the tropics. Curr. Landscape Ecol. Rep. 2, 51–60.

Goldewijk, K.K., Beusen, A., van Drecht, G., de Vos, M., 2011. The HYDE 3.1 spatially explicit database of human-induced global land-use change over the past 12,000 years. Glob. Ecol. Biogeogr. 20, 73–86.

Gonzalez, A., Cardinale, B.J., Allington, G.R.H., Byrnes, J., Endsley, K.A., et al., 2016. Estimating local biodiversity change: a critique of papers claiming no net loss of local diversity. Ecology 97, 1949–1960.

Gray, C.L., Hill, S.L.L., Newbold, T., Hudson, L.N., Borger, L., et al., 2016. Local biodiversity is higher inside than outside terrestrial protected areas worldwide. Nat. Commun. 7.

Guggenheim, D., 2006. An Inconvenient Truth. Paramount Classics.

Harfoot, M., Tittensor, D.P., Newbold, T., McInerny, G., Smith, M.J., Scharlemann, J.P.W., 2014. Integrated assessment models for ecologists: the present and the future. Glob. Ecol. Biogeogr. 23, 124–143.

Harrison, X.A., 2014. Using observation-level random effects to model overdispersion in count data in ecology and evolution. PeerJ 2, e616.

Hedges, L.V., Gurevitch, J., Curtis, P.S., 1999. The meta-analysis of response ratios in experimental ecology. Ecology 80, 1150–1156.

Heidorn, P.B., 2008. Shedding light on the dark data in the long tail of science. Libr. Trends 57, 280–299.

Hillebrand, H., Blasius, B., Borer, E.T., Chase, J.M., Downing, J., et al., 2017. Biodiversity change is uncoupled from species richness trends: consequences for conservation and monitoring. J. Appl. Ecol. 55, 169–184. https://doi.org/10.1111/1365-2664.12959.

Holling, C.S., 1959. The functional response of predators to prey density and its role in mimicry and population regulation. Mem. Entomol. Soc. Can. 97, 5–60.

Hooper, D.U., Chapin, F.S., Ewel, J.J., Hector, A., Inchausti, P., et al., 2005. Effects of biodiversity on ecosystem functioning: a consensus of current knowledge. Ecol. Monogr. 75, 3–35.

Hooper, D.U., Adair, E.C., Cardinale, B.J., Byrnes, J.E.K., Hungate, B.A., et al., 2012. A global synthesis reveals biodiversity loss as a major driver of ecosystem change. Nature 486, 105–108.

Hortal, J., de Bello, F., Diniz, J.A.F., Lewinsohn, T.M., Lobo, J.M., Ladle, R.J., 2015. Seven shortfalls that beset large-scale knowledge of biodiversity. Annu. Rev. Ecol. Evol. Syst. 46, 523–552.

Hoskins, A.J., Bush, A., Gilmore, J., Harwood, T., Hudson, L.N., et al., 2016. Downscaling land-use data to provide global 30″ estimates of five land-use classes. Ecol. Evol. 6, 3040–3055.

Houghton, J.T., Jenkins, G.J., Ephraums, J.J. (Eds.), 1990. Report Prepared for Intergovern-mental Panel on Climate Change by Working Group I. Australia Cambridge University Press, Cambridge, Great Britain, New York, NY, USA and Melbourne.

Hudson, L.N., Newbold, T., Contu, S., Hill, S.L.L., Lysenko, I., et al., 2014. The PRE-DICTS database: a global database of how local terrestrial biodiversity responds to human impacts. Ecol. Evol. 4, 4701–4735.

Hudson, L.N., Newbold, T., Contu, S., Hill, S.L.L., Lysenko, I., et al., 2016. Dataset: The 2016 Release of the PREDICTS Database. Natural History Museum Data Portal (data. nhm.ac.uk). https://doi.org/10.5519/0066354.

Hudson, L.N., Newbold, T., Contu, S., Hill, S.L.L., Lysenko, I., et al., 2017. The database of the PREDICTS (projecting responses of ecological diversity in changing terrestrial sys-tems) project. Ecol. Evol. 7, 145–188.

Hurtt, G.C., Chini, L.P., Frolking, S., Betts, R.A., Feddema, J., et al., 2011. Harmonization of land-use scenarios for the period 1500-2100: 600 years of global gridded annual land-use transitions, wood harvest, and resulting secondary lands. Clim. Change 109, 117–161.

IPBES, 2017. Update on the Work on Knowledge and Data (Deliverables 1 (d) and 4 (b)). United Nations, UNEP, UNESCO, Food and Agriculture Organization of the United Nations, UNDP. http://www.ipbes.net/sites/default/files/downloads/pdf/ipbes-5-inf-5.pdf.

IPCC, 2013. Summary for Policymakers. In: Climate Change 2013: The Physical Science Basis. Contribution of Working Group I to the Fifth Assessment Report of the Intergov-ernmental Panel on Climate Change Cambridge. Cambridge University Press, United Kingdom and New York, NY, USA.

Isbell, F., Calcagno, V., Hector, A., Connolly, J., Harpole, W.S., et al., 2011. High plant diversity is needed to maintain ecosystem services. Nature 477, 199–202.

IUCN, UNEP, 2014. The World Database on Protected Areas (WDPA). UNEP-WCMC. http://www.protectedplanet.net.

Joppa, L.N., O'Connor, B., Visconti, P., Smith, C., Geldmann, J., et al., 2016. Filling in biodiversity threat gaps. Science 352, 416–418.

Jung, M., Hill, S.L.L., Platts, P.J., Marchant, R., Siebert, S., et al., 2017. Local factors mediate the response of biodiversity to land use on two African mountains. Anim. Conserv. 20, 370–381.

Koricheva, J., Gurevitch, J., 2014. Uses and misuses of meta-analysis in plant ecology. J. Ecol. 102, 828–844.

Lawton, J.H., Bignell, D.E., Bolton, B., Bloemers, G.F., Eggleton, P., et al., 1998. Biodi-versity inventories, indicator taxa and effects of habitat modification in tropical forest. Nature 391, 72–76.

Leadley, P.W., Krug, C.B., Alkemade, R., Pereira, H.M., Sumaila, U.R., et al., 2014. Pro-gress towards the Aichi Biodiversity Targets: An Assessment of Biodiversity Trends, Pol-icy Scenarios and Key Actions. Montreal, Canada.

Lichstein, J.W., 2007. Multiple regression on distance matrices: a multivariate spatial analysis tool. Plant Ecol. 188, 117–131.

Loeys, T., Moerkerke, B., De Smet, O., Buysse, A., 2012. The analysis of zero-inflated count data: beyond zero-inflated Poisson regression. Br. J. Math. Stat. Psychol. 65, 163–180.

Loh, J., Randers, J., MacGillivray, A., Kapos, V., Groombridge, B., Jenkins, M., 1998. Liv-ing Planet Report 1998: Overconsumption is Driving the Rapid Decline of the world's Natural Environments. Gland, Switzerland, WWF International.

Loh, J., Green, R.E., Ricketts, T., Lamoreux, J., Jenkins, M., et al., 2005. The Living Planet Index: using species population time series to track trends in biodiversity. Philos. Trans. R. Soc. Lond. B Biol. Sci. 360, 289–295.

Mace, G., Masundire, H., Baillie, J., Ricketts, T., Brooks, T., et al., 2005. Biodiversity. In: Hassan, R., Scholes, R., Ash, N. (Eds.), In: Ecosystems and Human Well-Being: Current State and Trends, vol. 1. Island Press, Washingto, DC.

Mace, G.M., Reyers, B., Alkemade, R., Biggs, R., Chapin, F.S., et al., 2014. Approaches to defining a planetary boundary for biodiversity. Glob. Environ. Chang. 28, 289–297.

Magurran, A.E., 2004. Measuring Biological Diversity. Blackwell, Oxford.

Mann, M.E., Bradley, R.S., Hughes, M.K., 1998. Global-scale temperature patterns and climate forcing over the past six centuries. Nature 392, 779–787.

Martin, T.G., Wintle, B.A., Rhodes, J.R., Kuhnert, P.M., Field, S.A., et al., 2005. Zero tolerance ecology: improving ecological inference by modelling the source of zero observations. Ecol. Lett. 8, 1235–1246.

Maxwell, S., Fuller, R.A., Brooks, T.M., Watson, J.E.M., 2016. The ravages of guns, nets and bulldozers. Nature 536, 143–145.

Mayfield, M.M., Dwyer, J.M., Chalmandrier, L., Wells, J.A., Bonser, S.P., et al., 2013. Differences in forest plant functional trait distributions across land-use and productivity gradients. Am. J. Bot. 100, 1356–1368.

Mayor, S.J., Boutin, S., He, F.L., Cahill, J.F., 2015. Limited impacts of extensive human land use on dominance, specialization, and biotic homogenization in boreal plant communities. BMC Ecol. 15, 5.

McGill, B.J., Enquist, B.J., Weiher, E., Westoby, M., 2006. Rebuilding community ecology from functional traits. Trends Ecol. Evol. 21, 178–185.

McGill, B.J., Dornelas, M., Gotelli, N.J., Magurran, A.E., 2015. Fifteen forms of biodiversity trend in the Anthropocene. Trends Ecol. Evol. 30, 104–113.

McKinney, M.L., 1997. Extinction vulnerability and selectivity: combining ecological and paleontological views. Annu. Rev. Ecol. Syst. 28, 495–516.

Meyer, C., Jetz, W., Guralnick, R.P., Fritz, S.A., Kreft, H., 2016. Range geometry and socio-economics dominate species-level biases in occurrence information. Glob. Ecol. Biogeogr. 25, 1181–1193.

Millar, R.B., Anderson, M.J., Tolimieri, N., 2011. Much ado about nothings: using zero similarity points in distance-decay curves. Ecology 92, 1717–1722.

Millennium Ecosystem Assessment, 2005. Ecosystems and Human Well-Being: Synthesis. Island Press, Washington, DC.

Mora, C., Tittensor, D.P., Adl, S., Simpson, A.G.B., Worm, B., 2011. How many species are there on earth and in the ocean? PLoS Biol. 9, 8.

Murphy, G.E.P., Romanuk, T.N., 2014. A meta-analysis of declines in local species richness from human disturbances. Ecol. Evol. 4, 91–103.

Naeem, S., Prager, C., Weeks, B., Varga, A., Flynn, D.F.B., et al., 2016. Biodiversity as a multidimensional construct: a review, framework and case study of herbivory's impact on plant biodiversity. Proc. R. Soc. B Biol. Sci. 283, 20153005.

Nellemann, C., Kullerud, L., Vistnes, I., Forbes, B.C., Husby, E., et al., 2001. GLOBIO Global Methodology for Mapping Human Impacts on the Biosphere. Nairobi, UNEP Division of Early Warning and Assessment.

Newbold, T., Scharlemann, J.P.W., Butchart, S.H.M., Sekercioglu, C.H., Alkemade, R., et al., 2013. Ecological traits affect the response of tropical forest bird species to land-use intensity. Proc. R. Soc. B Biol. Sci. 280, 20122131.

Newbold, T., Hudson, L.N., Phillips, H.R.P., Hill, S.L.L., Contu, S., et al., 2014. A global model of the response of tropical and sub-tropical forest biodiversity to anthropogenic pressures. Proc. R. Soc. B Biol. Sci. 281, 20141371.

Newbold, T., Hudson, L.N., Hill, S.L.L., Contu, S., Lysenko, I., et al., 2015. Global effects of land use on local terrestrial biodiversity. Nature 520, 45–50.

Newbold, T., Hudson, L.N., Arnell, A.P., Contu, S., De Palma, A., et al., 2016a. Has land use pushed terrestrial biodiversity beyond the planetary boundary? A global assessment. Science 353, 288–291.

Newbold, T., Hudson, L.N., Hill, S.L.L., Contu, S., Gray, C.L., et al., 2016b. Global patterns of terrestrial assemblage turnover within and among land uses. Ecography 39, 1151–1163.

Norfolk, O., Eichhorn, M.P., Gilbert, F.S., 2015. Contrasting patterns of turnover between plants, pollinators and their interactions. Divers. Distrib. 21, 405–415.

Noss, R.F., 1990. Indicators for monitoring biodiversity: a hierarchical approach. Conserv. Biol. 4, 355–364.

O'Hara, R.B., Kotze, D.J., 2010. Do not log-transform count data. Methods Ecol. Evol. 1, 118–122.

Oliver, T.H., 2016. How much biodiversity loss is too much? Science 353, 220–221.

O'Neill, B.C., Kriegler, E., Riahi, K., Ebi, K.L., Hallegatte, S., et al., 2014. A new scenario framework for climate change research: the concept of shared socioeconomic pathways. Clim. Change 122, 387–400.

Petchey, O.L., Pontarp, M., Massie, T.M., Kefi, S., Ozgul, A., et al., 2015. The ecological forecast horizon, and examples of its uses and determinants. Ecol. Lett. 18, 597–611.

Peters, S.E., Zhang, C., Livny, M., Re, C., 2014. A machine reading system for assembling synthetic paleontological databases. PLoS One 9, e113523.

Phalan, B., Onial, M., Balmford, A., Green, R.E., 2011. Reconciling food production and biodiversity conservation: land sharing and land sparing compared. Science 333, 1289–1291.

Phillips, H.R.P., Halley, J.M., Urbina-Cardona, J., Purvis, A., 2017a. The effect of fragment area on site-level biodiversity. Ecography. https://doi.org/10.1111/ecog.02956.

Phillips, H.R.P., Newbold, T., Purvis, A., 2017b. Land-use effects on local biodiversity vary between continents. Biodivers. Conserv. 26, 2251–2270.

PREDICTS, 2017. Projecting Responses of Ecological Diversity in Changing Terrestrial Systems: Data collection guidelines, 26 November 2017. PREDICTS Project, London.

Pulliam, H.R., 1988. Sources, sinks and population regulation. Am. Nat. 132, 652–661.

Purvis, A., 2008. Phylogenetic approaches to the study of extinction. Annu. Rev. Ecol. Evol. Syst. 39, 301–319.

Purvis, A., Hector, A., 2000. Getting the measure of biodiversity. Nature 405, 212–219.

Roberts, D.R., Bahn, V., Ciuti, S., Boyce, M.S., Elith, J., et al., 2017. Cross-validation strategies for data with temporal, spatial, hierarchical, or phylogenetic structure. Ecography 40, 913–929.

Rockstrom, J., Steffen, W., Noone, K., Persson, A., Chapin, F.S., et al., 2009. A safe operating space for humanity. Nature 461, 472–475.

Rouget, M., Cowling, R.M., Vlok, J., Thompson, M., Balmford, A., 2006. Getting the biodiversity intactness index right: the importance of habitat degradation data. Glob. Chang. Biol. 12, 2032–2036.

Sala, O.E., Vuuren, D.v., Pereira, H.M., Lodge, D., Alder, J., et al., 2005. Biodiversity across scenarios. In: Carpenter, S.R., Pingali, P.L., Bennett, E.M., Zurek, M.B. (Eds.), In: Ecosystems and Human Well-Being: Scenarios, vol. 2. Island Press, Washington, DC.

Schipper, A.M., Belmaker, J., de Miranda, M.D., Navarro, L.M., Bohning-Gaese, K., et al., 2016. Contrasting changes in the abundance and diversity of North American bird assemblages from 1971 to 2010. Glob. Chang. Biol. 22, 3948–3959.

Schleuter, D., Daufresne, M., Massol, F., Argillier, C., 2010. A user's guide to functional diversity indices. Ecol. Monogr. 80, 469–484.

Scholes, R.J., Biggs, R., 2005. A biodiversity intactness index. Nature 434, 45–49.

Scholes, R.J., Biggs, R., Palm, C., Duralappah, A., 2010. Assessing state and trends in ecosystem services and human wellbeing. In: Ash, N., Blanco, H., Brown, C.D., Garcia, K., Henrichs, T., Lucas, N., Raudsepp-Hearne, C., Simpson, R.D., Scholes, R.J., Tomich, T.P., Vira, B., Zurek, M. (Eds.), Ecosystems and Human Well-Neing: A Manual for Assessment Practitioners. Island Press, Washington, DC.

Secretariat of the Convention on Biological Diversity, 2010. Global Biodiversity Outlook 3. Montreal, Secretariat of the Convention on Biological Diversity.

Secretariat of the Convention on Biological Diversity, 2014. Global Biodiversity Outlook 4. Montreal, Canada.

Shipley, B., Vile, D., Garnier, E., 2006. From plant traits to plant communities: a statistical mechanistic approach to biodiversity. Science 314, 812–814.

Sobral, F.L., Lees, A.C., Cianciaruso, M.V., 2016. Introductions do not compensate for functional and phylogenetic losses following extinctions in insular bird assemblages. Ecol. Lett. 19, 1091–1100.

Soininen, J., McDonald, R., Hillebrand, H., 2007. The distance decay of similarity in ecological communities. Ecography 30, 3–12.

Steffen, W., Richardson, K., Rockstrom, J., Cornell, S.E., Fetzer, I., et al., 2015. Planetary boundaries: guiding human development on a changing planet. Science 347, 1259855.

Stocker, T.F., Qin, D., Plattner, G.-K., Tignor, M., Allen, S.K., et al., (Eds.), 2013. Climate Change 2013: The Physical Science Basis. Contribution of Working Group I to the Fifth Assessment Report of the Intergovernmental Panel on Climate Change. Cambridge University Press, Cambridge, United Kingdom and New York, NY, USA.

Suding, K.N., Lavorel, S., Chapin, F.S., Cornelissen, J.H.C., Diaz, S., et al., 2008. Scaling environmental change through the community-level: a trait-based response-and-effect framework for plants. Glob. Chang. Biol. 14, 1125–1140.

Tittensor, D.P., Walpole, M., Hill, S.L.L., Boyce, D.G., Britten, G.L., et al., 2014. A midterm analysis of progress toward international biodiversity targets. Science 346, 241–244.

Tscharntke, T., Sekercioglu, C.H., Dietsch, T.V., Sodhi, N.S., Hoehn, P., Tylianakis, J.M., 2008. Landscape constraints on functional diversity of birds and insects in tropical agroecosystems. Ecology 89, 944–951.

Tuck, S.L., Winqvist, C., Mota, F., Ahnstrom, J., Turnbull, L.A., Bengtsson, J., 2014. Land-use intensity and the effects of organic farming on biodiversity: a hierarchical meta-analysis. J. Appl. Ecol. 51, 746–755.

Tucker, C.M., Cadotte, M.W., Carvalho, S.B., Davies, T.J., Ferrier, S., et al., 2017. A guide to phylogenetic metrics for conservation, community ecology and macroecology. Biol. Rev. 92, 698–715.

Tylianakis, J.M., Klein, A.M., Tscharntke, T., 2005. Spatiotemporal variation in the diversity of hymenoptera across a tropical habitat gradient. Ecology 86, 3296–3302.

Tylianakis, J.M., Klein, A.M., Lozada, T., Tscharntke, T., 2006. Spatial scale of observation affects alpha, beta and gamma diversity of cavity-nesting bees and wasps across a tropical land-use gradient. J. Biogeogr. 33, 1295–1304.

UNEP, 2002. Global Environment Outlook 3: Past, Present and Future Perspectives. Earthscan, London.

UNEP, 2007. Global Environment Outlook 4: Environment for Development. Earthscan, London.

United Nations, 1992. Convention on Biological Diversity. United Nations.

Vackar, D., ten Brink, B., Loh, J., Baillie, J.E.M., Reyers, B., 2012. Review of multispecies indices for monitoring human impacts on biodiversity. Ecol. Indic. 17, 58–67.

van Vuuren, D.P., Edmonds, J., Kainuma, M., Riahi, K., Thomson, A., et al., 2011. The representative concentration pathways: an overview. Clim. Change 109, 5–31.

Vellend, M., Verheyen, K., Flinn, K.M., Jacquemyn, H., Kolb, A., et al., 2007. Homogenization of forest plant communities and weakening of species-environment relationships via agricultural land use. J. Ecol. 95, 565–573.

Vellend, M., Baeten, L., Myers-Smith, I.H., Elmendorf, S.C., Beausejour, R., et al., 2013. Global meta-analysis reveals no net change in local-scale plant biodiversity over time. Proc. Natl. Acad. Sci. U.S.A. 110, 19456–19459.

Vellend, M., Dornelas, M., Baeten, L., Beausejour, R., Brown, C.D., et al., 2017. Estimates of local biodiversity change over time stand up to scrutiny. Ecology 98, 583–590.

Vetter, D., Hansbauer, M.M., Vegvari, Z., Storch, I., 2011. Predictors of forest fragmentation sensitivity in Neotropical vertebrates: a quantitative review. Ecography 34, 1–8.

Violle, C., Navas, M.L., Vile, D., Kazakou, E., Fortunel, C., et al., 2007. Let the concept of trait be functional!. Oikos 116, 882–892.

Warton, D.I., Hui, F.K.C., 2011. The arcsine is asinine: the analysis of proportions in ecology. Ecology 92, 3–10.

Watson, J.E.M., Shanahan, D.F., Di Marco, M., Allan, J., Laurance, W.F., et al., 2016. Catastrophic declines in wilderness areas undermine global environment targets. Curr. Biol. 26, 2929–2934.

Wilson, K.A., Auerbach, N.A., Sam, K., Magini, A.G., Moss, A.S., et al., 2016. Conservation research is not happening where it is most needed. PLoS Biol. 14, e1002413.

Zuur, A.F., Ieno, E.N., Walker, N.J., Saveliev, A.A., Smith, G.M., 2009. Mixed Effects Models and Extensions in Ecology with R. Springer, New York.

CHAPTER SIX

Mapping Mediterranean Wetlands With Remote Sensing: A Good-Looking Map Is Not Always a Good Map

Christian Perennou*,[1], Anis Guelmami*, Marc Paganini[†],
Petra Philipson[‡], Brigitte Poulin*, Adrian Strauch[§], Christian Tottrup[¶],
John Truckenbrodt[‖], Ilse R. Geijzendorffer*

*Tour du Valat, Research Institute for the Conservation of Mediterranean Wetlands, Arles, France
[†]European Space Agency, Frascati, Italy
[‡]Brockmann Geomatics Sweden AB, Stockholm, Sweden
[§]University of Bonn, Center for Remote Sensing of Land Surfaces (ZFL), Bonn, Germany
[¶]DHI GRAS, Hoersholm, Denmark
[‖]Friedrich-Schiller-University Jena, Institute of Geography, Jena, Germany
[1]Corresponding author: e-mail address: perennou@tourduvalat.org

Contents

Advances in Ecological Research, Volume 58
ISSN 0065-2504
https://doi.org/10.1016/bs.aecr.2017.12.002

Abstract

Wetlands are a key habitat within the Mediterranean biodiversity hotspot and provide important ecosystem services for human well-being. Remote sensing (RS) has significantly boosted our ability to monitor changes in Mediterranean wetlands, especially in areas where little information is being collected. However, its application to wetlands has sometimes been flawed with uncertainties and unrecognized errors, to a large extent due to the inherent and specific ecological characteristics of Mediterranean wetlands. We present here an overview of the state of the art on RS techniques for mapping and monitoring Mediterranean wetlands, and the remaining challenges: delineating and separating wetland habitat types; mapping water dynamics inside wetlands; and detecting actual wetland trends over time in a context of high, natural variability. The most important lessons learned are that ecologists' knowledge need to be integrated with RS expertise to achieve a valuable monitoring approach of these ecosystems.

1. INTRODUCTION: THE CHALLENGES OF MONITORING WETLANDS STATUS AND TRENDS WITH REMOTE SENSING (RS) DATA

Mediterranean wetlands are part of a global biodiversity hotspot, hosting many endemic species (e.g. Darwall et al., 2014). Their global importance stretches further as they produce a global share of the ecosystem services that is greater than their relative habitat extent (Zedler and Kercher, 2005). With current decreasing trends in natural habitat extent and regionally increasing human population numbers (MWO, 2012a), the importance of the remaining, increasingly threatened Mediterranean wetlands will only further increase. The importance of Mediterranean wetlands is acknowledged in multiple Multilateral Environmental Agreements (MEAs), especially the Ramsar Convention (e.g. Gardner et al., 2015; Ramsar Convention Secretariat, 2015), one of the most influential agreements for the conservation of wetlands globally. The monitoring of changes in Mediterranean wetland state and extent therefore provides crucial information for decisions makers and feeds into a diversity of policy reporting activities at different spatial scales (e.g. Beltrame et al., 2015; MWO, 2012a, b, 2014; Perennou et al., 2016; Plan Bleu, 2009).

The use of satellite imagery for wetland inventories and monitoring offers a great potential because of repeated, homogeneous coverage of large areas (e.g. Dadaser-Celik et al., 2008; Özesmi and Bauer, 2002; Rebelo et al., 2009; Rosenqvist et al., 2007). The production and interpretation of wetland habitat maps have gone through a steep learning curve, leading to a wide diversity of available products (Guo et al., 2017). The emergence of new satellites offers possibilities to further improve our understanding of changes in wetlands. The Sentinels of the European Copernicus Program form a recent initiative, which provides free optical and radar observations at a high spatial resolution (10–20 m) and short revisiting time. All technical improvements will potentially allow us to better distinguish different wetland habitats.

However, this is also a moment where progress already made can be easily lost as new RS tools and new mapping nomenclatures have to be developed for the new satellite data. International cooperation and collaborative development of RS methods, guidelines and best practices are required to avoid duplication of efforts, foster progress and innovation, and provide long-term access to the developed products, methods and tools. The new GEO-Wetlands initiative of the Group on Earth Observations (GEO) addresses this requirement and several of the projects contributing to this chapter form a part of this initiative.

This chapter presents the improvements in the monitoring of Mediterranean wetlands using RS, starting from the end of the 80s, when the first maps based on CORINE land cover (CLC) (Bossard et al., 2000) were produced, up to the most recent monitoring efforts of 2016. It should be emphasized that many of the issues that will be covered are not Mediterranean specific, but rather are specific to (semi)-arid regions. Ephemeral wetlands, rice paddies, artificial wetlands, irregular precipitations, all occur across the globe, and our Mediterranean review should thus be seen as a regional case study with lessons that are applicable globally. We use this overview to raise awareness of the challenges of using RS to monitor wetland habitats, some of which have still to be embraced. The most crucial recommendation for advancing our capacity to use RS for the monitoring of Mediterranean wetlands is that iterative exchanges between wetland ecologists, hydrologists and RS experts are necessary for obtaining credible results (e.g. Skidmore et al., 2015).

To be able to reflect on our capacity to monitor Mediterranean wetlands, we first present three of their characteristics that are drivers of their rich biodiversity, but that also generate challenges for monitoring change. Of course there are common general challenges related to the mapping of wetland habitats using RS data, but here we focus on those that particularly affect wetlands in arid and semiarid areas such as the Mediterranean. The true specificities of this region, e.g., how wetlands are classified for management purposes under local, regional and international legislation on wetlands, or how nomenclatures have led to over- or underestimation of regional habitats, do not lessen the applicability of most lessons to other bioclimatic regions too. Finally, we emphasize that many of the challenges reviewed below have either been addressed recently, or will be discussed further in Sections 2–4, with suggestions on how to address them.

1.1 Characteristic 1: Wetland Habitats and Surface Water Dynamics

The Ramsar Convention takes a broad approach in determining the wetlands which come under its mandate. In the text of the convention (Article 1.1), wetlands are defined as: "areas of marsh, fen, peatland or water, whether natural or artificial, permanent or temporary, with water that is static or flowing, fresh, brackish or salt, including areas of marine water the depth of which at low tide does not exceed six meters" (Ramsar Convention Secretariat, 2013).

These different wetland types have a variable level of detectability using RS data and, therefore, their identification, delineation and monitoring are challenging from a pure RS point of view. In particular, there is a frequent confusion between flooded areas (or surface waters) and wetlands in RS studies, despite their major ecological difference: some wetlands are only rarely and partly flooded, whereas many nonwetland habitats (e.g. agricultural or forest) can occasionally be flooded. For example, the GIEMS dataset (Global Inundation Extent from Multi-Satellites) and other measurements of surface water are frequently assimilated to wetland extents (Aires et al., 2013; Fluet-Chouinard et al., 2015; Papa et al., 2006; Prigent et al., 2001, 2012). These two essential metrics are in fact quite distinct, as national statistics for Mediterranean countries show (Fig. 1), and confusing them may lead to underestimations or overestimations of the total amount of wetland habitats.

One of the main reasons why RS tools often underestimate the extent of wetland habitats is that the spatiotemporal dynamics of flooded areas are difficult to tackle, even with good time series. For instance, some ephemeral wetlands are rarely flooded and are often missed by RS datasets recording "surface waters." Conversely, identifying and delineating flooded areas

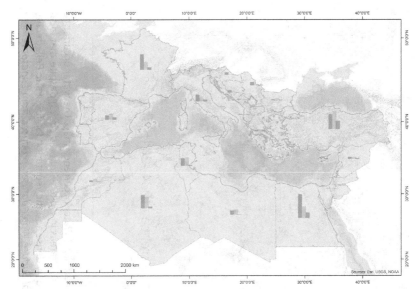

Fig. 1 Comparison between national estimates of total wetland surface for Mediterranean countries (Mediterranean Wetland Observatory, *green*; after Perennou et al., 2012; Global Lakes and Wetlands Database, *brown*; after Lehner and Döll, 2004) and surface waters (Global Surface Water, *blue*; after Pekel et al., 2016).

under a dense vegetation canopy are crucial for the monitoring of several Red List habitats, but still difficult to assess using optical RS data, where the presence of water is not easily detected under dense emergent vegetation (see Section 3.2 for some possibilities offered by new Synthetic Aperture Radar (SAR) data). As a result, what is often mapped as "flooded areas" are in fact only the "open water areas," leading to an underestimation of the real flooded areas. Fig. 2 illustrates a lake in Algeria mapped using the Global Surface Water dataset, a Landsat-based approach using optical data (Pekel et al., 2016) where parts that are temporally flooded are indicated as dry for more than 30 years (from 1984 to 2015). This is contrary to local ecological knowledge of this well-known Ramsar site (Saifouni and Bellatreche, 2014). The reason is simple: dense aquatic vegetation hinders the detection of truly inundated areas.

1.2 Characteristic 2: Artificial vs Natural Wetland Habitats and Their Relevance for Biodiversity

Wetland habitats, hydrology and water quality have been thoroughly manipulated by humans for centuries so wetlands that are strictly "natural" in their functioning are rare. It is therefore challenging to draw

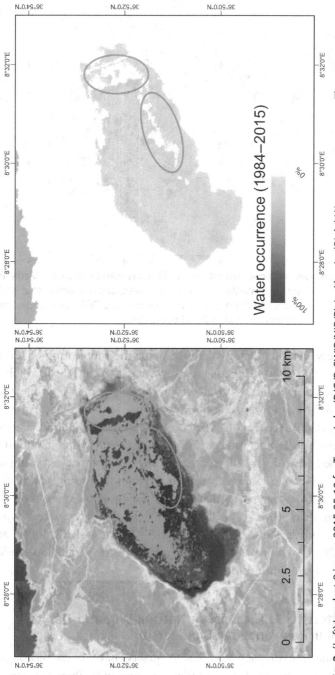

Fig. 2 (*Left*) Landsat-8 image 2015-05-12 for Tonga Lake (R/G/B: SWIR/NIR/R), in Algeria. (*Right*) Water occurrence map. The water occurrence map would suggest that part of the wetland was never flooded between 1984 and 2015 (areas in *white* within *red circles*) despite that local experts identified these areas as temporarily inundated during this period and covered by dense aquatic vegetation (Saifouni and Bellatreche, 2014). *Left: Landsat 8 image: Courtesy of the U.S. Geological Survey. Right: The Global Surface Water dataset, produced from Landsat remote sensing data (after Pekel et al. 2016).*

the line between "natural wetlands" and "artificial wetlands," given that many intermediate situations exist, e.g., very old modified wetlands that have reverted to a "near-natural" state in terms of habitats, while still having a strongly man-modified hydrology and water quality. Despite this difficulty, ecologists routinely use the distinction "natural" and "artificial" (e.g. Sebastián-González and Green, 2016).

One of the main reasons for the monitoring of Mediterranean wetlands is to obtain estimates of changes in habitat extent and related impacts on biodiversity. Of all the species that Mediterranean wetlands host, an important part is endemic and many Red List species depend on natural wetland habitats (MWO, 2012b; Riservato et al., 2009). Artificial wetlands host a significantly lower biodiversity (e.g. Sebastián-González and Green, 2016), and it is therefore important to be able to separate them reliably from natural wetlands (e.g. distinguishing a natural lake from a reservoir; a natural lagoon from a salina). But, in practice, telling the difference based on RS data alone is not always possible. Use of object-based classifications or of a predefined matrix to discriminate natural from man-made wetlands have been proposed (Camilleri et al., 2017). However, natural and artificial wetlands are often similar in shapes and updating an artificial wetland matrix can pose a problem for routine monitoring over large geographical scale comprising several hundreds of sample sites. Finally, and related to artificialization, water quality too has a profound impact on biodiversity, and RS monitoring methods are making progress (e.g. Brezonik et al., 2005; Ritchie et al., 2003; Sandström et al., 2016; Tyler et al., 2006; Vihervaara et al., 2017).

1.3 Characteristic 3: Strong Inter- and Intraannual Variability

The rate with which changes take place and the variability of water availability in wetlands in arid or semiarid regions such as the Mediterranean is a challenge for the detection of long-term trends. Although some changes take place rapidly and are relatively easy to detect (e.g. construction of a dam), others are slower and show greater variation over time (e.g. reduced water availability due to climate change, or agricultural land abandonment transforming it back gradually into a "natural" wetland), making it difficult to distinguish when a site has definitely changed or is "only" temporarily affected.

Mediterranean climate conditions vary irregularly within and between years. In the driest areas of North Africa, large wetlands (e.g. chotts and sebkhas) can remain virtually devoid of water for years, and refill irregularly depending on erratic rains. Assessing the boundaries of such wetlands through RS poses challenges, since long time series of images may be required to capture the full, potential extent of a given wetland, independently from its highly variable level of filling. Images that by chance focus on a dry period may be prone to misinterpretation as "severe reduction in the wetland size". Long-term, slow changes like the impact of declining precipitation are difficult to detect under such highly variable conditions and a reduction in wetland extent even more so. Clouds, which bring precipitation and refill wetlands, are an additional, frequent obstacle to obtaining good time series of satellite images, and may hinder the detection of flood extent in crucial periods.

Based on these characteristics, many of which are common to other arid and semiarid regions beyond the Mediterranean basin, we can identify three challenges for the use of RS to monitor trends in Mediterranean wetlands: (1) the delineation and separation of habitat types; (2) the mapping of the water dynamics inside wetlands; and (3) the detection of trends over time with respect to a naturally occurring variability.

In the next sections, we demonstrate how these challenges have previously led to errors in the interpretation of RS data for Mediterranean wetlands monitoring, how some challenges have been recently overcome and which other ones remain to be addressed. Table 1 presents a schematic guide to how these issues will be covered.

2. DELINEATION AND SEPARATION OF HABITAT TYPES

Temporal detection of changes in habitat type and extent of Mediterranean wetlands first requires that the delineation and separation of habitat types is consistent over time. Second, to determine the nature of the changes, the methods and nomenclature need to be constructed coherently. Key elements for delineating and mapping wetlands using RS are quality images; rules for detection of significant changes in wetland habitats; a wetland nomenclature; procedures for interpreting habitat classes; and robust validation procedures. Despite precautions, land cover misclassifications always occur (Kleindl et al., 2015), and a known margin of error is usually accepted by specialists. However, known and unknown type errors can easily outweigh the credibility of habitat maps if the mentioned key elements are not carefully developed and implemented. Known and unknown errors can be caused by intrinsic technical limitations, human errors, or an interaction of both. Misclassifications can be significantly reduced when local knowledge and ecological expertise of Mediterranean wetlands are combined with technical RS expertise.

Table 1 Schematic of the Challenges Posed by Wetland Remote Sensing and of Some of the Solutions Currently Applied

Inherent Characteristics of Wetlands in Arid Regions Such as the Mediterranean That Render the Interpretation of RS Information Difficult	Progress Made to Date	Remaining Challenges
(i) Intertwined wetland habitat and surface water dynamics	– Increased knowledge on uncertainties – Increased number of images per year – Improved habitat classification at local scale	– Quantification of uncertainties for habitat changes – **Retroactive studies (to periods with fewer images)** – Decision rules for habitat classification at large, multisites scale – **Integration of ecological knowledge with technical know-how** – Improved nomenclature based on dynamics as characteristics (trade-off between more classes and more confusion risks) – Improving semantics: wetlands vs flooded areas vs open waters – **Application of existing validation procedures**
(ii) Artificial vs natural wetland habitats and their relevance for biodiversity	– Object-based classifications	– Application to large wetland samples – Water quality assessment beyond a few pollutants
(iii) Strong inter- and intraannual variability	– Increased number of images per year	– Separating long-term trends vs annual variability in flooding – **Detection of water under vegetation** – Detection of ephemeral wetlands

Remaining challenges highlighted in bold are applicable to more than one of the characteristics in left column, but are only mentioned once.

2.1 Wetland Habitat Nomenclature

To produce maps for habitat monitoring derived from optical RS data, a nomenclature is required to identify and delineate separate habitats. Ideally the nomenclature consists of classes that are both ecologically relevant and distinguishable with optical RS data. In the absence of a satisfying standardized nomenclature for Mediterranean wetlands, most initiatives developed their own, which has led to a multitude of nomenclatures with varying degrees of applicability at different scales and geographic regions (Tomaselli et al., 2013). For instance, the GlobWetland-II (GW-II) ESA project (GlobWetland-II, 2012) developed a hybrid hierarchical typology between the European Union's CLC (European Commission/JRC and EEA, 1997) and the Ramsar Convention's nomenclature (Ramsar Convention Secretariat, 2010). Whereas CLC encompasses all classes of land cover to be found in Europe with only 11 classes being predominantly wetlands, the Ramsar typology (Ramsar Convention Secretariat, 2010) provides ecologically relevant and detailed descriptions of wetlands comprising 42 classes, but does not cover other habitats, which are out of the scope of the Ramsar Convention. Another example of hybrid classification developed for specific purposes is from the Horizon-2020 SWOS (Satellite-based Wetlands Observation Service) project that combines the hierarchical Mapping and Assessment of Ecosystem Services (MAES) nomenclature with wetland classes to monitor their potential for ecosystem services.

When nomenclatures do not contain many wetland-relevant classes, such as the much used CLC in European countries, a significant part of wetland habitats can go undetected as they are merged with larger (e.g. agricultural) classes. This was for instance the case in France where wet meadows, nationally one of the most important wetland types in terms of area, ended up being lumped with dry meadows in CLC, and therefore not identified as "wetlands" in the final maps (Perennou et al., 2012).

From an ecological point of view, a more detailed nomenclature is appealing because its application may provide better estimates of biodiversity and ecosystem services. However, use of ancillary data is then necessary to produce reliable maps from RS data. Ancillary data may include preexisting local land cover maps, in situ data, literature, VHR images and contextual data (e.g. topography, hydrology, precipitations). Collecting this information may be realistic when working at the scale of a site level or for a limited number of habitats, but less feasible at large scales. For instance, of the 103 different classes developed in GW-II, the 55 classes for wetlands comprised many classes that eventually proved difficult or even impossible to segregate using RS data alone. In theory, the hierarchical structure should have allowed mapping at higher level only, when the information available did not allow separating finer (lower-rank) classes. However, in practice, the lack of ancillary data often translated into detailed maps of high uncertainty although confusion matrices (see below) were developed. Unfortunately, this uncertainty is rarely estimated when mapping biodiversity and ecosystem services (e.g. Rocchini et al., 2011), resulting in visual representations that may be powerful, but include lots of unacknowledged errors (Hauck et al., 2013). This should eventually be overcome by not placing on the mapping operators any pressure for always having final results using the most precise habitat type level, rather accepting maps with less precise classes that are more accurately mapped. This could be done by aggregating habitat types into broader classes (higher levels in hierarchical nomenclatures), when dealing with regional assessments involving many sites and limited ancillary data. Aggregating land cover categories into less numerous classes has been shown to increase thematic map accuracy (Kleindl et al., 2015), thereby reducing the classification errors and increasing time efficiency for multiple sites assessments (e.g. MWO, 2014). However, it also reduces the capacity to monitor specific habitat transformations at wide scales, which is the main interest of using RS data. In addition, coarse land cover categories are likely to provide insufficient information if the maps are to be used to inform management decisions. Clearly, an overarching, hierarchical nomenclature is needed to make both local and broad-scale assessments, but deciding on the most relevant level of detail to use needs to be carefully set depending on the scale of the work (local vs regional) and the purpose for which the map will be used, and taking into account the necessary trade-off between more classes and more confusion risks.

Another approach that is particularly useful to document habitat transformation consists of using the Earth Observation Data for Habitat Monitoring (EODHaM) system developed by Lucas et al. (2015). Using the hierarchical land cover classification system (LCCS) from the Food and Agriculture Organization (FAO), this approach uses a combination of pixel- and object-based procedures. The first four levels of the FAO LCCS can be obtained based on simple rules that can be quantified by RS: vegetated vs nonvegetated areas, herbs vs trees, terrestrial vs aquatic, cultivated vs natural, etc. Combined with expert local knowledge (e.g. available land cover land use maps), these methods are also applicable to other nomenclatures (e.g. EUNIS, Annex I) for generating habitat maps. An additional module quantifies changes in the LCCS classes and their components, being particularly useful to monitor ecosystem evolution and support decisions relating to the use and conservation of protected areas, including wetlands.

2.2 Quality Images

Distinguishing different wetland types or wetlands from other habitat types based on satellite images often requires several scenes from contrasting seasons. This is the case for instance for separating ricefields (which are artificial wetlands) from dry crops or from reedbeds (natural wetlands), since the seasonal flooding regime or plant phenology are key criteria to distinguish these habitats. The initial Landsat Thematic Mapper (TM) and Enhanced Thematic Mapper (ETM) data had intrinsic limitations in temporal resolution, which limited the detection and interpretation of wetlands. In some areas, a too low image frequency for previous time periods (e.g. during the 80s and before) increased the probability that over a given year, no cloud-free images would be available which is a challenge especially in the rainy season. This leads to incorrect habitat mapping caused by omission errors (when a habitat is left out of the category being evaluated) or commission errors (when a habitat is incorrectly included in the category being evaluated).

Several solutions have been tested to overcome these misclassifications, each with their own limitations. For instance, periods over which RS images are used were extended, sometimes up to 2 years around official dates, to acquire enough satellite scenes at different seasons (GlobWetland-II, 2011). This however reduces the variability that can be detected over shorter time scales. Another approach, which has become available recently, is the inclusion of higher frequency images, e.g., SAR data from Sentinel-1 and optical data from Sentinel-2, which will certainly be useful for reducing the error rates due to image availability.

To reduce the unknowns when not enough ancillary data are available, mapping can focus on broad habitat classes encompassing all those that cannot be separated (e.g. "water bodies" instead of "natural lakes" vs "man-made reservoirs") (Mediterranean Wetlands Observatory internal protocol). This approach, however, will not permit the detection of all habitat transformations that take place.

To improve the accuracy of habitat delineation, especially for retrospective analysis, the number of images can be increased. New technologies, such as Sentinel can provide a higher number of images in a shorter time span, but to have a similar number of images for a date in the past requires searching for images at such a very large range around the intended date (e.g. sometimes ±2 years in the case of GlobWetland–II, 2012) that inter- and intraannual variation can no longer be distinguished. While this approach is reasonable for detecting and delineating natural habitat types that are unlikely to change much over a few years, it leads to systematic overestimations of specific habitat classes in agricultural landscapes with a high crop rotation frequency, such as rice. Flooded fields that are used for rice production are counted as wetlands and by aggregating images of multiple years, the total surface covered with (flooded) rice fields in any year is summed up, rather than averaged. This can lead to an overestimation of rice fields up to three times their actual surface, e.g., in CLC (Perennou et al., 2012). However, the availability of new optical and SAR RS data with very high temporal resolutions (Landsat-8, Sentinel-1 and Sentinel-2) allows to better capture interannual dynamics of habitats such as ricefields (Fig. 3).

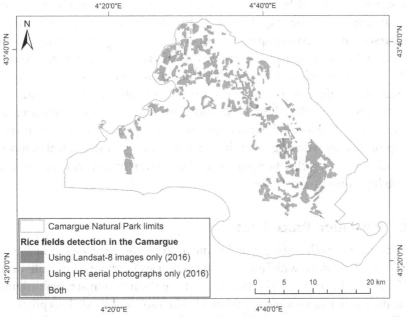

Fig. 3 Comparison between two different methods delineating rice fields in the Camargue in 2016 an object-based classification of 8 Landsat-8 (L8) images covering the whole hydroperiod and using the GEOclassifier software developed by the SWOS project (total rice fields area = 11,201 ha); and a "field reality" consisting of a land cover map provided by the Regional Natural Park management body, based on aerial photographs interpretation (total rice fields area = 10,694 ha). Rice field areas are almost the same, with a little overestimation for the L8-derived map (90% of existing rice fields are detected using L8 images).

2.3 Procedures for Interpreting Habitat Classes

Technical experts that create RS maps are not necessarily familiar with all the (wetland) habitat types and wetland ecology. Therefore, field data are crucial for training and validating maps, as are explicit guidelines for identifying and reducing supervised classification omission and commission errors, and obtaining comparable results from different operators. These guidelines should include explicit decision rules on how particular habitat types should be separated, their flooding calendar, vegetation phenology and how/when/what ancillary data would best assist the mapping, considering which habitat types are commonly confounded in the Mediterranean. They should also assist with what to do in borderline cases, i.e., situations where the habitat delineation and identification can lead to two different habitat type interpretations, both being valid from different wetland perspectives. For instance, decreasing water levels in a man-made reservoir will expose banks that, when covered by aquatic vegetation, appear very similar to natural marshes and could be mapped as such. In a dry year, the falling water levels in a reservoir may consequently be misinterpreted as a decrease in man-made reservoir habitat coupled with an increase in (natural) marshes (e.g. MWO, 2014). Such decision rules are clearly a much needed avenue for future research, and developments are ongoing but remain so far unpublished.

In summary, when human decisions are required to produce accurate maps, clear, detailed and explicit guidelines and training data will enhance map quality by reducing both the variability between individual mappers, as well as reducing classification errors. This in turn greatly increases the replicability of results.

2.4 Validation Procedures

Despite having diligently applied nomenclatures using the best available RS and ancillary data with interpretation protocols, errors can still occur (Kleindl et al., 2015; see also Box 1). To keep them within acceptable limits, the produced maps should always be checked and validated. This process of validation can rely on comparing the produced maps with "reality" by using spatially and temporally specific reference material, combined with a critical, independent assessment based on wetland ecology expertise.

For land cover maps, the ideal situation is an in situ (field) information verification at a date sufficiently close to the satellite image(s) used, but RS-derived maps can also be compared with other independent and more accurate maps, such as often produced by local management bodies, usually based on a higher spatial resolution data and integrating more complex thematic details, or with some regional wetland inventories (e.g. Congalton, 1991; Fluet-Chouinard et al., 2015; Sanchez et al., 2015). Known borderline cases and clusters of often confused habitat types can be reviewed by wetland experts to further decrease uncertainties in habitat identifications. In addition, wetland experts can easily detect some habitat identification errors by comparing produced maps or their trends over time. For instance, after a dam is built on a river and the reservoir fills up, the habitat behind the dam should be identified as a "human-made reservoir/lake" and should not be identified as an (expanding) "river habitat," i.e., a natural wetland type (Fig. 4).

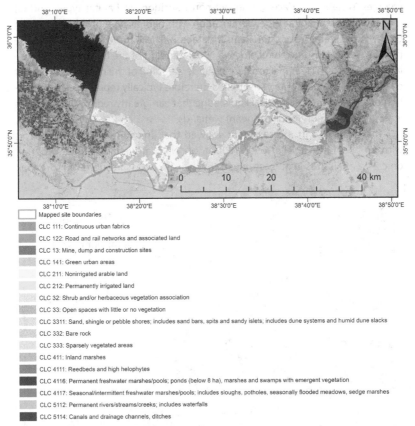

Fig. 4 A typical error affecting the largest man-made wetland in Syria: Al-Assad reservoir on the Euphrates, mapped as a "Permanent river." Background: Landsat TM 2006-06-24 (R/G/B: SWIR/NIR/R). *Data from GlobWetland-II, 2014. GlobWetland-II, a regional pilot project of the Ramsar Convention on Wetlands: handbook. GW-II project documentation. JenaOptronik, Jena, Germany. 110 p. http://www.globwetland.org/.*

BOX 1 A Quality Control Exercise of Mediterranean Wetland Mapping Using RS Data—The GlobWetland-II Case Study

The GlobWetland-II (GW-II) project (ESA DUE project, 2010–14) aimed at producing a homogeneous assessment of 284 wetlands spread all over the Mediterranean region (Fig. 5) and their change over time (1975–1990–2005), using Landsat imagery. The approach followed in the project accumulated uncertainties due to many of the challenges presented in this chapter.

In this box, we present an analysis of the Mediterranean Wetlands Observatory (internal document) in which we quantified the errors of the habitats mapped in the GlobWetland-II project. The quality control consisted of the scrutiny of the database on three possible inconsistencies: (1) unrealistically large changes in habitat surface; (2) uncommon transitions in habitat types; and (3) mismatch of the habitat type identification with habitat type observations from the field, the literature and the Ramsar database. These three inconsistencies are partly based on insufficient ecological understanding of the Mediterranean wetlands and partly based on a too general understanding of the mapping of habitat changes. The quality control we applied therefore typically represents an integration of technical and ecological knowledge that can greatly reduce uncertainties in RS monitoring of (Mediterranean) wetlands.

The quality control assessment showed that, in the initial analysis, errors with an absolute value of over 1000 ha of "Natural wetland areas" occurred on c.24% of all sites in at least one of the years; misclassifications of more than

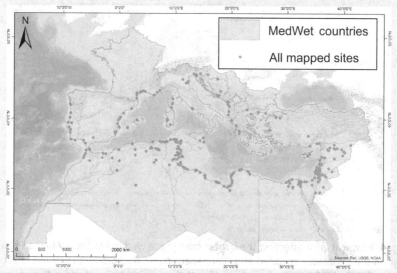

Fig. 5 Distribution of the GlobWetland-II project sites.

BOX 1 A Quality Control Exercise of Mediterranean Wetland Mapping Using RS Data—The GlobWetland-II Case Study— cont'd

10,000 ha of natural wetlands were found on 5% of the sites, and two sites had more than 100,000 ha of natural wetland areas incorrectly classified. To put these numbers in perspective, a total of 1.97 million ha of natural wetland habitat was mapped in the Mediterranean region across the 284 sites in 2005. The misclassifications generated both under- as well as overestimations (Fig. 6) and were not systematic. This means that their effect compensated each other to a variable extent in different years for the pan-Mediterranean assessment.

Overall, these errors lead to a distortion and an overall underestimation of the actual loss of natural wetlands, which proved to be 30% higher than initially estimated (i.e. a natural wetland habitat loss of 13% instead of 10% in 30 years). The errors also caused an underestimation of the overall gain in human-made wetlands which turned out to be +159% instead of the originally reported +54%. A reanalysis of the whole dataset had to be undertaken, using new internal decision rules of the Mediterranean Wetlands Observatory based upon the lessons learnt from the first, flawed analysis (e.g. Beltrame et al., 2015; MWO, 2014).

The steps followed can be viewed as a quality control for end products, i.e., both for maps and for key indicator values, such as the natural and artificial wetland surface at each date. This example of a posteriori quality control clearly demonstrates the need for wetland expertise and data to validate the remotely sensed produced maps.

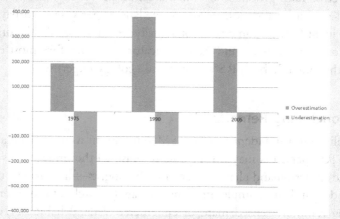

Fig. 6 Cumulated errors (in hectares) for the surface areas of natural wetland habitats for the 284 sites per studied period.

A confusion (or error) matrix is an important requirement (e.g. Camilleri et al., 2017). It should be developed using ancillary data to quantify for each habitat class, or aggregation of classes, the risk of confusion with other classes and the overall error rate (both omission and commission). This will in turn allow any user to understand the limit of the end products and their margins of error, whether for decision making or for use in ecological modelling. Accepted error margins in wetland assessments at a single site are typically in the order of 10%–20% (e.g. Dadaser-Celik et al., 2008; Guo et al., 2017; Rapinel et al., 2015). In the GW-II project, the initially reported error rate across multiple sites was 12.4% for all habitat classes and 10.6% for wetland habitat classes (GlobWetland-II, 2014), although it later proved to be significantly higher. In two ongoing projects, SWOS (http://swos-service.eu/) and GlobWetland-Africa (http://globwetland-africa.org/), error margins of 15% are considered acceptable.

Similarly, when monitoring at wide geographical scales using standard RS-based approaches, results on status and trends (such as total areas of the different habitats at different dates) can also be checked for any unlikely and implausible changes. Wetland experts play a crucial role in detecting unlikely habitat changes which require a closer look. For instance, the transition of a freshwater lake into a coastal brackish lagoon, or transitions of thousands of hectares from lagoons into marshes are very unlikely over a 6–15 years period (see e.g. Ernoul et al., 2012). To increase the reliability of the error rates produced by the validation protocol, when designing a ground-truthing validation protocol, wetland experts need to pay special care to identify what can or cannot actually be validated. Validation procedures for outputs (maps and trend indicators) are therefore most effective when they include both RS and wetland ecology expertise.

2.5 Concluding Remarks

The delineation and identification of Mediterranean wetland habitats can be tricky. RS specialists may not be able to grasp all the implications of how wetlands function and change over time. Ecologists and managers may also not understand the complexities and caveats (miss-classification errors) of creating remotely sensed maps of ephemeral wetland habitats. Technical development should therefore be coupled with ecological knowledge notably in deciding upon the required detail of nomenclatures, availability of ancillary data, accessibility to reliable information for the selection of training data (field information, ancillary data or local knowledge) and image frequencies, as well as development of explicit interpretation guidelines and validation procedures. An integrated approach can significantly improve the quality and accuracy of produced maps and indicators.

3. MAPPING THE WATER DYNAMICS OF WETLANDS

Beyond identifying wetland habitats, mapping flooded areas is essential, since wetland biodiversity reacts to flooding regimes, both in terms of quantity (flooded surfaces) and timing (hydrological cycles). The flood regime influences the distribution of aquatic plant species and communities as well as wetland primary productivity (Díaz-Delgado et al., 2016; Tamisier and Grillas, 1994). The key elements for delineating and mapping flood extent and water dynamics using RS include ability to distinguish "inundated/flooded" vs "open/surface water" areas; good seasonal coverage of the whole hydroperiod; and a validation procedure involving ground-truth data and/or expertise on wetland ecology/hydrology.

3.1 Mapping Inundated or Open Water Areas

In this discussion, we define flooded (or inundated) areas as being covered by water irrespective of vegetation presence, and open (or surface) waters as flooded areas free of vegetation. This distinction is ecologically relevant, as key biodiversity features differ between these two zones. For instance, large flocks of wintering waterbirds will usually favour open waters, but not flooded areas with dense emergent vegetation (Tamisier and Dehorter, 1999). Yet, flooding regimes are important for the maintenance of emergent vegetation, which is typically used by waterfowl for nesting activities. As already discussed, imprecise terms may lead to ambiguous maps and/or interpretations (e.g. Aires et al., 2014). A more precise use of terminology would significantly increase the consistency and coherence between maps of flooding regimes.

Mapping flooded areas, especially under vegetation is a particular challenge. Although some long SAR wavelengths can penetrate vegetation (Hess et al., 2003), dense, emergent aquatic vegetation hinders the detection of water when using multispectral optical or radar sensors (Cazals et al., 2016; Horritt and Mason, 2001), in a variable way depending on the dominant species (Davranche et al., 2013). SAR sensors generally perform slightly better although they do not penetrate the vegetation completely. As a consequence, "open water" areas which are easier to detect are often considered to represent "inundated/flooded areas" (e.g. Aires et al., 2014) causing the true extent of flooded areas to be underestimated (Fluet-Chouinard et al., 2015; Smith, 1997).

Currently, a combination of optical and SAR images offers the most reliable results (Töyrä et al., 2002), and the Sentinel-1 and -2 developments are likely to make this approach feasible at larger and finer scales. Additionally, some tools using the mid-infrared band of SPOT-5 images have shown promising results for the mapping of surface water dynamics independently of vegetation type and density in shallow marshes (Davranche et al., 2013). A newly developed index, the Automated Water Extraction Index (AWEI; Feyisa et al., 2014) significantly increases the detection of surface water, by reducing the risk of confusion with other dark surfaces. Such accuracy problems remained frequent until recently (Feyisa et al., 2014; Sanchez et al., 2015). The capacity of mapping water dynamics even under dense vegetation using Landsat-8 data was also improved by combining water indices like the Modified Normalized Difference Water Index (MNDWI: Xu, 2006) with vegetation indices like the Temperature Vegetation Dryness Index (TVDI: Gastal, 2016; Sandholt et al., 2002). However, in all cases, the use of ancillary data is strongly recommended to estimate mapping accuracy of flooded areas.

3.2 Images Covering the Whole Hydroperiod

Mapping the water dynamics of wetlands requires that the flood extent is mapped at different dates throughout the annual hydrological cycle (e.g. Camilleri et al., 2017). This information is often overlaid to produce a single map providing flooding duration of different wetlands according to classes, e.g., as "Never," "Seasonally" and "Permanently flooded" areas (GlobWetland-II, 2011) over a given cycle, or as flood/submersion frequencies, i.e., percentages of the analysed images in which a given pixel was flooded (e.g. Davranche et al., 2013; Pekel et al., 2016; also see Fig. 7), or in terms of days/months (e.g. Díaz-Delgado et al., 2016). Obviously, increasing the number of scenes will increase the precision of flooding duration estimation for highly seasonal wetlands. In the past, due to limited availability of cloud-free optical images, "water dynamics" maps have been produced with as little as two images from a hydrological year (GlobWetland-II, 2011), which is not sufficient for monitoring short-term dynamics of wetland hydrology having a seasonal water regime.

SAR satellites provide an improvement for mapping open water dynamics throughout the year due to their independence of daylight and cloud cover. Furthermore, some sensors offer the ability to detect water overgrown by vegetation. Although several SAR satellite missions have been launched in the past decades, ERS, ENVISAT ASAR, TerraSAR-X and ALOS PALSAR to name a few, water body monitoring was hindered by infrequent data acquisition schemes and acquisition cost. General overviews on the use of SAR for wetland monitoring and water mapping are provided in Brisco (2015) and White et al. (2015).

A rather large and consistent historical image archive is available from the ASAR C-Band sensor aboard ESA's ENVISAT satellite, which was operated from 2002 to 2012. In particular its Wide Swath Mode (WSM), although not optimal due to its limited VV polarization and low spatial resolution (150 m), has been successfully used in several water mapping studies (e.g. Bartsch et al., 2008; Kuenzer et al., 2013; Matgen et al., 2011; Schlaffer et al., 2015) and also specifically to characterize wetlands by Schlaffer et al. (2016).

With ESA's Sentinel-1 mission, currently consisting of two satellites with a combined repeat rate of down to 6 days and data provided at no cost, water dynamics mapping at high resolution and temporal frequency now becomes easier and cheaper. At the time of writing still only few publications can be found on Sentinel-1 water mapping, but it has been used for flood detection by, e.g., Boni et al. (2016) and Twele et al. (2016). Like ASAR WSM data, Sentinel-1 images are acquired in VV polarization. Thus, the imagery is very sensitive to waves, which can easily be mistaken for land surfaces (Brisco, 2015), as shown for instance in Fig. 7.

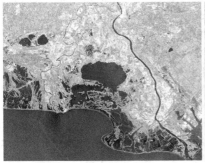

Fig. 7 Backscatter characteristics of water surfaces under different environmental conditions. Smooth water surfaces (*left*) appear as *dark*, while wind-induced waves generate a much brighter signal (*right*). The reduced contrast between water and land surfaces in the right image will result in large areas of water to be mistaken for land in automated classifications. Camargue, Southern France, Sentinel-1 IW VV, acquired on January 31, 2015 and February 05, 2015, respectively (approx. $50 \times 42\,km^2$). *Produced from ESA remote sensing data.*

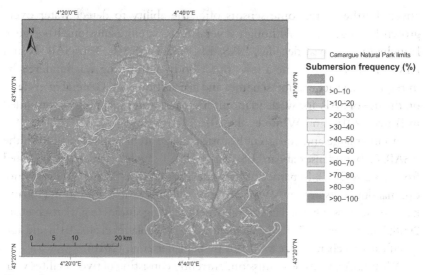

Fig. 8 Camargue, France: 2015 intraannual open water dynamics based on SAR data from Sentinel-1 Interferometric Wide Swath mode (IW). All available images acquired on 75 days throughout the year were included. Though an improvement from previous maps, some inconsistencies remain. *Produced from ESA remote sensing data.*

One opportunity to circumvent this is by analysing the dense time series of images of individual pixels instead of the mostly used per image classification. For instance, Schlaffer et al. (2015, 2016) used harmonic models to estimate seasonal behaviour of SAR backscatter and detect outliers for flood detection. Fig. 8 shows an example of a time-series analysis technique, which assesses not only a pixel's intensity but also its class likeliness based on the temporal stability of the observed pixel. This way, images acquired under suboptimal conditions (e.g. heavy clouds or wind-induced waves) can be reliably classified, although the single scene shows no contrast between land and sea surfaces. This ongoing research will be extended to historic ASAR and ERS data once the approach has been refined to a satisfactory degree of accuracy. In the meantime, care has to be taken for trend analysis where flood regime maps are compared over time, notably for past trend analysis going back further than 1980.

3.3 Validation of Flood Regime Maps

Flood regime mapping has its own challenges and a thorough comparison with ground-truth data is required to validate final maps. Validation is less frequently performed for the flooding regime than for habitat classification, but Davranche et al. (2013) and Thomas et al. (2015) found overall accuracy rates of 83% in the Camargue, France, and 93%–95% for the Macquarie marshes, Australia, respectively. A particularly robust field validation covering 6000 points in 31 ground-truth field campaigns from 2003 to 2013 was implemented in the Doñana marshes, Spain, with an accuracy estimate of 94% (Díaz-Delgado et al., 2016).

A situation where careful validation is required is where hydrological cycles do not show regular annual patterns, e.g., in large temporary wetlands in arid regions which do not fill up completely every year, and whose outermost parts often remain dry. These margins should nevertheless not be identified as switching from dry (e.g. steppe) habitat to wetland habitat, depending on their seasonal or annual flood conditions (e.g. Fig. 9). To avoid these misclassifications which can lead to large errors in wetland area estimation, interpretation procedures may recommend to carefully reevaluate all rapid and unlikely back-and-forth shifts between dry land and wetland habitats.

3.4 Concluding Remarks

The hydroperiod is a crucial indicator for monitoring trends in ecology, functions and services of Mediterranean wetlands. Unfortunately, the mapping of flooding regime is also technically difficult. With increasing availability of multiple images per year, hydrological cycles as well as water extent under vegetation will potentially be better captured. However, for the interpretation of trends detected by RS, ecological knowledge is required to distinguish unpredictable variations typical of Mediterranean wetlands from actual habitat transitions.

4. DETECTION OF TRENDS OVER TIME

Opportunities and difficulties to identify habitat and flood extent of Mediterranean wetlands given the uncertainties in land cover maps and the irregularity of flooding regimes have been addressed in the two preceding sections. Use of RS observation to estimate long-term trends in ecosystem quality and biodiversity, infer some complementary challenges.

4.1 Uncertainty of Detecting Trends

A major question when it comes to detecting changes in Mediterranean wetlands is how to quantify the actual rates of change when they are in

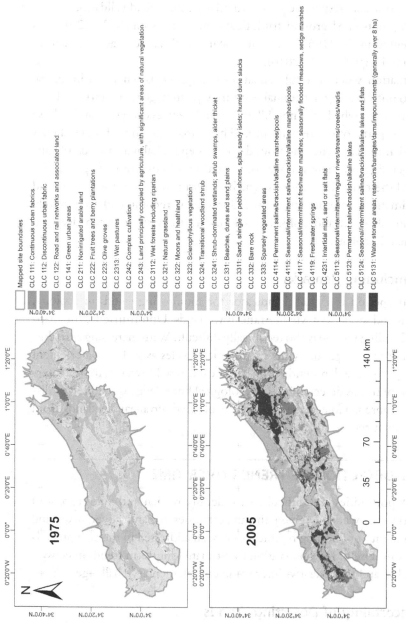

Fig. 9 A typical confusion affecting large wetlands in arid zones (Chott Ech-Chergui, Algeria) in 1975 (*upper map*) and 2005 (*lower*): a higher inundation level in 2005 due to higher rainfalls in previous years was mapped as "large increase in marshes," i.e. natural wetland area" (+147,000ha), and conversely as a "decrease in terrestrial habitats such as Steppes/Pastures." In reality the chott area did not vary; only its level of flooding did. At the pan-Mediterranean scale, this lead initially to underestimating the true natural wetland loss, since in the total figures for the 284 sites, this apparent "increase" offset 147,000ha of actual loss elsewhere (GlobWetland-II, 2014).

Legend entries:

Mapped site boundaries
CLC 111: Continuous urban fabrics
CLC 112: Discontinuous urban fabric
CLC 122: Road and rail networks and associated land
CLC 141: Green urban areas
CLC 211: Nonirrigated arable land
CLC 222: Fruit trees and berry plantations
CLC 223: Olive groves
CLC 2313: Wet pastures
CLC 242: Complex cultivation
CLC 243: Land principally occupied by agriculture, with significant areas of natural vegetation
CLC 3112: Wet forests including riparian
CLC 321: Natural grassland
CLC 322: Moors and heathland
CLC 323: Sclerophyllous vegetation
CLC 324: Transitional woodland shrub
CLC 3241: Shrub-dominated wetlands; shrub swamps, alder thicket
CLC 331: Beaches, dunes and sand plains
CLC 3311: Sand, shingle or pebble shores, spits, sandy islets; humid dune slacks
CLC 332: Bare rock
CLC 333: Sparsely vegetated areas
CLC 4114: Permanent saline/brackish/alkaline marshes/pools
CLC 4115: Seasonal/intermittent saline/brackish/alkaline marshes/pools
CLC 4117: Seasonal/intermittent freshwater marshes; seasonally flooded meadows, sedge marshes
CLC 4119: Freshwater springs
CLC 4231: Intertidal mud, sand or salt flats
CLC 5113: Seasonal/intermittent/irregular rivers/streams/creeks/wadis
CLC 5123: Permanent saline/brackish/alkaline lakes
CLC 5124: Seasonal/intermittent saline/brackish/alkaline lakes and flats
CLC 5131: Water storage areas; reservoirs/barrages/dams/impoundments (generally over 8 ha)

practice often inferior to, or of the same order as the error estimation. For instance, MWO (2014) detected a loss of 10% in natural wetlands in a sample of 214 sites between 1975 and 2005. However, the habitat classification error rate was estimated at 12.3% for any given year. In such cases, because the estimate falls within the confidence interval, a cautionary approach would be to refrain from estimating any quantitative loss; on the other hand, a qualitative systematic trend (e.g. an overall loss in wetland habitat) can still be tested rigorously, e.g., through nonparametric statistics applied to the large sample of sites. For a quantitative approach, the recent improvements in image resolution and frequency still need to be translated into lower error rates (i.e. narrower confidence intervals) for habitat mapping, so as to allow the detection of wetland habitat trends of, e.g., 5%–15%. Uncertainty and error rates differ between habitat classes; for instance fewer errors are made when identifying sand and beaches than wet meadows. This means that the uncertainty of detected transformations of habitats depends on the habitat types involved, as reflected in the confusion matrices. Error rates for habitat identification are likely to decrease over time with the increasing availability of high-quality images, ancillary data and Mediterranean wetland expertise. To date, there is no estimation of the impact of the combined uncertainties on the detection of long-term trends (especially retrospectively) in Mediterranean wetlands, mainly due to a lack of validation data for older maps and this is unlikely to change.

4.2 Detecting Long-Term Changes When Flooding Extent Varies Interannually

Seasonal wetlands under dry climates (e.g. chotts and sebkhas) pose a particular challenge due to their extreme hydrological variability. Under such climates, rainfall is erratic and unpredictable, and so is the resulting extent of flooding in these typically large wetlands (see example of Chott Ech-Chergui above, Fig. 9). Comparing two maps at a 15–20 years interval may result, by chance alone, in comparing a very dry vs a relatively wet year—or the reverse—and a superficial analysis would conclude to a large increase (or conversely, decrease) in the flooding extent in a given site.

Detecting accurate long-term trends in wetland flooding remains an important issue, given the increasing human (e.g. MWO, 2012a) and climate (Giorgi and Lionello, 2008) pressure on freshwater in arid areas. One way to address this issue is to consider that trends in flooding regimes can only be assessed by using multiple seasonal maps over several years (e.g. the GSW produced by Pekel et al. (2016) and covering 32 years from 1984 to 2015). Although this approach is more time consuming, it will likely provide higher accuracy in trend estimations.

RS methods could also potentially borrow from ecological methods which aim to identify trends despite having very noisy data, e.g., occupancy or abundance modelling and other demographic modelling techniques.

4.3 Biodiversity: Changes in Ecosystem Quality

Tracking changes in wetland types and extent is important but distilling indications on ecosystem quality bring us closer to the monitoring of trends in biodiversity. Here we zoom into efforts on identifying habitat fragmentation, time lags in responses of Mediterranean biodiversity to water shortage and measures of water quality.

Many studies and projects have developed habitat fragmentation indices (e.g. Liu et al., 2014; Tomaselli et al., 2012), most of which are based on the assumption that a reduction in (natural) habitat extent should be interpreted as an increase of habitat fragmentation. Some of these indices are based on the evaluation of the landscape connectivity using specific metrics related to the size of habitat patches and the distance between them (Minor and Urban, 2008). However in the case of Mediterranean wetlands characterized by temporary water coverage, the variability in intra- and interannual surface water is not necessarily linked to these fragmentation metrics. For instance, temporary ponds may be connected during times of high water availability and separated in periods of water shortages—without any implication in terms of habitat fragmentation. This dynamic is natural for Mediterranean wetlands and occurs with a frequency which is not always predictable. The detection of trends in habitat fragmentation therefore has to take into account the ecological reality of what defines a temporary wetland habitat in the Mediterranean region (Perennou et al., 2013). Assessing long-term trends in wetland fragmentation will therefore require data series analysed over a long period to distinguish natural variability in habitat delineation from a long-term degradation of habitat extent. This should be used carefully, and only when long time series of data are available.

Long-term data series are also required for predicting the impacts of changes in flooding regimes on Mediterranean biodiversity. Many taxa and species are adapted to unpredictable availability of water and long periods of water shortages. For instance, the flamingo *Phoenicopterus ruber* is a long-lived species with an optional nesting behaviour adapted to fluctuating conditions; the mosquito *Ochlerotatus caspius* lays quiescent eggs on the ground which can survive to long periods of drought (Balenghien et al., 2010); damselflies found in brackish temporary marshes in the Camargue exhibit strong interannual variations in abundance related to marsh flooding duration (Aguesse, 1961); abundance of breeding passerines in temporary reedbeds depends on the duration of flooding from June through December in the preceding year, which determines food level during the following nesting season (Poulin et al., 2002). This means that the impact of one dry spell on overall species richness and abundances is hard to predict, and that species richness and abundances may react with a time lag relative to long-term trends in water shortages: longer time series are therefore required.

A key component of ecosystem quality is water quality. It is a long-term driver for biodiversity in Mediterranean wetlands, especially because nutrient runoff and pesticides from agriculture, as well as wastewater from settlements often end up in wetland habitats (EEA, 2012). Beyond pollutants, other substances like suspended particulate materials (SPMs) originating from soil erosion can affect biodiversity. Eutrophic (nutrient-rich) waters are usually characterized by a high productivity and a poor plant and animal species richness. Nutrients cannot be directly estimated by RS, but high nutrient availability often leads to massive development of algae. Since the concentration of chlorophyll a, the main pigment in algae, can be estimated from RS data (e.g. Brezonik et al., 2005; Matthews, 2011; Odermatt et al., 2012; Ritchie et al., 2003), it often serves as an index for algal biomass and as a proxy for eutrophication.

Turbidity is another common water quality parameter that can be estimated from RS data (Dogliotti et al., 2015; Ritchie et al., 2003). Turbidity is related to the concentration of SPM. At this moment, it is not possible to separate artificial from natural turbidity based on RS data alone. The suspended particles will affect the transparency of the water and therefore the transmittance of sunlight through the water, which can result in low plant productivity. In addition, the dissolved organic matter (DOM) present in natural waters is known to absorb light and therefore also affects water transparency. The coloured fraction of the dissolved organic matter (CDOM) is an optically active substance that affects the reflectance measured by the satellite sensor, and which thereby can be estimated based on RS data (Beltrán-Abaunza et al., 2014). These substances bind metals as well as organic contaminants, and organic substances from discharges are one major cause of pollution in surface waters, which can have a strong effect on biodiversity.

Especially adapted high-frequency images do exist for estimating water quality parameters (e.g. MERIS and Sentinel-3), but the spatial resolution is only 300 m, which limits the applicability to larger water bodies. In addition, a proper estimation of the water quality requires that the water is optically deep, i.e., that the bottom substrate cannot be seen from the surface. Many shallow wetlands are not of sufficient depth for RS to provide useful information and indicators on these issues. Improvements are expected from new images of high spatial resolution provided by Sentinel-2A (launched in 2015) and 2B (launched in 2017), which should enhance our capability to monitor water quality from space in smaller water bodies. Sentinel-2 applications are especially relevant for small lakes and for patchy waters where few or no information exists. In these habitats, Sentinel-2 can contribute to a first assessment of the water condition, in terms of eutrophication, transport of suspended sediments and distribution of invasive floating plants such as the water hyacinth *Eichhornia crassipes*.

4.4 Concluding Remarks

The ultimate objective of detecting and quantifying trends in Mediterranean wetlands and their biodiversity over time is a challenge that largely remains unsolved. New Sentinel-3 data could provide promising advances and potential on estimates of water quality. More work is needed to address how we can relate computations of flood regimes to changes in quality of ecosystems and their related species richness and abundances.

5. CONCLUSIONS

The use of RS data for the mapping and monitoring of wetlands of arid and semiarid areas such as the Mediterranean has improved since 1980. To allow for the progress made to be included in the RS tools and interpretation procedures that are being developed for new types of RS data, we here summarize the most important progress made in the form of recommendations, as well as the challenges that still need to be addressed. By identifying current best methods as well as gaps, this chapter contributes to the development of a repository for best practices of wetland RS under the framework of GEO Wetlands, and to applications beyond the Mediterranean ecoregion.

In general, the uncertainties that come with any method applied to RS data can be greatly reduced by integrating ecological expertise and ground-truth data in the different steps of the technical process. This chapter uses several ways to do this. To increase accuracy of RS products, ecological understanding of the habitats and their dynamics is required. This can be achieved by developing RS products through active collaboration with ecologists and site managers. This would allow RS experts to improve monitoring tools and ecology experts to recognize both the value and the limitations of RS data, as recommended by Skidmore et al. (2015).

The development of a robust habitat nomenclature for Mediterranean wetlands that would fulfil the needs of scientists, managers and decision makers is a tricky task, especially for an application to areas where we can only rely on RS data for monitoring these habitats. Habitat maps using typologies with many classes require enhanced validation procedures with ancillary data or expert knowledge. To ensure low uncertainties in large-scale mapping exercises, the number of habitat classes can be reduced, but this also means that some of the changes, i.e., those occurring within one of the enlarged habitat classes, will go undetected. In addition, if the purpose of a map is to inform management decisions, a more detailed classification may be required.

Uncertainties can be further reduced and consistency across maps increased, through the development of protocols for habitat identification based on integrated knowledge of wetland ecology and hydrology and RS techniques. In particular, interpretation protocols which stipulate explicit decision rules for habitat identification as well as rigorous validation procedures, can both help reduce and quantify the error estimates for each habitat class. Any mapping activity should be complemented by a systematic validation phase that should be developed and continuously adapted to include unlikely habitat transformations, recurrent interpretation errors, risks of confusion between habitats and flood extent, and focused efforts on difficult clusters of habitats.

To date, a number of challenges have been solved at the local scale of one site through the use of local experts or available ancillary data, and for assessing one-off, recent situations through improved satellite imagery. However, that does not solve the double challenge for larger-scale assessments or multisite assessments (e.g. the Mediterranean region), and for retroactive studies, i.e., comparisons with the earlier periods of RS. Progress is still needed to better segregate particular habitats, such as wet meadows from other meadows, lagoons and lakes from their peripheral marshes, and ricefields from other crops. Estimates of changes in Mediterranean wetland biodiversity would be greatly improved if we could better separate similar-looking natural vs man-made habitats, e.g., lakes vs reservoirs or fish ponds, or lagoons vs salinas. Citizen science could be used for these, for instance by getting people to send their photos, or using online georeferenced photos to confirm classifications. Changes in the flood regimes of Mediterranean wetlands still requires better detection of flooding under dense and/or emergent vegetation, as well as the development of procedures for distinguishing the natural variability in hydrological cycles from long-term trends. And, last but not least, the ability to develop reliable tools for monitoring long-term trends in wetlands of arid and semiarid areas such as the Mediterranean will require the quantification of uncertainties, both for individual maps and for changes derived from comparison of maps over time.

ACKNOWLEDGEMENTS

This work was partly supported by a suite of completed and ongoing projects, namely GlobWetland2 (GW-II), GlobWetland-Africa, SWOS and ECOPOTENTIAL. GW-II was a European Space Agency (ESA-DUE) project, in partnership with the Ramsar Convention on Wetlands and the Mediterranean Wetlands Observatory. GlobWetland Africa is an Earth Observation application project funded by the European Space Agency (ESA) in partnership with the African Team of the Ramsar convention. SWOS (Contract No. 642088) and ECOPOTENTIAL (Contract No. 641762) are ongoing projects under the Horizon 2020 Programme funded by the European Commission. SWOS and GlobWetland Africa are partners to the GEO-Wetlands Initiative, to which this study contributes. The authors are grateful to the Editor and two anonymous reviewers for suggesting substantial improvements.

REFERENCES

Aguesse, P., 1961. Contribution à l'étude écologique des Zygoptères de Camargue. Thèse de doctorat, Université de Paris. 156 pp.

Aires, F., Papa, F., Prigent, C., 2013. A long-term, high-resolution wetland dataset over the Amazon basin, downscaled from a multi-wavelength retrieval using SAR data. J. Hydrometeorol. 14, 594–607. https://doi.org/10.1175/JHM-D-12-093.1.

Aires, F., Papa, F., Prigent, C., Cretaux, J.F., Berge-Nguyen, M., 2014. Characterization and space–time downscaling of the inundation extent over the inner Niger delta using GIEMS and MODIS data. J. Hydrometeorol. 15, 171–192. https://doi.org/10.1175/JHM-D-13-032.1.

Balenghien, T., Carron, A., Sinègre, G., Bicout, D.J., 2010. Mosquito density forecast from flooding: population dynamics model for *Aedes caspius* (Pallas). Bull. Entomol. Res. 100, 247–254.

Bartsch, A., Pathe, C., Wagner, W., Scipal, K., 2008. Detection of permanent open water surfaces in central Siberia with ENVISAT ASAR wide swath data with special emphasis on the estimation of methane fluxes from tundra wetlands. Hydrol. Res. 39, 89.

Beltrame, C., Perennou, C., Guelmami, A., 2015. Évolution de l'occupation du sol dans les zones humides littorales du bassin méditerranéen de 1975 à 2005. Méditerranée (125), 97–111.

Beltrán-Abaunza, J.M., Kratzer, S., Brockmann, C., 2014. Evaluation of MERIS products from Baltic Sea coastal waters rich in CDOM. Ocean Sci. 10, 377–396. https://doi.org/10.5194/os-10-377-2014.

Boni, G., Ferraris, L., Pulvirenti, L., Squicciarino, G., Pierdicca, N., Candela, L., Pisani, A.R., Zoffoli, S., Onori, R., Proietti, C., Pagliara, P., 2016. A prototype system for flood monitoring based on flood forecast combined with COSMO-SkyMed and Sentinel-1 data. IEEE J. Sel. Top. Appl. Earth Obs. Remote Sens. 9, 2794–2805.

Bossard, M., Feranec, J., Otahel, J., 2000. CORINE land cover technical guide—addendum 2000. Technical Report No 40, European Environment Agency, Copenhagen, Denmark.

Brezonik, P., Menken, K.D., Bauer, M., 2005. Landsat-based remote sensing of lake water quality characteristics, including chlorophyll and colored dissolved organic matter (CDOM). Lake Reserv. Manage. 21, 373–382.

Brisco, B., 2015. Mapping and monitoring surface water and wetlands with synthetic aperture radar. In: Tiner, R.W., Lang, M.W., Klemas, V.V. (Eds.), Remote Sensing of Wetlands: Applications and Advances. CRC Press, Boca Raton, pp. 119–135.

Camilleri, S., De Giglio, M., Stecchi, F., Perez-Hurtado, A., 2017. Land use and land cover change analysis in predominantly man-made coastal wetlands: towards a methodological framework. Wetlands Ecol. Manage. 25, 23–43. https://doi.org/10.1007/s11273-016-9500-4.

Cazals, C., Rapinel, S., Frison, P.L., Bonis, A., Mercier, G., Mallet, C., Corgne, S., Rudant, J.P., 2016. Mapping and characterization of hydrological dynamics in a coastal marsh using high temporal resolution sentinel-1A images. Remote Sens. (Basel) 8, 570. https://doi.org/10.3390/rs8070570.

Congalton, R.G., 1991. A review of assessing the accuracy of classifications of remotely sensed data. Remote Sens. Environ. 37, 35–46.

Dadaser-Celik, F., Bauer, M.E., Brezonik, P.L., Stefan, H.G., 2008. Changes in the Sultan marshes ecosystem (Turkey) in satellite images 1980–2003. Wetlands 28, 852–865.

Darwall, W., Carrizo, S., Numa, C., Barrios, V., Freyhof, J., Smith, K., 2014. Freshwater Key Biodiversity Areas in the Mediterranean Basin Hotspot: Informing Species Conservation and Development Planning in Freshwater Ecosystems. Cambridge, UK and Malaga, Spain, IUCN. https://portals.iucn.org/library/sites/library/files/documents/SSC-OP-052.pdf.

Davranche, A., Poulin, B., Lefebvre, G., 2013. Mapping flooding regimes in Camargue wetlands using seasonal multispectral data. Remote Sens. Environ. 138, 165–171.

Díaz-Delgado, R., Aragonés, D., Afán, I., Bustamante, J., 2016. Long-term monitoring of the flooding regime and Hydroperiod of Doñana marshes with Landsat time series (1974–2014). Remote Sens. (Basel) 8, 775. https://doi.org/10.3390/rs8090775.

Dogliotti, A.I., Ruddick, K., Nechad, B., Doxaran, D., Knaeps, E., 2015. A single algorithm to retrieve turbidity from remotely-sensed data in all coastal and estuarine waters. Remote Sens. Environ. 156, 157–168.

EEA (European Environment Agency), 2012. European waters—assessment of status and pressures. EEA Report no 8/2012, EEA, Copenhagen, DK. 100 p.

Ernoul, L., Sandoz, A., Fellague, A., 2012. The evolution of two great Mediterranean deltas: remote sensing to visualize the evolution of habitats and land use in the Gediz and Rhone deltas. Ocean Coast. Manag. 69, 111–117.

European Commission/JRC and EEA, 1997. CORINE Land Cover Technical Guide. 130 p, http://image2000.jrc.ec.europa.eu/reports/technical_guide.pdf.

Feyisa, G.L., Meilby, H., Fensholt, R., Proud, S.R., 2014. Automated water extraction index: a new technique for surface water mapping using Landsat imagery. Remote Sens. Environ. 140, 23–35. https://doi.org/10.1016/j.rse.2013.08.029.

Fluet-Chouinard, E., Lehner, B., Rebelo, L.M., Papa, F., Hamilton, S.K., 2015. Development of a global inundation map at high spatial resolution from topographic downscaling of coarse-scale remote sensing data. Remote Sens. Environ. 158, 348–361.

Gardner, R.C., Barchiesi, S., Beltrame, C., Finlayson, C.M., Galewski, T., Harrison, I., Paganini, M., Perennou, C., Pritchard, D.E., Rosenqvist, A., Walpole, M., 2015. State of the World's Wetlands and their services to people: a compilation of recent analyses. Ramsar Briefing Note no. 7, Ramsar Secretariat, Gland, CH.

Gastal, V., 2016. Méthodologie appliquée de cartographie des dynamiques d'inondation des zones humides méditerranéennes. M.Sc. thesis, University of Orléans/Tour du Valat, Arles, France, 104 p.

Giorgi, F., Lionello, P., 2008. Climate change projections for the Mediterranean region. Global Planet. Change 63, 90–104.

GlobWetland-II, 2011. GlobWetland-II, a regional pilot project of the Ramsar Convention on Wetlands: technical specifications. GW-II project documentation, JenaOptronik, Jena, Germany. 115 p. http://www.globwetland.org/.

GlobWetland-II, 2012. GlobWetland-II, a regional pilot project of the Ramsar Convention on Wetlands: legend handbook. GW-II project documentation, JenaOptronik, Jena, Germany. 52 p. http://www.globwetland.org/.

GlobWetland-II, 2014. GlobWetland-II, a regional pilot project of the Ramsar Convention on Wetlands: handbook. GW-II project documentation, JenaOptronik, Jena, Germany. 110 p. http://www.globwetland.org/.

Guo, M., Li, J., Sheng, C., Xu, J., Wu, L., 2017. A review of wetland remote sensing. Sensors 17, 777. https://doi.org/10.3390/s17040777.

Hauck, J., Görg, C., Varjopuro, R., Ratamäki, O., Maes, J., Wittmer, H., Jax, K., 2013. "Maps have an air of authority": potential benefits and challenges of ecosystem service maps at different levels of decision making. Ecosyst. Serv. 4, 25–32. https://doi.org/10.1016/j.ecoser.2012.11.003.

Hess, L.L., Melack, J.M., Novo, E.M., Barbosa, C., Gastil, M., 2003. Dual-season mapping of wetland inundation and vegetation for the central Amazon basin. Remote Sens. Environ. 87, 404–428.

Horritt, M.S., Mason, D.C., 2001. Flood boundary delineation from synthetic aperture radar imagery using a statistical active contour model. Int. J. Remote Sens. 22, 2489–2507.

Kleindl, W.J., Powell, S.L., Hauer, F.R., 2015. Effect of thematic map misclassification on landscape multi-metric assessment. Ecol. Monit. Assess. 187, 321. https://doi.org/10.1007/s10661-015-4546-y.

Kuenzer, C., Guo, H.D., Huth, J., Leinenkugel, P., Li, X.W., Dech, S., 2013. Flood mapping and flood dynamics of the Mekong Delta: ENVISAT-ASAR-WSM based time series analyses. Remote Sens. (Basel) 5, 687–715.

Lehner, B., Döll, P., 2004. Development and validation of a global database of lakes, reservoirs and wetlands. J. Hydrol. 296, 1–22.

Liu, G., Zhang, L., Zhang, Q., Musyimi, Z., Jiang, Q., 2014. Spatiotemporal dynamics of wetland landscape patterns based on remote sensing in Yellow River Delta, China. Wetlands 34 (4), 787–801.

Lucas, R.M., Blonda, P., Bunting, P., Jones, G., Inglada, J., Arias, M., Kosmidou, V., Manakos, I., Adamo, M., Charmock, R., Tarantino, C., Mücher, C.A., Jongman, R.H.G., Kramer, H., Arvor, D., Honrado, J.P., Mairota, P., 2015. The earth observation data for habitat monitoring (EODHaM) system. Int. J. Appl. Earth Observ. Geoinform. 37, 1–6.

Matgen, P., Hostache, R., Schumann, G., Pfister, L., Hoffmann, L., Savenije, H.H.G., 2011. Towards an automated SAR-based flood monitoring system: lessons learned from two case studies. Phys. Chem. Earth 36, 241–252.

Matthews, M.W., 2011. A current review of empirical procedures of remote sensing in inland and near-coastal transitional waters. Int. J. Remote Sens. 32, 6855–6899.

Minor, E.S., Urban, D.L., 2008. A graph-theory framework for evaluating landscape connectivity and conservation planning. Conserv. Biol. 22, 297–307. https://doi.org/10.1111/j.1523-1739.2007.00871.x.

MWO (Mediterranean Wetlands Observatory), 2012a. Mediterranean Wetlands Outlook 2012—Technical report. Tour du Valat, Arles, France. 126 p.

MWO (Mediterranean Wetlands Observatory), 2012b. Biodiversity—status and trends of species in Mediterranean Wetlands. MWO Thematic collection, 1. Tour du Valat Research Institute for the Conservation of Mediterranean Wetlands, Arles, France.

MWO (Mediterranean Wetlands Observatory), 2014. Land cover—spatial dynamics in Mediterranean Coastal Wetlands From 1975 to 2005. MWO Thematic collection, 2. Tour du Valat Research Institute for the Conservation of Mediterranean Wetlands, Arles, France.

Odermatt, D., Gitelson, A., Brando, V.E., Schaepman, M., 2012. Review of constituent retrieval in optically deep and complex waters from satellite imagery. Remote Sens. Environ. 118, 116–126.

Özesmi, S.I., Bauer, M.E., 2002. Satellite remote sensing of wetlands. Wetlands Ecol. Manage. 10, 381–402.

Papa, F., Prigent, C., Durand, F., Rossow, B., 2006. Wetland dynamics using a suite of satellite observations: a case study of application and evaluation for the Indian subcontinent. Geophys. Res. Lett. 33. L08401, https://doi.org/10.1029/2006GL025767.

Pekel, J.F., Cottam, A., Gorelick, N., Belward, A.S., 2016. High-resolution mapping of global surface water and its long-term changes. Nature 540, 418–422. https://doi.org/10.1038/nature20584.

Perennou, C., Beltrame, C., Guelmami, A., Tomas Vives, P., Caessteker, P., 2012. Existing areas and past changes of wetland extent in the Mediterranean region: an overview. Ecol. Mediterr. 38, 53–66.

Perennou, C., Guelmami, A., Alleaume, S., Isenmann, M., Abdulmalak, D., Sanchez, A., RhoMéO Axe B., 2013. Rapport final de la 1ère phase (2011–2012), Technical report Agence de l'Eau Rhône-Méditerranée/Tour du Valat Research Institute for the Conservation of Mediterranean Wetlands, Arles, France. 190 p. + Annexes. http://doc-oai.eaurmc.fr/cindocoai/download/4931/1/rapport%20Final%20RhomeO%20Axe%20B.pdf_35408Ko.

Perennou, C., Guelmami, A., Gaget, E., 2016. Les milieux humides remarquables, des espaces naturels menacés. Quelle occupation du sol au sein des sites Ramsar de France métropolitaine? Rétrospective 1975–2005. French Ministry of Environment, Tour du Valat. Technical Report, 53 p. http://www.naturefrance.fr/sites/default/files/fichiers/ressources/pdf/161003_brochure_ramsar_occ-sol_tome_1_complet.pdf.

Plan Bleu, 2009. State of the Environment and Development in the Mediterranean. UNEP-MAP Plan Bleu, Athens, 200 p.

Poulin, B., Lefebvre, G., Mauchamp, A., 2002. Habitat requirements of passerines and reedbed management in southern France. Biol. Conserv. 107, 315–325.

Prigent, C., Matthews, E., Aires,F., Rossow, W.B., 2001. Remote sensing of global wetland dynamics with multiple satellite data sets. Geophys. Res. Lett. 28, 4631–4634, https://doi.org/10.1029/2001GL013263.

Prigent, C., Papa, F., Aires, F., Jimenez, C., Rossow, W.B., Matthews, E., 2012. Changes in land surface water dynamics since the 1990's and relation to population pressure. Geophys. Res. Lett. 39. L08403, https://doi.org/10.1029/2012GL051276.

Ramsar Convention Secretariat, (Ed), 2010. Designating Ramsar Sites, Handbook 17, fourth ed., Ramsar Convention Secretariat; Gland, CH. http://www.ramsar.org/sites/default/files/documents/pdf/lib/hbk4-17.pdf.

Ramsar Convention Secretariat, (Ed.), 2013. The Ramsar Convention Manual, sixth ed. Ramsar Convention Secretariat, Gland, CH http://www.ramsar.org/sites/default/files/documents/library/manual6-2013-e.pdf.

Ramsar Convention Secretariat, 2015. In: Regional overview of the implementation of the Convention and its Strategic Plan in Europe. Report to the 12th Meeting of the Conference of the Parties to the Convention on Wetlands, Punta del Este, Uruguay, 1–9 June 2015—Ramsar COP12 Doc.11. http://www.ramsar.org/sites/default/files/documents/library/cop12_doc11_summary_europe_e.pdf.

Rapinel, S., Bouzillé, J.B., Oszwald, J., Bonis, A., 2015. Use of bi-seasonal Landsat-8 imagery for mapping marshland plant community combinations at the regional scale. Wetlands 35, 1043–1054. https://doi.org/10.1007/s13157-015-0693-8.

Rebelo, L.M., Finlayson, C.M., Nagabhatla, N., 2009. Remote sensing and GIS for wetland inventory, mapping and change analysis. J. Environ. Manage. 90, 2144–2153.

Riservato, E., Boudot, J.P., Ferreira, S., Jović, M., Kalkman, V.J., Schneider, W., Samraoui, B., Cuttelod, A., 2009. The Status and Distribution of Dragonflies of the Mediterranean Basin. IUCN, Gland, Switzerland/Malaga, Spain.

Ritchie, J.C., Zimba, P.V., Everitt, J.H., 2003. Remote sensing techniques to assess water quality. Photogramm. Eng. Remote Sensing 69, 695–704.

Rocchini, D., Hortal, J., Lengyel, S., Lobo, J.M., Jimenez-Valverde, A., Ricotta, C., Bacaro, G., Chiarucci, A. 2011. Accounting for uncertainty when mapping species distributions: the need for maps of ignorance. *Prog. Phys. Geogr.* 35, 211–226. https://doi.org/10.1177/0309133311399491.

Rosenqvist, A., Finlayson, C.M., Lowry, J., Taylor, D.M., 2007. The potential of spaceborne (L-band) radar to support wetland applications and the Ramsar Convention. Aquat. Conserv. Mar. Freshwater Ecosyst. 17, 229–244.

Saifouni, A., Bellatreche, M., 2014. Cartographie numérique des habitats de reproduction de l'avifaune nicheuse du lac Tonga, parc national d'El-Kala (Nord Est Algérien). Medit. Ser. Estud. Biol II (25), 10–52. https://doi.org/10.14198/MDTRRA2014.25.01.

Sanchez, A., Abdulmalak, D., Guelmami, A., Perennou, C., 2015. Development of an indicator to monitor Mediterranean wetlands. PLoS One 31, 2015. https://doi.org/10.1371/journal.pone.0122694.

Sandholt, I., Rasmussen, K., Andersen, J., 2002. A simple interpretation of the surface temperature/vegetation index space for assessment of surface moisture status. Remote Sens. Environ. 79, 213–224.

Sandström, A., Philipson, P., Asp, A., Axenrot, T., Kinnerbäck, A., Ragnarsso-Stabo, H., Holmgren, K., 2016. Assessing the potential of remote sensing-derived water quality data to explain variations in fish assemblages and to support fish status assessments in large lakes. Hydrobiologia 780 (1), 71–84. https://doi.org/10.1007/s10750-016-2784-9.

Schlaffer, S., Matgen, P., Hollaus, M., Wagner, W., 2015. Flood detection from multitemporal SAR data using harmonic analysis and change detection. Int. J. Appl. Earth Observ. Geoinform. 38, 15–24.

Schlaffer, S., Chini, M., Dettmering, D., Wagner, W., 2016. Mapping wetlands in Zambia using seasonal backscatter signatures derived from ENVISAT ASAR time series. Remote Sens. (Basel) 8, 402.

Sebastián-González, E., Green, A.J., 2016. Reduction of avian diversity in created versus natural and restored wetlands. Ecography 39, 1176–1184.

Skidmore, A.K., Pettorelli, N., Coops, N.C., Geller, G.N., Hansen, M., Lucas, R., Mücher, C.A., O'Connor, B., Paganini, M., Pereira, H.M., Schaepman, M.E., Turner, W., Wang, T., Wegmann, M., 2015. Agree on biodiversity metrics to track from space. Nature 523, 403–405.

Smith, L.C., 1997. Satellite remote sensing of river inundation area, stage, and discharge: a review. Hydrol. Process. 11, 1427–1439.

Tamisier, A., Dehorter, O., 1999. Camargue—Canards et foulques. Fonctionnement et devenir d'un prestigieux quartier d'hiver. Centre ornithologique du Gard, Nîmes, France.

Tamisier, A., Grillas, P., 1994. A review of habitat changes in the Camargue: an assessment of the effects of the loss of biological diversity on the wintering waterfowl community. Biol. Conserv. 70, 39–47.

Thomas, R.F., Kingsford, R.T., Lu, Y., Cox, S.J., Sims, N.C., Hunter, S.J., 2015. Mapping inundation in the heterogeneous floodplain wetlands of the Macquarie marshes, using Landsat thematic mapper. J. Hydrol. 524, 194–213.

Tomaselli, V., Tenerelli, P., Sciandrello, S., 2012. Mapping and quantifying habitat fragmentation in small coastal areas: a case study of three protected wetlands in Apulia (Italy). Environ. Monit. Assess. 184, 693–713.

Tomaselli, V., Dimopoulos, P., Marangi, C., Kallimanis, A.S., Adamo, M., Tarantino, C., Panitsa, M., Terzi, M., Veronico, G., Lovergine, F., Nagendra, H., Lucas, R., Mairota, P., Mücher, C.A., Blonda, P., 2013. Translating land cover/land use classifications to habitat taxonomies for landscape monitoring: a Mediterranean assessment. Landsc. Ecol. 28, 905. https://doi.org/10.1007/s10980-013-9863-3.

Töyrä, J., Pietroniro, A., Martz, L.W., Prowse, T.D., 2002. A multi-sensor approach to wetland flood monitoring. Hydrol. Process. 16, 1569–1581.

Twele, A., Cao, W., Plank, S., Martinis, S., 2016. Sentinel-1-based flood mapping: a fully automated processing chain. Int. J. Remote Sens. 37, 2990–3004.

Tyler, A.N., Svab, E., Preston, T., Présing, M., Kovács, W.A., 2006. Remote sensing of the water quality of shallow lakes: a mixture modelling approach to quantifying phytoplankton in water characterized by high-suspended sediment. Int. J. Remote Sens. 27, 1521–1537.

Vihervaara, P., Auvinen, A., Mononen, L., Törmä, M., Ahlroth, P., Antilla, S., Böttcher, K., Forsius, M., Heina, J., Heliölä, J., Koskelainen, M., Kuussaari, M., Meissner, K., Ojala, O., Tuominen, S., Viitasalo, M., Virkkala, R., 2017. How essential biodiversity variables and remote sensing can help national biodiversity monitoring. Global Ecol. Conserv. 10, 43–59. https://doi.org/10.1016/j.gecco.2017.01.007.

White, L., Brisco, B., Dabboor, M., Schmitt, A., Pratt, A., 2015. A collection of SAR methodologies for monitoring wetlands. Remote Sens. (Basel) 7, 7615–7645.

Xu, H., 2006. Modification of normalised difference water index (NDWI) to enhance open water features in remotely sensed imagery. Int. J. Remote Sens. 27, 3025–3033. http://www.tandfonline.com/doi/abs/10.1080/01431160600589179.

Zedler, J.B., Kercher, S., 2005. Wetland resources: status, trends, ecosystem services, and restorability. Annu. Rev. Env. Resour. 30, 39–74.

ADVANCES IN ECOLOGICAL RESEARCH VOLUME 1–58

CUMULATIVE LIST OF TITLES

Aerial heavy metal pollution and terrestrial ecosystems, **11**, 218

Advances in monitoring and modelling climate at ecologically relevant scales, **58**, 101

Age determination and growth of Baikal seals (*Phoca sibirica*), **31**, 449

Age-related decline in forest productivity: pattern and process, **27**, 213

Allometry of body size and abundance in 166 food webs, **41**, 1

Analysis and interpretation of long-term studies investigating responses to climate change, **35**, 111

Analysis of processes involved in the natural control of insects, **2**, 1

Ancient Lake Pennon and its endemic molluscan faun (Central Europe; Mio-Pliocene), **31**, 463

Ant-plant-homopteran interactions, **16**, 53

Anthropogenic impacts on litter decomposition and soil organic matter, **38**, 263

Arctic climate and climate change with a focus on Greenland, **40**, 13

Arrival and departure dates, **35**, 1

Assessing the contribution of micro-organisms and macrofauna to biodiversity-ecosystem functioning relationships in freshwater microcosms, **43**, 151

A belowground perspective on Dutch agroecosystems: how soil organisms interact to support ecosystem services, **44**, 277

The benthic invertebrates of Lake Khubsugul, Mongolia, **31**, 97

Big data and ecosystem research programmes, **51**, 41

Biodiversity, species interactions and ecological networks in a fragmented world **46**, 89

Biogeography and species diversity of diatoms in the northern basin of Lake Tanganyika, **31**, 115

Biological strategies of nutrient cycling in soil systems, **13**, 1

Biomanipulation as a restoration tool to combat eutrophication: recent advances and future challenges, **47**, 411

Biomonitoring of human impacts in freshwater ecosystems: the good, the bad and the ugly, **44**, 1

Edwards Brothers Inc.
Ann Arbor MI. USA
February 20, 2018